基于文献学的澳洲龙纹斑研究与应用进展

◎ 罗 钦 等 编著

中国农业科学技术出版社

图书在版编目（CIP）数据

基于文献学的澳洲龙纹斑研究与应用进展／罗钦等编著．—北京：中国农业科学技术出版社，2018.9
ISBN 978-7-5116-3748-2

Ⅰ.①基⋯ Ⅱ.①罗⋯ Ⅲ.①鲈形目-淡水养殖 Ⅳ.①S965.211

中国版本图书馆 CIP 数据核字（2018）第 129179 号

责任编辑	李 雪 徐定娜
责任校对	贾海霞

出 版 者	中国农业科学技术出版社
	北京市中关村南大街 12 号 邮编：100081
电 话	（010）82109707（编辑室） （010）82109702（发行部）
	（010）82109709（读者服务部）
传 真	（010）82106626
网 址	http：//www.castp.cn
经 销 者	各地新华书店
印 刷 者	北京建宏印刷有限公司
开 本	787 mm×1 092 mm 1/16
印 张	19
字 数	370 千字
版 次	2018 年 9 月第 1 版 2018 年 9 月第 1 次印刷
定 价	58.00 元

◁版权所有·翻印必究▷

《基于文献学的澳洲龙纹斑研究与应用进展》

编著委员会

主编著 罗 钦　李冬梅　罗土炎

副编著 潘 葳　林 虬　黄惠珍

编著人（以姓氏笔画为序）

　　　　　刘 洋　张 斌　李 巍

　　　　　陈红珊　柯文辉　饶秋华

　　　　　徐庆贤　涂杰峰　黄敏敏

顾 问 翁伯琦

前 言

我国幅员辽阔，水产养殖业发展迅猛，近十几年来，已经成为世界上水产养殖产量最大的国家，水产养殖种类丰富多样，除了虾、蟹、贝类等多种水产养殖之外，名优鱼类与海参等特种水产养殖产业也在蓬勃发展，为丰富城乡居民的"菜篮子"与老百姓的小康生活做出了积极的贡献。我国是水产养殖大国，同时也是水产品消费大国，随着社会经济的快速发展，人民生活水平日益提高，城乡居民对优质鱼类品种需求水平也在不断提高，人们渴望有更多的优质鱼类品种上市。然而，市场上的水产养殖新品种供不应求，多数的高端水产品主要依靠捕捞或国外输入进行补充供给。尤其是近年来境内外市场对于高端水产品的需求量不断增加，但国内高质量的名优鱼类养殖规模比较小，产量不高，难以满足市场不断增长的需求。面对国际国内市场对优质水产品的巨大需求，有效开发品质优良且适合规模养殖及市场前景广阔的新兴鱼类品种，不仅有利于发挥我国南方水产资源及地方区位优势，而且有利于促进山区特种水产养殖业的持续发展。

2018年中共中央国务院关于实施乡村振兴战略的意见（中央1号文件），提出加快发展现代水产业。建设现代水产业，要把提升效益放在优先位置，效益决定收入，效益决定投入，效益决定竞争力。通过降低生产成本、促进第一、第二、第三产业融合发展、拓展农业功能、推进适度规模经营等多种途径提高农业效益、增加农民收入。2018年农业部关于大力实施乡村振兴战略加快推进农业转型升级的意见，指出要以提质减量、改善养殖生态环境为重点推进渔业结构调整。加快实施绿色水产养殖行动，优化养殖区域布局，编制发布养殖水域滩涂规划，划定禁养区、限养区、养殖区。压减水库、湖泊、近海网箱密度，稳定基本养殖水域。继续开展健康养殖示范创建，新认定健康养殖示范场500个以上、渔业健康养殖示范县10个以上。在内陆推广池塘、工厂化、大水面生态养殖、稻渔综合种养、盐碱水养殖，在近海推广贝藻间养、立体生态养殖，在深远海推广深水网箱、大型智能平台养殖。开展循环水养殖和零用药等养殖技术示范。

开展水产苗种产地检疫试点，强化水生动物疫病防控。大力发展休闲渔业，建设美丽渔村。加强现代渔港和渔港经济区建设，强化渔业安全生产监管，提高远洋渔业发展质量。完善海洋伏季休渔制度，启动实施黄河禁渔期制度。全面实施海洋渔业资源总量管理制度，进一步压减内陆水域捕捞，加快捕捞渔民减船转产，实现捕捞产量负增长。持续开展"绝户网"清理整治和涉渔"三无"船舶取缔行动，严厉打击电鱼行为。推进现代化海洋牧场建设，合理规划空间布局，新创建国家级海洋牧场示范区20个以上。加快实施国家海洋渔业生物种质资源库、3000吨级渔业资源调查船等重大工程。大力推进水产业走出去。深化与"一带一路"沿线国家和地区水产品贸易、水产业投资关系。

一、澳洲龙纹斑主要特征与开发意义

澳洲龙纹斑是生长于澳大利亚墨瑞河的一种品质优良及种群类型较大的淡水鱼类。1990年以来，由于部分流域的环境污染等问题，澳洲龙纹斑逐渐被列为极度濒危物种，澳大利亚相关生态科研单位与水产研究所开展了澳洲龙纹斑人工繁殖研究，以有效恢复因受水利工程设施及环境污染等影响而衰退的鱼类资源。通过多年的治理与防控，取得良好成效。近几年，在自然水系中澳洲龙纹斑已能够在人工控制的环境中产卵孵化，一部分苗放流于自然水域，一部分苗作为人工养殖的苗源，也有部分苗出口国外。

澳洲龙纹斑的营养和经济价值极高，加上数量稀少及在澳大利亚的饮食文化中独有的地位，有着澳洲"国宝鱼"的美称。澳洲龙纹斑不仅肉质结实细嫩、刺少、味道鲜美、无土腥味，且有一股淡淡的独特香味，口感、卖相俱佳；而且，鱼体高蛋白、低脂肪，同时富含DHA和EPA等不饱和脂肪酸等，适合现代高蛋白低脂肪的健康餐饮要求；此外，鱼肉中还含丰富的维生素、矿物质及多种生物活性物质，有助于心脏、脑部发育及眼睛视力的健康，有效促进关节健康，降低患心脑血管系统疾病的风险，增强免疫系统的功效。养殖实践表明，澳洲龙纹斑具有生长速度快、抗病能力强、对水温的耐受范围较大、人工配合饲料转化率高等特点，适合不同养殖区域集约化养殖。自1990年起，澳洲龙纹斑引进台湾驯养，此后在大陆有多家企业进行小规模饲养，如青岛七好生物科技股份有限公司、江苏中洋集团股份有限公司、浙江港龙渔业股份有限公司等多家水产养殖企业，产品深受消费者的喜爱，供不应求。目前，澳洲龙纹斑已成为我国重要的优质鱼类之一而备受关注。可以预测，其消费市场的前景广阔。

澳洲龙纹斑虽然是澳大利亚的原生鱼种，通过品种引进与技术改进，可望成为我国水产养殖的一个新兴产业。澳洲龙纹斑是一种肉食性淡水鱼，对其进行规模化产业开发之前，首先要对其养殖技术与方法进行深入研究，力求保证其获得高产、优质与高效益；其次，要着力于鱼苗繁育与本土化规模养殖技术的模式创新，力求做到人工规模养殖依然保持其肉质结实、白而细嫩、味道鲜美、无土腥味，依然保持有一股淡淡的独特香味，肉嫩鲜美，同时依然含有丰富的4种香味氨基酸、Ω-3多不饱和脂肪酸，以及EPA与DHA等功能成分。也就是说：不仅要食之美味，而且要富有营养，尽可能保持原产地的风味和品质。专家认为，引进与发展澳洲龙纹斑产业，也是立足资源与市场的重要实践，不仅有益于农业科技交流，而且有利于新兴产业开发。

二、澳洲龙纹斑养殖技术与攻关进展

澳洲龙纹斑作为顶级的白肉鱼，受其成熟周期长，以及养殖技术、养殖条件及澳洲出口控制的制约，养殖澳洲龙纹斑在我国尚属起步阶段，在人工繁育与科学养殖方面依然有许多制约环节，目前澳洲龙纹斑仅在浙江、福建有少数企业试验养殖，少量销售。福建省地处亚热带季风气候，全年温暖湿润，雨量丰富，气温适宜，气候环境优越，水资源丰富，非常适合澳洲龙纹斑养殖。福建省农业科学院农业质量标准与检测技术研究所科研人员已在福建邵武基地成功规模化繁育出澳洲龙纹斑鱼苗，成为福建省唯一一家具有澳洲龙纹斑繁育能力的单位，同时掌握了澳洲龙纹斑疾病防控、饲料配制和鱼苗培育等关键技术，为澳洲龙纹斑在福建省区域产业化发展奠定了基础。澳洲龙纹斑作为我国新引进的优质高值珍稀鱼类，是一个值得在福建山区大力推广且具有广阔市场前景的新兴养殖品种。

由于自然水域中捕捞强度不断加大，野生资源日益衰减，鱼苗的进口难以充分保障。澳洲龙纹斑在福建省农业科学院邵武养殖基地的繁育及养殖成功，意味着将大大增加苗种供应量，有望解决市场对鱼苗的需求，优化了福建省淡水鱼类种质资源，同时有利于培育壮大新兴淡水养殖产业，具有巨大的经济效益。尽管福建省并不是最早引进澳洲龙纹斑的省份，但有一支科研团队一直专注于澳洲龙纹斑开发技术的系统研究，福建省农业科学院从2012年成立联合攻关课题组开展前期研究工作，于2014年将澳洲龙纹斑项目推荐列为"福建省种业创新与产业化工程"的十大重要品种之一。获得正式立项之后，项目组深入开展了一系列科技攻关工作：包括建设良种培育技术体系、苗种繁育基地和苗种越冬基地，创建规模化苗种繁育、养殖及疾病防控等关键技术和科技服务平台，建立精养池养殖模式、水库网箱养殖模式、循环水设施养殖模式示范场，开展病害防控、专用饲料、水质调控、安全用药等方面的研究。项目组人员经过4年的辛勤努力、刻苦攻关，取得了良好成效。

三、以科技创新服务新兴产业的发展

2014年国内水产养殖研究团队陆续构建澳洲龙纹斑繁育体系与改进相关设施，开展澳洲龙纹斑繁育技术攻关，同时深入开展规模化养殖技术与病害防控等科研工作，取得良好的创新进展与推广成效。通过一系列的科研攻关和推广实践，使我们深刻体会到科技成果要有效转化为现实生产力，必须把握五个重要环节：一是要结合产业发展实际，有效开展科技创新工作；二是要发挥专业组合优势，有的放矢进行联合攻关；三是要注重科研团队建设，实行上下链接分工负责；四是要立足规模养殖需求，重点解决鱼苗繁育技术；五是要重视过程技术配套，实现科企共建示范基地。只有通过单项攻关与技术集成的有机结合，才有助于理论与实践的结合，有助于创新与创业的结合，有助于科技与经济的结合，使科技创新与科技创业有效链接。

很显然，澳洲龙纹斑繁育与养殖技术体系的优化构建与集成推广，可为淡水鱼类新兴养殖开发提供强有力技术支撑。在现代渔业转型升级过程中，不断培育与壮大新兴产业，有助于产业结构优化，有助于乡村振兴发展，有助于农民增收致富。近年来，福建省农业科学院相关单位承担了福建省种业创新与产业化工程项目"澳洲龙纹斑设施种苗工厂化繁育技术产业化工程"（2014S1477-2）、福建省发改委五新推广项目"澳洲龙纹斑新品种产业化繁育与设施健康养殖技术集成推广"（闽发改投资〔2016〕482号）、福建省属公益类科研专项"澳洲鳕鱼白点病的发生与防控技术研究"（2015R1025-3）、福建省属公益类科研专项"小瓜虫浸染澳洲龙纹斑细胞的超微结构研究"（2015R1025-6）、福建省属公益类科研专项"海藻寡糖对澳洲龙纹斑肠道菌群及肠道超微结构的影响"（2014R1025-8）、福建省地方标准"澳洲龙纹斑人工繁殖技术"（闽质监〔2017〕113号）、福建省属公益类科研专项"澳洲龙纹斑主要细菌病原菌鉴定与组织病理学研究"（2016R1024-2）、福建省农业科学院学术著作出版基金项目"基于文献学的澳洲龙纹斑研究与应用进展"（CBZX2017-26）、福建省农业科学院科技下乡"双百"行动项目"澳洲龙纹斑受精卵孵化过程中水霉病的防治及种苗推广"（sbmb1624）等项目。进行协同创新与集成推广，取得良好成效。

在研究过程中，我们虽然取得了显著的进展与良好的成效，但依然还有许多技术需要深化研究与不断探索，力求更加完善，也更加系统，进而为新兴养殖产业发展提供更为有效的保障。项目组的科技人员扩大收集了全球澳洲龙纹斑的研究资料，专门总结梳理了研究进展与推广经验，分门别类进行综述与编著，同时将主要研究论文摘要与相关文献观点罗列出来，供从事澳洲龙纹斑研发的科技人员参考与借鉴。本书包括前言（罗钦、罗土

炎、翁伯琦撰写），其他具体内容共分七章，第一章主要内容为澳洲龙纹斑种质资源及生物特性分析（罗钦、罗土炎、柯文辉撰写）；第二章主要内容为澳洲龙纹斑繁育与养殖（罗钦、涂杰峰、陈红珊撰写）；第三章主要内容为澳洲龙纹斑疾病与防控（罗钦、刘洋、饶秋华撰写）；第四章主要内容为澳洲龙纹斑营养与品质（罗钦、潘葳、黄敏敏撰写）；第五章主要内容为澳洲龙纹斑饲料与制备（罗钦、林虬、李巍撰写）；第六章主要内容为澳洲龙纹斑引种与开发（罗钦、李冬梅、张斌撰写）；第七章主要内容为澳洲龙纹斑产业与展望（罗钦、黄惠珍、徐庆贤撰写）。翁伯琦研究员负责全书的章节设计与具体统稿，罗钦、罗土炎、黄惠珍等同志为本书的具体编辑与组织修订付出辛勤劳动，同时本书在编写过程得到了福建师范大学协和学院的李冬梅老师的帮助，在此表示衷心感谢。本书罗钦作为主编著完成22万字以上，罗土炎等其他人完成13万字以上。

我们组织科技人员编著《基于文献学的澳洲龙纹斑研究与应用进展》一书，其主要目的在于集中报道项目组科技人员跟踪和收集国内外有关澳洲龙纹斑的文献资料并对研究成果进行系统综述。本书的编写是跨学科大协作的一个新的尝试，既有面上调研，点中剖析；又有文献征集，分类总结。既有经验分享，技术提升；又有进展分析，对策研究。内容丰富，有序归类，便于从事科研与推广工作者查询及应用。福建省农业科学院农业质量标准与检测技术研究所、福建师范大学协和学院、福建省农业科学院农业经济与科技信息研究所、福建省农业科学院农业工程技术研究所的科技人员进行分工协作，广泛收集文献资料，认真翻译外文摘要，深入总结研究内容，进行优化集成创作。本书对115篇英文和39篇中文报道澳洲龙纹斑的研究论文进行总结归纳，按专题进行综述，同时为了便于读者节约时间与快速查阅，专门梳理了相关重点论文的主要内容与作者观点，进行摘编并附在各个章节之后，有利于读者交替阅读。本书编写过程充分体现了协同创新的优势和成效，我们从中得到深刻的启示：协同创新是十分具体的，必须合理规划，做好优化设计；系统工作要做到精细化，必须合理分工，做到优化链接。这是我们的目标，更是我们的愿望。全书收录了1990年1月至2017年12月国内外有关澳洲龙纹斑研究的已经发表的文献报道154篇，按专题分类，并进行观点分析归纳，同时附上研究进展摘编以及索引。本书的出版可供我国乃至全球从事澳洲龙纹斑生产与研究的的科研人员、工程技术人员和管理人员参考查阅。由于时间仓促及编著者水平有限，书中的欠妥、不足甚至错误之处在所难免，敬请广大读者谅解并不吝斧正。

【罗钦、罗土炎、翁伯琦】

2018年3月31日

目 录

第一章 澳洲龙纹斑种质资源及生物特性分析 ………………………… (1)
    第一节 专题研究概况 ………………………………………………… (1)
    第二节 研究论文摘编 ………………………………………………… (9)

第二章 澳洲龙纹斑繁育与养殖 ………………………………………… (38)
    第一节 专题研究概况 ………………………………………………… (38)
    第二节 研究论文摘编 ………………………………………………… (53)

第三章 澳洲龙纹斑疾病与防控 ………………………………………… (87)
    第一节 专题研究概况 ………………………………………………… (87)
    第二节 研究论文摘编 ………………………………………………… (116)

第四章 澳洲龙纹斑营养与品质 ………………………………………… (138)
    第一节 专题研究概况 ………………………………………………… (138)
    第二节 研究论文摘编 ………………………………………………… (153)

第五章 澳洲龙纹斑饲料与制备 ………………………………………… (167)
    第一节 专题研究概况 ………………………………………………… (167)
    第二节 研究论文摘编 ………………………………………………… (191)

第六章 澳洲龙纹斑引种与开发 ………………………………………… (221)
    第一节 专题研究概况 ………………………………………………… (221)
    第二节 研究论文摘编 ………………………………………………… (229)

第七章 澳洲龙纹斑产业与展望 ………………………………………… (264)
    第一节 产业发展对策与建议 ………………………………………… (264)
    第二节 产业发展趋势与展望 ………………………………………… (270)

附录 研究进展摘编索引 ………………………………………………… (274)

第一章 澳洲龙纹斑种质资源及生物特性分析

第一节 专题研究概况

澳洲龙纹斑（*Macculochella peelii peelii*）隶属硬骨鱼纲（Osteichthyes）鲈形目（Perciformes）鮨鲈科（Percichthyidae）亦译为肖鲈科、真鲈科或暖鲈科，鳕鲈属（Maccullochella）亦译为麦鳕鲈属。英文名 Murray cod，中文俗称河鳕、鳕鲈、墨瑞鳕、墨累鳕、东洋鳕、澳洲鳕鲈、澳洲石斑、虫纹石斑、虫纹鳕鲈、墨瑞河真鲈、虫纹麦鳕鲈、澳洲淡水鳕、墨累河鳕鱼、澳洲淡水鳕鲈等。在澳大利亚素有"国宝鱼"之美称。因其体表布有黄褐色的虫纹斑点又名虫纹石斑，正统的称谓应是虫纹鳕鲈。

一、澳洲龙纹斑形态特征

（一）外部形态

澳洲龙纹斑鱼体体长为体高的 4.1~4.8 倍，为头长的 2.9~3.2 倍。头长为吻长的 3.1~3.5 倍，为眼径的 6.7~7.3 倍，为眼间距的 3.4~4.7 倍；尾柄长为尾柄高的 1.9~2.2 倍。澳洲龙纹斑身体延长，稍侧扁，鱼体呈纺锤形（图 1-1），左右对称，眼后微凹，头后部稍隆起，头长为全长的 1/3；口端位，口裂较大，口裂末端基本与眼前缘在一条线上，下颌略长于上颌。上、下颌骨上密布细齿，犁骨、颚骨、上咽骨、下咽骨上均有许多小齿。上下颌齿细小，犁骨及颚骨上有

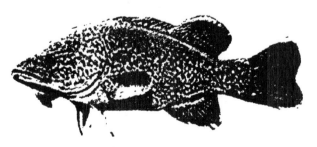

图 1-1 澳洲龙纹斑（韩茂森，2003）

排列成绒毛状的牙带。鳃弓4对，鳃耙短疏、棘状。第一鳃弓外鳃耙数为17。头部腹面具小孔，左右各6个，司感觉作用，颏部左右各3个小孔。侧线完全，在体侧上半部与背缘相平行，向后延伸至鳍基部。背鳍2个，基部相连。第一背鳍由11根鳍棘组成，第1、2根背鳍硬棘等长，第5、6根最长，余渐短；第二背鳍由13~14根鳍条组成，鳍条排列整齐，末端呈圆弧状。背鳍起点在胸鳍长的近1/2处，第二背鳍起点相对臀鳍起点稍后。胸鳍位低，后缘扇形，有19根鳍条，其长度达第一背鳍的第4硬棘处。腹鳍胸位，起点稍前于胸鳍起点，由1根硬棘5根鳍条组成，其中第一根鳍条延长。臀鳍起点在第一背鳍第11硬棘的垂线处，由3根硬棘13根鳍条组成。尾鳍呈圆形，由17~18根鳍条组成。肉眼观体表似无鳞，以手触之光滑，实际体披细小而密的栉鳞，侧线鳞为65~81。澳洲龙纹斑个体很大，体色有2种，一种是黑色加橄榄绿，另一种是黑绿色。鱼体两侧上半部褐色，腹部淡黄至白色，无斑点，背面为黄绿色，密布不规则黑色斑点，从背部至腹部越往下体色越浅。体表遍布似虫纹状的花斑，在背鳍、臀鳍、尾鳍等处亦有分布，有黑色斑点，体型及体色极其美丽（曹凯德，2001；韩茂森，2003；王波，2003）。

（二）内部器官

澳洲龙纹斑胃发达，呈"Y"形；肠道短，其长度约为体长的1/2，幽门盲囊2~4个（见图1-2）。胃壁厚，肌肉层极为发达，食道及胃内壁多皱褶。肠较短，占体长的60.2%~63.8%。鳔一室，壁薄，明显可见，呈大的囊袋状，约占腹腔体积的1/3。肝脏分三叶；胆囊卵圆形，脾脏紧贴胃，呈"松子"状。性腺一对，雌雄异体（王波，2003）。

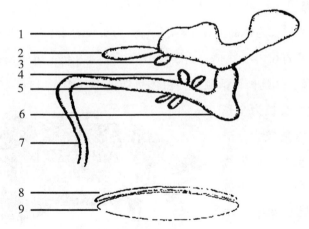

1. 肝；2. 胆囊；3. 脾；4. 幽门盲囊；5. 肠；6. 胃；7. 肠；8. 肾；9. 鳔

图1-2 澳洲龙纹斑的内部器官（王波，2003）

二、澳洲龙纹斑地理分布

澳洲龙纹斑属于鲐鲈科，这一族鱼类在世界上共有30多个品种，分布在热带、温带的海水、咸淡水和淡水水域中，在澳大利亚水域中有7个品种。澳洲龙纹斑是澳大利亚的原生鱼种（陈杰，2002），澳洲龙纹斑是澳大利亚迄今发现的最大的淡水鱼类。澳洲龙纹斑是大洋洲极为古老的鱼类之一，同大多数淡水鱼类类似，其祖先来自海洋。在新南韦尔斯的一溯到2600万年前土层中出土的墨列鳕鱼化石，解剖上与现代的墨瑞鳕完全一致。澳洲龙纹斑原生长在江河、湖泊水域中，原栖生地是澳大利亚东南部的墨瑞河（Murray river）。它的自然分布范围贯穿墨瑞-达令盆地（图1-3），西部划分范围从昆士兰东南，经过新南威尔士州到维多利亚和澳大利亚南部。澳洲龙纹斑栖息在墨瑞-达令水系的低洼河流中，喜欢微流水和较深的洞穴（曹凯德，2001；陈杰，2002；韩茂森，2003；杨小玉，2013；杨宝琴，2014；赵志玉，2014）。

Llewellyn, L. C.（2014）在1959年至1970年期间，在马兰比季河和墨累河上对大型淡水鱼类进行了标记和重新捕获计划，低数量的墨瑞鳕鱼和澳洲银鲈被重新捕获（Llewellyn, L. C., 2014）。

澳洲龙纹斑目前已能够在人工控制的环境中产卵孵化，一部分苗放流于自然水域，一部分苗作为人工养殖的苗源，也有一部分苗出口国外。澳洲龙纹斑已被引入我国，1999年首次由澳洲引进台湾，随后引进福建、浙江、江苏、广东等地，目前我国青岛、广州、济南等地已经开始引进养殖。澳洲龙纹斑也向港澳台地区及东南亚出口。国内已有几家引养了澳洲龙纹斑，见表1-1（王波，2003；韩茂森，2003；李娴，2013；罗土炎，2015）。

表1-1 中国养殖澳洲龙纹斑的企业名单

序号	地点	公司名称	备注
1	广东省	东莞市安业水族有限公司	（杨宝琴，2014；吕华当，2017；欧小华，2017）
2	浙江省	乐清市港龙渔业有限公司	（张善发，2009）
3	江苏省	江苏中洋集团江苏中润农业发展有限公司	（郭正龙，2012）
4	山东省	青岛晟华种苗公司	（韩茂森，2003）
5	江苏省	江苏省淡水水产研究所	（韩茂森，2003）
6	山东省	山东省淡水水产研究所	（张龙岗，2012）
7	福建省	福建省农业科学院	（罗土炎，2015）

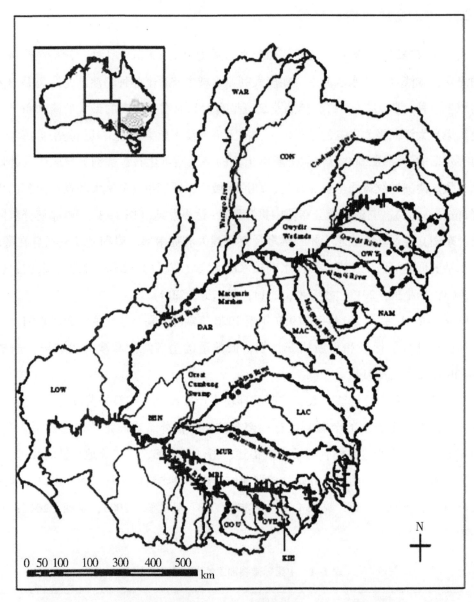

图1-3 墨瑞-达令盆地的采样点（●），采样流域（...），主要河流（_）和终端湿地．流域：WAR（Warrego河）；CON（Condamine河）；BOR（Border河）；GWY（Gwydir河）；NAM（Namoi河）；DAR（Darling河）；MAC（Maquarie河）；LAC（Lachlan河）；BEN（Benanee河）；MUR（Murrumbidgee河）；MRI（墨瑞瑞福利纳）；LOW（墨瑞河下游）；GOU（Goulburn河）；OVE（Ovens河）；KIE（Kiewa河）．所有的河流都流向了南部的墨瑞河，然后流到墨瑞河下游，最后到达了大海（M. L. Rourke，2011）

三、澳洲龙纹斑摄食习性

Whiterod, N. S. (2013) 研究了澳洲龙纹斑的游泳能力，确定其是一名游泳通才，在所研究的游泳能力参数的各个方面保持适中的能力。澳洲龙纹斑能量需求低，维持较低的质量特异性标准（21.3~140.3mg/h/kg）和最大活性代谢率（75.5~563.8mg/h/kg），导致活动范围小（最大活动代谢率-标准代谢率：1.4~5.9）。他们是相当有效率的游泳运动员（绝对和相对最佳游泳速度，分别为 0.17~0.61m/s 和 0.77~1.93BL/s）并且能够重复长时间运动（恢复比=0.99）。游泳能力方面的异速变化是通过体重来实现的，而广泛的游泳能力在很宽的温度范围内保持不变。澳洲龙纹斑所展示的游泳能力反映了一种坐稳等待的觅食策略。澳洲龙纹斑属典型肉食类型，是非常凶猛的鱼类，并具有高度拓疆性和攻击性，对于入侵其领地者给以猛烈反击（陈杰，2002；Whiterod, N. S., 2013）。

Tonkin, Z. D. (2006) 研究了澳洲龙纹斑的形态学和摄食行为的遗传变化。在开始摄食时，澳洲龙纹斑游离胚胎体积大且发育良好，饲喂小型哲水蚤目桡足类，大型哲水蚤目桡足类，小型水蚤和大型水蚤后，经测试幼鱼偏爱中型猎物（宽 300~500μm）。即使它们的嘴巴大大超过所提供的最大猎物，情况也是如此。水蚤比相似大小的桡足类消耗得更多。这项研究的结果表明，如果在幼年期没有宽度<500 的可捕食猎物，那么澳洲龙纹斑物种的存活将受到威胁。他们还认为可能会发生种间竞争猎物，尤其是当幼体非常小的时候（Tonkin, Z. D. 2006）。

澳洲龙纹斑喜食活饵，不同生长发育阶段，摄食对象有所不同，5~8mm 的仔、稚鱼主要摄食浮游动物，包括轮虫、桡足类、昆虫幼体、摇蚊幼虫及大型蚤等；15~20mm 的鱼苗即可捕食水中昆虫；成鱼主要捕食小型鱼类、虾蟹类、甲壳类，也捕食蛙类、水禽、乌龟、水老鼠等。在饥饿无其他饲料时，会互相残食。澳洲龙纹斑主要食物有软体动物、甲壳类、小鱼和其他半水生脊椎动物（曹凯德，2001；郭松，2012）。

澳洲龙纹斑经过人工驯化可喂食人工配合饲料。人工养殖的澳洲龙纹斑，受精卵孵化后，8~10d 后卵黄囊吸收完毕，开始摄食小型浮游动物，再过 4~6 周鱼苗发育至 15~20mm 时，转为摄食水生昆虫、线虫、小型甲壳类等。鱼苗的规格达 25~30mm 时，投喂人工配合饲料。每天投饵 2 次，因澳洲龙纹斑不喜强光，投饵时间为清晨和傍晚（韩茂森，2003；李娴，2013）。

四、澳洲龙纹斑栖息习性

澳洲龙纹斑喜栖于阴暗处，白天多在水域的边缘、草丛、树根、石缝中，不大游动，活动较少，特别喜欢藏于水草丛中躲避强光；夜间出来活动觅食，每年夏季尤其夏季的夜晚最为活跃。澳洲龙纹斑是温水性鱼类，对温度有较强的适应性，在7~30℃均可生存。适合生长的温度为7~32℃，最佳温度为16~26℃。最佳生长水温为18~22℃，秋冬低温季节，潜于深水处越冬，春季逐渐游到食物丰富的沿岸水草丛中觅食。pH6.5~8.5，最低溶氧为2mL/L（左瑞华，2001；韩茂森，2003；郭松，2012；李娴，2013）。

在澳大利亚墨瑞-达令盆地洪水和关联的缺氧黑水事件期间，利用无线电调查了36尾在主要渠道和墨瑞河下游栖息地中的澳洲龙纹斑的活动。在洪峰期间大约93000mL/d，溶解氧降至1.2mg/L。澳洲龙纹斑集中出现在支流栖息地上，但也在支流和主渠道之间大面积活动。鱼一直位于主渠道或永久支流，这表明河漫滩栖息地的短暂利用是有限的，突出了渠道外和主渠道栖息地连通的重要性。溶解氧低浓度导致无线电标记鱼的死亡率相当大（25%），表明缺氧黑水环境可能对墨瑞河下游中澳洲龙纹斑的种群有重大影响。2004年2月，在达令河下游的下梅宁迪和蓬卡里之间澳洲龙纹斑出现大量死亡。在延绵超过160km的流中出现1000~5000尾死亡，主要是大的澳洲龙纹斑。就在死亡事件发生之前，达令河下游已经断流了，只剩下残留的水池。鱼群死亡的那一周，最高气温超过40℃，水温超过30℃。这些死亡是缺氧造成的（最可能结合极端温度，藻类耗氧量高和有机污染物负荷大），伴随着可能增加的来自池塘底层还原性沉积物中释放出来的有毒硫化物和/或氨。澳洲龙纹斑从来没有在澳洲布罗肯河下游的死水中发现过（Ellis, I.M., 2004; A. J. King, 2010; Leigh, S. J., 2013）。

在一个大型的澳洲旱河巴望-达令河中，澳洲龙纹斑与大木材栖息地密切相关，具有很高的栖息地专属性。Koehn, J.D.（2009）研究了澳洲龙纹斑在澳大利亚东南部的Ovens和Murray河上选择栖息地的信息。澳洲龙纹斑成鱼（总长度>450mm, LT）和年龄0岁（总长度<150mm）的澳洲龙纹斑都选择了基于结构变量的宏观环境和微生境。在宏观环境中，成鱼选择了河流中的河道栖息地，在高洪水区的漫滩海峡以及能避免洪泛区的Mulwala湖内。成年和年龄0岁的鱼选择类似的微生境，无论地点或水文条件如何，选择主要受结构木质栖息地较高负荷的影响，包括深水区较高的c.v.、更多的悬垂植被、较浅的深度、更低的水流速度，更靠近河岸。0岁的澳洲龙纹斑似乎选择了较多结构性木本的较浅栖息地，比成年鱼更靠近河岸。（Koehn, J. D., 2013）澳洲龙纹斑

采用较深的栖息地（分别为2.8m和2.9m），地表水速度较高（分别为0.37m/s和0.49m/s），距河岸较远的地方（Boys，C.A.，2006；Koehn，J.D.，2009）。

五、澳洲龙纹斑繁殖习性

澳洲龙纹斑性成熟需要3~6年。雄鱼比雌鱼成熟得早。雌鱼每年产卵1次。澳洲龙纹斑产卵最适水温为18~22℃。不同纬度地区，繁殖时间不同，通常5—8月为繁殖盛期。一般在春季或夏初水温达到18℃时开始产卵。河道及湖泊入口为澳洲龙纹斑产卵的理想场所，繁殖季节通常是一尾雌鱼与多尾雄鱼在安全隐蔽处或有遮盖物的水域交配。其卵为黏性，卵径3.0~3.5mm，卵底栖，受精卵黏附在浅水处的树洞、水管孔内或其他固体如树木、礁石及泥土的表面。产卵后，雄鱼驱散雌鱼，担负护卵重任，雄鱼守护在卵附近以防被其他水生动物侵害和吃掉，在卵受精发育阶段，时而游来游去，以使水流畅通，确保孵化顺利进行直至出苗（陈永乐，1999；曹凯德，2001；郭松，2012）。

通常情况下，4~5岁的澳洲龙纹斑体重为2.5~4.0kg/尾时可达性成熟，通常雌雄性征不明显，近性成熟的雌鱼，生殖孔较大、突出、紫红色，腹部膨胀。雌鱼一般怀卵量为1万~4万粒，随个体的增大而增加，体重2.5~3.0kg的雌鱼可产卵10000粒，体重5.0kg的雌体可产卵14000~30000粒，体重23kg的雌鱼可产卵90000粒。其卵径为3.0~3.5mm，受精卵孵化温度为21~30℃，最适温度为25~28℃。在20~22℃水温条件下，受精卵经5~7d孵化出膜，刚出膜的苗体长5.0~8.0mm，卵黄囊清晰可见，过8~10d后卵黄囊吸收完毕，再过4~6周鱼苗发育至15~20mm（陈永乐，1999；郭松，2012）。

六、澳洲龙纹斑年龄和生长

Disspain，M.C.F.（2013）分析了墨瑞河下游挖掘出的耳石，这些耳石的年代可以追溯到全新世中晚期。大部分鱼类（确定为澳洲龙纹斑$n=22$和圆尾麦氏鲈$n=2$）在淡水环境中被捕获，并且已经成长到性成熟的年龄和大小。数据表明，澳洲龙纹斑能生长到比现在大得多的尺寸；历史数据表明，这种鱼体大小的减少可能是欧洲捕捞活动、引入物种和栖息地退化的结果（Disspain，M.C.F.，2013）。

澳洲龙纹斑的生长速度较快，当年可达到200g、体长约23cm，翌年可达800g、体长达35cm，第3年可长至2kg、体长达50cm，之后生长相对缓慢。4~5岁澳洲龙纹斑体长分别是56cm和64cm，检查23kg个体的耳石，其年龄约为15岁。在自然状态下，澳洲龙纹斑生长很大，文献记载1902年在沃格特（Waigett）附近的河流中人们曾捕获的最大的体重达113.5kg，长180cm，估计年龄有80岁。在澳大利亚的河流中时有游钓者钓到20~40kg的个体，2~3kg的个体比比皆是（陈永乐，1999；陈杰，2002；韩茂森，2003；郭松，2012）。

澳洲龙纹斑生长速度与水温、栖息地和饵料资源情况有关。养殖时间为2008年4月12日到10月12日，投放鱼苗27000尾，平均重1.15g/尾，总重31kg，6个月后剩余鱼苗24354尾，成活率90.2%，总重600kg，平均每尾重24.6g。其中45g左右的鱼有7568尾、占总数的31.07%，20g左右的10358尾、占总数的42.53%，20g以下的占26.4%。江苏中洋集团工厂化循环水养殖中的一组作为研究组，测量放养时平均规格为0.502kg/尾、放养密度6.275kg/m。经过6个月的饲养，在2012年4月25日再次测量平均规格为1.596kg/尾、放养密度约为20kg/m。平均体重为（1.29±0.23）g，平均全长为（4.56±0.30）cm的澳洲龙纹斑鱼苗，经97d培育，平均体重和全长分别达12.84±1.30g和10.03±0.36cm。澳洲龙纹斑鱼种体重和全长特定生长率分别为2.37%day^{-1}和0.81%d^{-1}；该鱼种体重与全长的最优回归方程为幂函数方程$W = 0.016L^{2.884}$。以Gompertz非线性生长模型算得的澳洲龙纹斑鱼种体重和全长生长拐点分别为11.8g和8.5cm，拐点日龄分别为170d和149d；全长生长拐点日龄的出现先于体重生长拐点（曹凯德，2001；张善发，2009；郭正龙，2012；罗钦，2016）。

七、澳洲龙纹斑遗传多样性

Meaghan，R.（2010）从澳洲龙纹斑DNA文库中分离出102个多态微卫星基因座，并评估了它们在13种本地和6种引入的淡水鱼种中的扩增成功率。该基因座为了达到评估野生群体遗传结构和未来保存工作的双重目的，以及为水产养殖识别关键数量性状基因座的标记。从部分基因组测序数据中回收澳洲龙纹斑完整线粒体基因组，其由16442bp（58%A+T含量）组成，含有13个蛋白质编码基因，2个核糖体亚基基因，22个转移RNA和768bp非编码富含AT-富集区组成。这是鳕鲈属的第一线粒体基因组序列，是暖鲈科属的第四线粒体基因组序列。在澳洲龙纹斑、东部淡水鳕鱼和鳟鱼鳕鱼的野生种群上测试了来自东部淡水鳕鱼的7个新型微卫星基因座和来自澳洲龙纹斑的6个先前公开的基因座。多态性水平在不同物种之间有不同的变化，澳洲龙纹斑有13个多态性基因座，鳟鱼鳕鱼有9个多态性基因座，东部鳕鱼有7个多态性基因座。观察到的杂合度范围为0.053~0.842（Nock，C.J.，2008；Meaghan，R.，2010；Austin，C.M.，2016）。

澳洲龙纹斑引进群体的单倍型多样性为0.300，核苷酸多样性（Pi）为0.00073，其单倍型多样性以及核苷酸多样性均处于较低水平。从遗传学和进化角度看，一个物种的遗传多样性高低与其适应能力、生存能力和进化潜力密切相关，贫乏的遗传多样性给物种生存、进化及种质资源的保护和利用带来许多不利影响。澳洲龙纹斑引进群体的遗传变异较低可能与其分布范围狭窄，种群数量有限，以及生存环境逐渐恶化有关。澳洲龙纹斑分布在澳大利亚墨瑞河及达令河流域，分布范围狭窄，该地区属于典型的地中海式气候。该流域有效集水面积只有 40 万 km^2。在其缓慢流程中，河水蒸发量渐次增加，河流水量不大，一些河道经常干涸，致使澳洲龙纹斑种群数量有限，面临逐渐恶化的水域环境，可能导致其遗传多样性逐渐降低。另外，近亲交配和环境的双重压力造成的瓶颈效应以及遗传漂变作用，可能使澳洲龙纹斑基因型变异较为单一，群体遗传多样性贫乏。再者，本试验的样品来源于同一批次引进的个体，可能为同一自然群体产生的后代，所以导致遗传变异较小。澳洲龙纹斑 COI 基因序列 NJ 系统树表明：该引进群体的 25 个个体明显聚为两个分支。作为真核生物胞质遗传的重要组成部分，mtDNA 是由卵细胞传递给后代，属于典型的母系遗传。因此，可初步推断该群体的 25 个个体可能起源于两个具有明显遗传差异的不同母系祖先（张龙岗，2013）。

（罗钦、罗土炎、柯文辉）

第二节　研究论文摘编

1. Identifying environmental correlates of intraspecific genetic variation.

［英文摘要］：Genetic variation is critical to the persistence of populations and their capacity to adapt to environmental change. The distribution of genetic variation across a species' range can reveal critical information that is not necessarily represented in species occurrence or abundance patterns. We identified environmental factors associated with the amount of intraspecific, individual-based genetic variation across the range of a widespread freshwater fish species, the Murray cod Maccullochella peelii. We used two different approaches to statistically quantify the relative importance of predictor variables, allowing for nonlinear relationships: a random forest model and a Bayesian approach. The latter also accounted for population history. Both approaches identified associations between homozygosity by locus and both disturbance to the natural flow regime and mean annual flow. Homozygosity by locus was negatively associated with disturbance to the natural flow regime, suggesting that river reaches with more disturbed flow regimes may support larger, more genetically diverse populations. Our findings are consistent with the hypothesis that artificially induced perennial flows in regulated channels may provide greater and more consistent habitat and reduce the frequency of population bottlenecks that can occur frequently under the highly variable and unpredictable natural flow regime of the system. Although extensive river regulation across eastern Australia has not had an overall positive effect on Murray cod numbers over the past century, regulation may not represent the primary threat to Murray cod survival. Instead, pressures other than flow regulation may be more critical to the persistence of Murray cod (for example, reduced frequency of large floods, overfishing and chemical pollution).

[译文]:
鉴定种内遗传变异的环境相关性

遗传变异对种群持续存在及其适应环境变化的能力至关重要。物种范围内遗传变异的分布可以揭示关键信息，这些信息不一定代表物种发生或丰度模式。我们确定了与广泛淡水鱼类物种（墨瑞鳕鱼）范围内的种内个体基因变异相关的环境因素。我们用两种不同的方法来统计量化预测变量的相对重要性，从而考虑非线性关系：随机森林模型和贝叶斯方法。后者也列入种群史。两种方法均确定了基因座纯合性与自然流态和平均年流量两者之间的关联。基因座纯合性与自然流动状态的干扰负相关，表明越混乱流态的河流可能支持更大、基因更多样的群体。我们的研究结果与这样一个假设是一致的，即人为诱导的常年流量在受管制渠道可能提供更大更一致的栖息地，并减少在高度可变且不可预测的自然流动规则体系下频繁发生的群体频率。在过去的一个世纪中，虽然澳大利亚东部地区广泛的河流整治并未对墨瑞鳕鱼数量产生整体积极影响，但河流的整治监管可能不是对墨瑞鳕鱼生存的主要威胁。相反，流量调节以外的其他压力（例如，大洪水频率减少，过度捕捞和化学污染）可能对墨瑞鳕鱼的持续存在更为重要。

参考文献: Harrisson, K. A., Yen, J. D., Pavlova, A., *et al.* 2016. Identifying environmental correlates of intraspecific genetic variation [J]. Heredity, 117 (3), 155.

（罗钦、李冬梅、黄惠珍 译）

2. The complete mitogenome of the Murray Cod, Maccullochella peelii (Mitchell, 1838) (Teleostei: Percichthyidae)

[英文摘要]: The complete mitochondrial genome of the iconic Australian freshwater fish, the Murray Cod, Maccullochella peelii, was recovered from partial genome sequencing data using the HiSeq platform (Illumina, San Diego, CA). The mitogenome consists of

16442bp (58%A+T content) containing 13 protein-coding genes, 2 ribosomal subunit genes, 22 transfer RNAs, and a 768bp non-coding AT-rich region. This is the first mitogenome sequence for the genus Maccullochella, and the fourth for the family Percichthyidae.

[译文]：

墨瑞鳕鱼的完整线粒体基因组（米切尔：1838）（真骨附类：暖鲈科）

使用 HiSeq 平台（Illumina, San Diego, CA）从部分基因组测序数据中回收标志性澳大利亚淡水鱼（墨瑞鳕鱼）的完整线粒体基因组。线粒体基因组由 16442bp（58%A+T 含量）组成，含有 13 个蛋白质编码基因，2 个核糖体亚基基因，22 个转移 RNA 和 768bp 非编码富含 AT-富集区组成。这是鳕鲈属的第一线粒体基因组序列，是暖鲈科属的第四线粒体基因组序列。

参考文献：Austin, C. M., Tan, M. H., Lee, Y. P.. 2016. The complete mitogenome of the murray cod, maccullochella peelii (mitchell, 1838) (teleostei: percichthyidae) [J]. Mitochondrial Dna Part A Dna Mapping Sequencing & Analysis, 27 (1), 729.

（罗钦、李冬梅、黄惠珍　译）

3. Comparative habitat use by large riverine fishes

[英文摘要]：The present radio-tracking study compared adult daytime microhabitat use by three large Australian native freshwater fishes (Murray cod, Maccullochella peelii, trout cod, M. macquariensis, golden perch, Maquaria ambigua) and introduced carp, Cyprinus carpio, in the Murray River, south-eastern Australia. The paper describes habitat patches used by all species and quantifies differences among species. All species were strongly associated with structural woody habitat (>68% cover), deeper (>2.4m), slower water (<0.2m s^{-1}) closer to the river bank, with variations in substrate. Murray cod and trout cod used deeper habitats (2.8m and 2.9m, respectively), with higher surface water velocities (0.37m s^{-1} and 0.49m s^{-1}, respectively) and further from the bank than the habitats of golden perch (2.6m; 0.31m s^{-1}) or carp (2.4m; 0.20m s^{-1}), the latter species using wood higher in the water column than did cod species. Trout cod used habitats furthest from the bank and carp those closest. These data provide support and direction for reintroduction of structural woody habitat patches for rehabilitation which, in general, should have >70% cover, be >1.5m high, located <15% of the river channel (width) closest to the bank, with surface water velocities of 0.3~0.6m s^{-1}.

[译文]：

比较大型河鱼的栖息地使用

在澳大利亚东南部的墨累河中，目前对三种澳大利亚原生淡水鱼（墨瑞鳕鱼，黄金鲈鱼，圆尾麦氏鲈）和引入的鲤鱼，采用无线电追踪研究比较了成鱼日间小环境利用。该论文描述了所有物种使用的栖息地斑块，并量化了物种间的差异。所有物种都与结构木质栖息地（>68%覆盖度），较深（>2.4m），较接近河岸的缓流（<0.2m/s）以及底物变化密切相关。瑞鳕鱼和鳟鱼鳕鱼采用较深的栖息地（分别为2.8m和2.9m），地表水速度较高（分别为0.37m/s和0.49m/s），距河岸较远的地方，而黄金鲈鱼栖息地（2.6m；0.31m/s）或鲤鱼栖息地（2.4m；0.20m/s），后两种鱼类使用木材的水柱比鳕鱼种类高。鳟鱼鳕鱼使用的栖息地远离岸边，并且鲤鱼离那些最近。这些数据为重新引入用于复原的结构性木本生境斑块提供了支持和指导，一般来说，结构性木本生境斑块覆盖率应>70%，高度>1.5m，位于离河岸最近的河道（宽度）在为15%以内，地表水速度为0.3~0.6m/s。

参考文献：Koehn, J. D., Nicol, S. J. 2013. Comparative habitat use by large riverine fishes [J]. Marine & Freshwater Research, 65 (2), 164-174.

（罗钦、李冬梅、黄惠珍　译）

4. Movements of golden perch Macquaria ambigua (Richardson) in the mid Murray and lower Murrumbidgee Rivers (New South Wales) with notes on other species

[英文摘要]：A tagging and recapture program was carried out between 1959 and 1970 on large freshwater fish species in the Murrumbidgee and Murray Rivers to determine the patterns of their movement. Tagging was carried out at four main sites Stoney Crossing, Hattah, Hay and Narrandera. At the time internal body tags were found to be the most reliable method of tagging. Adequate numbers of Golden Perch, Macquaria ambigua, and low numbers of Murray Cod Maccullochella peelii peelii and Silver Perch Bidyanus bidyanus were recaptured. It is concluded that up-and downstream movement of M. ambigua from these sites varied. The western two sites at Stoney Crossing and Hattah on the Murray River showed a marked net upstream movement; movement from Hay on the Murrumbidgee showed dispersal both up and downstream, while movement from Narrandera showed a net downstream movement Within the size range sampled 309~563mm, the larger fish moved significantly less than the smaller fish; when split into upstream and downstream movement, this was only significant for upstream movement B. Bidyanus, tagged in the upper two sites, showed a net downstream movement from Narrandera with a single upstream movement from Hay. M. peelii peelii showed only minor movements (up to 141km) up and downstream from Narrandera, with only one long distance movement from Balranald to Gundagai. Net downstream stream movement of M. ambigua declined as one moved downstream closer to the mouth of the Murray River.

[译文]：

在墨累河中部黄金鲈鱼和马兰比季河下游（新南威尔士州）的其他标记物种的运动

在1959年至1970年期间，在马兰比季河和墨累河上对大型淡水鱼类进行了标记和重新捕获计划，以确定它们的运动模式。在四个主要地点Stoney Crossing，Hattah，Hay和Narrandera进行了标记实验。当时体内标签被认为是最可靠的标记方法。足够数量的黄金鲈鱼，圆尾麦氏鲈，低数量的墨瑞鳕鱼和澳洲银鲈被重新捕获。得出的结论是，来自这些地点的圆尾麦氏鲈上下游运动各不相同。墨累河上Stoney Crossing和Hattah这两个西部地点显示出明显的纯上游运动；在马兰比季河上Hay地点的鱼群表现出向上和向下游的扩散，而在Narrandera的鱼群在309~563mm取样的尺寸范围内表现出纯下游运动，较大鱼的移动速度显著低于较小的鱼；当分成上游和下游运动时，只有澳洲银鲈的上游运动比较明显。上面提及的两个地点，标记显示了鱼群在Narrandera纯上游运动，以及鱼群在Hay的单次上游运动。墨瑞鳕鱼种群只显示了在Narrandera上下游的微小移动（长达141km），就只有这一个从Balranald到Gundagai的长距离移动。随着向下游靠近墨累河口，圆尾麦氏鲈种群纯下游运动减少。

参考文献：Llewellyn, L. C. 2014. Movements of golden perch macquaria ambigua (richardson) in the mid murray and lower murrumbidgee rivers (new south wales) with notes on other species [J]. Australian Zoologist, 37 (2), 139-156.

（罗钦、李冬梅、黄惠珍　译）

5. The swimming capacity of juvenile Murray cod (Maccullochella peelii): an ambush predator endemic to the Murray-Darling Basin, Australia

[英文摘要]：This study documented the swimming capacity of a large ambush predator, Murray cod Maccullochella peelii, endemic to the Murray-Darling Basin, Australia. It was evident that the species is a swimming generalist, maintaining moderate ability across all aspects of the swimming capacity parameters that were investigated. For instance, the species was capable of prolonged swimming performance (critical swimming speed, Ucrit：absolute, 0.26~

$0.60 \text{m} \cdot \text{s}^{-1}$, relative, $1.15 \sim 2.20 \text{BL} \cdot \text{s}^{-1}$) that was inferior to active fish species, but comparable with other ambush predators. The species had low energetic demands, maintaining a low mass-specific standard ($21.3 \sim 140.3 \text{mg} \cdot \text{h}^{-1} \text{kg}^{-1}$) and maximum active metabolic rate ($75.5 \sim 563.8 \text{mg} \cdot \text{h}^{-1} \text{kg}^{-1}$), which lead to a small scope for activity (maximum active metabolic rate-standard metabolic rate; $1.4 \sim 5.9$). They were reasonably efficient swimmers (absolute and relative optimal swimming speed, $0.17 \sim 0.61 \text{ m} \cdot \text{s}^{-1}$ and $0.77 \sim 1.93 \text{BL} \cdot \text{s}^{-1}$, respectively) and capable of repeat bouts of prolonged performance (recovery ratio = 0.99). Allometric changes in aspects of swimming capacity were realised with body mass, whereas broad swimming capacity was maintained across a wide range of temperatures. The swimming capacity demonstrated by M. peelii reflects a sit-and-wait foraging strategy that seeks to conserve energy characteristic of ambush predators, but with distinct features (e.g., lack of fast-start ability) that may reflect their evolution in some of the world's most hydrologically and thermally variable rivers.

[译文]：

墨瑞鳕幼鱼的游泳能力：在澳大利亚的墨累-达令盆地，一种伏击性的捕食者

这项研究记录了澳大利亚墨累-达令盆地特有的大型伏击捕食者墨瑞鳕鱼的游泳能力。很明显，该物种是一名游泳通才，在所研究的游泳能力参数的各个方面保持适中的能力。例如，该物种能够延长游泳性能（临界游泳速度，绝对值，$0.26 \sim 0.60 \text{m/s}$，相对值，$1.15 \sim 2.20 \text{BL/s}$），差于活性鱼种，但比得上其他伏击掠食者。该种群能量需求低，维持较低的质量特异性标准（$21.3 \sim 140.3 \text{mg/h/kg}$）和最大活性代谢率（$75.5 \sim 563.8 \text{mg/h/kg}$），导致活动范围小（最大活动代谢率-标准代谢率：$1.4 \sim 5.9$）。他们是相当有效率的游泳运动员（绝对和相对最佳游泳速度，分别为$0.17 \sim 0.61 \text{m/s}$和$0.77 \sim 1.93 \text{BL/s}$）并且能够重复长时间运动（恢复比=$0.99$）。游泳能力方面的异速变化是通过体重来实现的，而广泛的游泳能力在很宽的温度范围内保持不变。墨瑞鳕鱼所展示的游泳能力反映了一种坐稳等待的觅食策略，该策略旨在保护伏击掠食者的能量特征，但具有明显的特征（例如，缺乏快速启动能力），这可能反映它们在某些世界上最具水文和热量变化的河流中的演化。

参考文献：Whiterod, N. S. 2013. The swimming capacity of juvenile murray cod (maccullochella peelii): an ambush predator endemic to the murray-darling basin, australia [J]. Ecology of Freshwater Fish, 22 (1), 117-126.

（罗钦、李冬梅、黄惠珍 译）

6. Morphological and chemical analysis of archaeological fish otoliths from the Lower Murray River, South Australia.

[英文摘要]：We analysed otoliths from excavations along the Lower Murray River ($n=24$), dating from the mid-to late-Holocene period. We identified the species, and estimated the size and age of fish. The potential habitat that fish used throughout their life was estimated from chemical information in the otoliths. The majority of the fish (identified as Maccullochella peelii $n=22$ and Macquaria ambigua $n=2$) were caught in freshwater environments during the warm season, and had grown to an age and size indicative of sexual maturity. These observations accord with Ngarrindjeri oral tradition concerning sustainable management strategies. Data indicate that M. peelii grew to a significantly larger size than present fish; historical data suggests this size reduction may be the result of European fishing practices, introduced species and habitat degradation. The study demonstrates the unique nature of otoliths and their potential for investigating Indigenous subsistence strategies.

[译文]：

澳大利亚南部墨累河下游考古鱼类耳石的形态和化学分析

我们分析了墨累河下游挖掘出的耳石，这些耳石的年代可以追溯到全新世中晚期。我们确定了物种，并估计了鱼的大小和年龄。鱼类其一生的潜在栖息地是根据耳石中的化学信息估算出来的。在暖季期间，大部分鱼类（确定为墨瑞鳕鱼 $n=22$ 和圆尾麦氏鲈 $n=2$）在淡水环境中被捕获，并且已经成长到性成熟的年龄和大小。这些观测数据符合 Ngarrindjeri 关于可持续管理战略的口述。数据表明，墨瑞鳕鱼能生长到比现在大得多的尺寸；历史数据表明，这种鱼体大小的减少可能是欧洲捕捞活动、引入物种和栖息地退化的结果。该研究表明耳石的独特性质及其调查本土生存策略的潜力。

参考文献：Disspain, M. C. F., Wilson, C. J., Gillanders, B. M. 2013. Morphological and chemical analysis of archaeological fish otoliths from the lower murray river, south australia [J]. Archaeology in Oceania, 47 (3), 141–150.

（罗钦、李冬梅、黄惠珍 译）

7. Multi-scale habitat selection by Murray cod Maccullochella peelii peelii in two lowland rivers.

[英文摘要]：This study provides information on habitat selection by the threatened Murray cod Maccullochella peelii peelii at two spatial scales in the Ovens and Murray Rivers in south-eastern Australia. Both adult (>450mm total length, LT) and age 0 year (<150 mm LT) M. p. peelii selected macro and microhabitats based on structural variables. At the macrohabitat scale, adults selected channel habitats in the river, floodplain channels at high floods and within Lake Mulwala, whereas the floodplain proper was avoided. Adult and age 0 year fish selected similar microhabitats regardless of site or hydrologic conditions, and selection was primarily influenced by the presence of higher loadings of structural woody habitat, higher c. v. in depth, more overhanging vegetation, shallower comparative depths and lower water velocities, closer to the bank. Age 0 year M. p. peelii appeared to select shallower habitats with greater amounts of structural woody habitat, closer to the river bank than adult fish.

[译文]：

墨累-鳕鱼在两个平原河流的多尺度生境选择

这项研究提供了受威胁的墨瑞鳕鱼在澳大利亚东南部的Ovens和Murray河两个空间尺度上选择栖息地的信息。成鱼（总长度>450mm，LT）和年龄0岁（总长度<150mm）的墨瑞鳕鱼都选择了基于结构变量的宏观环境和微生境。在宏观环境中，成鱼选择了河流中的河道栖息地，在高洪水区的漫滩海峡以及能避免洪泛区的Mulwala湖内。成年和0岁年龄的鱼选择类似的微生境，无论地点或水文条件如何，选择主要受结构木质栖息地较高负荷的影响，包括深水区较高的c.v.、更多的悬垂植被、较浅的深度、更低的水流速度，更靠近河岸。Age 0 year M. p. 0岁的墨瑞鳕鱼似乎选择了较多结构性木本的较浅栖息地，比成年鱼更靠近河岸。

参考文献：Koehn, J. D. 2009. Multi-scale habitat selection by murray cod maccullochella peelii peelii, in two lowland rivers [J]. Journal of Fish Biology, 75 (1), 113.

（罗钦、李冬梅、黄惠珍 译）

8. Polymorphic microsatellite loci for species of Australian freshwater cod, Maccullochella

[英文摘要]：The Australian freshwater cod genus, Maccullochella is represented by three species: Murray cod, M. peelii peelii, eastern freshwater cod, M. ikei, and trout cod, M. macquariensis. Seven novel microsatellite loci from M. ikei and six previously published loci from M. peelii peelii were tested on wild populations of Murray, eastern and trout cod. Levels of polymorphism varied between species with 13 loci polymorphic in Murray cod, 9 in trout cod and 7 in eastern cod. Observed heterozygosities ranged from 0.053 to 0.842. This suite of microsatellite loci will facilitate future studies of the genetic status of wild and hatchery bred populations of Maccullochella.

[译文]：

澳大利亚淡水鳕鱼 Maccullochella 物种的多态微卫星基因座

澳大利亚淡水鳕鱼属 Maccullochella 由三种物种代表：墨瑞鳕鱼，东部淡水鳕鱼和鳟鱼鳕鱼。在墨瑞鳕鱼，东部淡水鳕鱼和鳟鱼鳕鱼的野生种群上测试了来自东部淡水鳕鱼的 7 个新型微卫星基因座和来自墨瑞鳕鱼的 6 个先前公开的基因座。多态性水平在不同物种之间有不同的变化，墨瑞鳕鱼有 13 个多态性基因座，鳟鱼鳕鱼有 9 个多态性基因座，东部鳕鱼有 7 个多态性基因座。观察到的杂合度范围为 0.053~0.842。这套微卫星基因座将有助于今后对鳟鱼鳕鱼野生种群和孵化种群遗传状态的研究。

参考文献：Nock, C. J., Baverstock, P. R. 2008. Polymorphic microsatellite loci for species of australian freshwater cod, maccullochella [J]. Conservation Genetics, 9 (5), 1353-1356.

（罗钦、李冬梅、黄惠珍 译）

9. Isolation and characterization of 102 new microsatellite loci in Murray cod, Maccullochella peelii peelii (Percichthyidae), and assessment of cross-amplification in 13 Australian native and six introduced freshwater species

[英文摘要]：We have isolated 102 polymorphic microsatellite loci from an enriched Murray cod DNA library and also assessed their amplification success in 13 native and six introduced freshwater fish species. The loci will serve the dual purpose of assessing wild population genetic structure for future conservation efforts, and for identifying markers for key quantitative trait loci important for aquaculture.

[译文]：

在墨瑞鳕鱼中分离和鉴定 102 个新的微卫星基因座，并评估了 13 个澳大利亚本土和 6 个引入的淡水鱼种的交叉扩增

我们从富集的墨瑞鳕鱼 DNA 文库中分离出 102 个多态微卫星基因座，并评估了它们在 13 种本地和 6 种引入的淡水鱼种中的扩增成功率。该基因座为了达到评估野生群体遗传结构和未来保存工作的双重目的，以及为水产养殖识别关键数量性状基因座的标记。

参考文献：Meaghan, R., Jenny, N., Hayley, M., *et al.* 2010. Isolation and characterization of 102 new microsatellite loci in murray cod, maccullochella peelii peelii (percichthyidae), and assessment of cross-amplification in 13 australian native and six introduced freshwater species [J]. Molecular Ecology Resources, 7 (6), 1258-1264.

（罗钦、李冬梅、黄惠珍　译）

10. Ontogeny of feeding in two native and one alien fish species from the Murray-Darling Basin, Australia.

[英文摘要]：Investigations into the feeding of the early stages of fishes can provide insights into processes influencing recruitment. In this study, we examined ontogenetic changes in morphology and feeding behaviour of two native Australian freshwater species, Murray cod, Maccullochella peelii peelii, and golden perch, Macquaria ambigua, and the alien species, common carp, Cyprinus carpio. Murray cod free embryos are large and well developed at the onset of feeding, whereas the other two species begin exogenous feeding much younger and are smaller and less-developed. Carp commence exogenous feeding 302days earlier than golden perch, and show more advanced development of the eyes and ingestive apparatus. We conducted feeding experiments, presenting larvae of the three species with a standardised prey mix (comprising equal numbers of small calanoid copepods, large calanoid copepods, small Daphnia, and large Daphnia). Larvae of most tested ages and species showed a preference for mid-sized prey (300~500μm wide). This was true even when their gapes substantially exceeded the largest prey offered. Daphnia were consumed more than similar-sized copepods. The results of this study suggest that survival through their larval period will be threatened in all three species if catchable prey<5000208m in width are not available throughout such time. They also suggest that interspecific competition for prey may occur, especially when larvae are very young. The precocious development of structures involved in feeding and the extended transition from endogenous to exogenous feeding of early carp larvae are likely to have contributed to the success of this species since its introduction to Australia.

[译文]：

澳大利亚墨累-达令盆地两种本地鱼类和一种外来鱼类的个体发育

对鱼类早期摄食的调查可以提供对增员过程的了解。在这项研究中，我们研究了两种澳大利亚原生淡水物种，墨瑞鳕鱼和黄金鲈鱼以及外来物种鲤鱼的形态学和摄食行为的遗传变化。在开始摄食时，墨瑞鳕鱼游离胚胎体积大且发育良好，而另外两种开始外源性摄食的物种则更年幼，体型更小，发育较差。鲤鱼开始外源性摄食比黄金鲈鱼早302d，并表现出更加发达的眼睛和摄食器官。我们进行了饲养实验，向三种物种的幼体提供了标准化的猎物混合物（包含相同数量的小型哲水蚤目桡足类，大型哲水蚤目桡足类，小型水蚤和大型水蚤）。经测试的大多数年龄和物种的幼体偏爱中型猎物（宽300~500μm）。即使它们的嘴巴大大超过所提供的最大猎物，情况也是如此。水蚤比相似大小的桡足类消耗得更多。这项研究的结果表明，如果在幼年期没有宽度<500的可捕食猎物，那么所有三种物种的存活将受到威胁。他们还认为可能会发生种间竞争猎物，尤其是当幼体非常小的时候。早期鲤鱼仔鱼摄食结构涉及的早熟发育，从内源性向外源性摄食的延伸转变，都可能是该物种自澳大利亚引进成功的原因。

参考文献：Tonkin, Z. D., Humphries, P., Pridmore, P. A. 2006. Ontogeny of feeding in two native and one alien fish species from the murray-darling basin, australia [J]. Environmental Biology of Fishes, 76 (2-4), 303-315.

（罗钦、李冬梅、黄惠珍　译）

11. Ontogenetic patterns of habitat use by fishes within the main channel of an Australian floodplain river

[英文摘要]：The ontogenetic patterns of habitat use by a community of fishes in the main channel of the Broken River, an Australian lowland river, was investigated. Stratified sampling was conducted fortnightly across six habitat types throughout the spring-summer period within the main channel. As predicted by the 'low flow recruitment hypothesis', backwaters and still littoral habitats were important nursery habitats for most species. These habitats were found to be used by some species throughout all stages of their life cycle, while other species showed clear ontogenetic shifts in habitat preference. Only one species, Murray cod Maccullochella peelii peelii, was never found in backwaters. This study confirms the significance of main channel habitats in the rearing of larvae of some riverine fish species, and emphasizes the importance of considering the habitat requirements of all stages of a fish's life cycle in the management and restoration of rivers and streams.

[译文]：

澳洲洪泛区河主要河道内鱼类栖息地的个体发生模式

调查了位于澳洲布罗肯河下游主要渠道的鱼类栖息地的个体发生模式。在主要渠道内的整个春夏期间，对6种类型栖息地进行每两周一次分层抽样。如"低水流量补充假设"所预测的那样，对大多数鱼类来说死水和沿岸静水栖息地是重要的育苗栖息地。这些栖息地被一些鱼类在其生命周期的各个阶段使用，而其他鱼类清楚地表现出栖息地的改变。只有一种鱼类，墨瑞鳕鱼，从来没在死水中发现过。本研究证实了主要渠道栖息地在养殖一些河流鱼类幼鱼的意义,并强调要考虑在河川的管理和恢复中鱼类生命周期各阶段的栖息地需求的重要性。

参考文献:A. J. King †. 2010. Ontogenetic patterns of habitat use by fishes within the main channel of an australian floodplain river [J]. Journal of Fish Biology, 65 (6), 1582-1603.

(罗钦、李冬梅、黄惠珍 译)

12. A Large-scale, Hierarchical Approach for Assessing Habitat Associations of Fish Assemblages in Large Dryland Rivers

[英文摘要]:Multiple-scale assessments of fish-habitat associations are limited despite the fact that riverine fish assemblages are influenced by factors operating over a range of spatial scales. A method for assessing fish-habitat assemblages at multiple scales is proposed and tested in a large Australian dryland river, the Barwon-Darling River. Six discrete mesohabitat types (large wood, smooth bank, irregular bank, matted bank, mid-channel and deep pool) nested within 10km long river reaches were sampled. Individual reaches were, in turn, nested within four larger geomorphological zones, previously identified along the river. Fish assemblages varied significantly between mesohabitat types and at different spatial scales. Golden perch (Macquaria ambigua), Murray cod (Maccullochella peelii peelii) and common carp (Cyprinus carpio) were strongly associated with large wood, but golden perch and Murray cod exhibited higher habitat specificity than carp. Bony herring (Nematalosa erebi) were more common in shallow edgewater habitats. At the river-scale, regional differences in the fish assemblage occurred at scales closely corresponding to geomorphological zones and these differences were associated with changes in the relative abundance of species rather than the addition or replacement of species. The proposed hierarchical framework improves the efficiency of fish surveys in large rivers by viewing meso-scale fish-habitat associations in the context of larger-scale geomorphological processes.

[译文]:

评估大型旱地河流中鱼群的栖息群落的一种大规模的分级方法

鱼类栖息群落的多尺度评估是有限的,尽管事实上河流鱼群受到一系列超空间尺度操作因素的影响。在一个大型的澳洲旱河巴望-达令河中,建议并测试了一种评估多尺度鱼类栖息地组合的方法。抽取10km河流中出现的分散的6种栖息地(大木材,平整的岸,不规则的岸,暗淡的岸,中间渠道和深水池)。在以前确定的沿着河流的四个较大地貌区内个别河段轮流被筑巢。鱼类的聚集在栖息地类型和不同空间尺度之间差异很大。黄金鲈鱼，墨瑞鳕鱼和鲤鱼与大木材栖息地密切相关，但黄金鲈鱼和墨瑞鳕鱼比鲤鱼具有更高的栖息地专属性。骨鲱鱼（北澳海鳉）在浅水滨栖息地更为常见。在河流尺度上，鱼类聚集的地区差异发生在密切相关的地貌区的尺度上，这些差异与鱼类相对丰度的变化相关，而不是鱼类的添加或替换。推荐的层次框架结构提高了大型河流鱼类调查的效率，通过在较大尺度地貌过程的环境中观察中尺度鱼类栖息地群落。

参考文献：Boys, C. A., Thoms, M. C. 2006. A large-scale, hierarchical approach for assessing habitat associations of fish assemblages in large dryland rivers. Hydrobiologia, 572 (1), 11-31.

（罗钦、李冬梅、黄惠珍　译）

13. Movement and mortality of Murray cod, Maccullochella peelii, during overbank flows in the lower River Murray, Australia

[英文摘要]：Conservation of Murray cod (Maccullochella peelii), a large endangered fish species of Australia's Murray-Darling Basin, relies on a detailed understanding of life history, including movement patterns and habitat use. We used radio-tracking to investigate the movement of 36 Murray cod in main channel and anabranch habitats of the lower River Murray during a flood and associated hypoxic blackwater event. During a flood peak of ~93000ML d^{-1}, dissolved oxygen decreased to 1.2mg L^{-1}. Four movement types were observed：(1) localised small-scale movement, (2) broad-scale movement within anabranch habitats, (3) movement between anabranch and main channel habitats, and (4) large-scale riverine movement. Murray cod exhibited high fidelity to anabranch habitats but also moved extensively between anabranches and the main channel. Fish were consistently located in the main channel or permanent anabranches, suggesting that use of ephemeral floodplain habitats is limited, and highlighting the importance of connectivity between off-channel and main channel habitats. Mortality of radio-tagged fish was considerable (25%) in association with low dissolved oxygen concentrations, indicating that hypoxic blackwater may have had a substantial impact on Murray cod populations in the lower River Murray.

[译文]：

澳洲墨瑞河下游漫滩水域中墨瑞鳕鱼的存活率和死亡率

保护澳洲墨瑞-达令盆地大型濒危鱼类墨瑞鳕鱼，是依靠对其生活史的详细理解，包括活动模式和栖息地的使用。在洪水和关联的缺氧黑水事件期间,我们利用无线电调查了36尾在主要渠道和墨瑞河下游栖息地中的墨瑞鳕鱼的活动。在洪峰期间大约93000ML/d,溶解氧降至1.2mg/L。观察到四种活动类型:(1)局部小规模的活动,(2)分布范围内大规模的活动,(3)支流与主要通道栖息地之间的活动,(4)大型河流的活动。墨瑞鳕鱼集中出现在支流栖息地上,但也在支流和主渠道之间大面积活动。鱼一直位于主渠道或永久支流,这表明河漫滩栖息地的短暂利用是有限的,突出了渠道外和主渠道栖息地连通的重要性。溶解氧低浓度导致无线电标记鱼的死亡率相当大(25%),表明缺氧黑水环境可能对墨瑞河下游中墨瑞鳕鱼的种群有重大影响。

参考文献:Leigh, S. J., Zampatti, B. P. 2013. Movement and mortality of murray cod, maccullochella peelii, during overbank flows in the lower river murray, australia [J]. Australian Journal of Zoology, 61 (2), 160–169.

(罗钦、李冬梅、黄惠珍 译)

14. An Independent Review of the February 2004 Lower Darling River Fish Deaths: Guidelines for Future Release Effects on Lower Darling River Fish Populations.

[英文摘要]:In February 2004 an extensive kill of Murray cod (Maccullochella peelii peelii) occurred on the Lower Darling River between Menindee and Pooncarie. The deaths, which stretched for over 160 km of river, involved predominantly large Murray cod. Landowner and newspaper estimates of numbers of dead Murray cod ranged from 2 dead fish per kilometre of river, to 30 dead per kilometre. An adjusted estimate over the whole affected stretch of river produced a total of 3000 Murray cod deaths (with minimum and maximum estimates of 1000 and 5000 deaths respectively). Immediately prior to the deaths the Lower Darling had ceased flowing and was reduced to a series of remnant pools. During the week of the fish kill maximum air temperature exceeded 40℃, with water temperatures over 30℃. Initial reports indicated large Murray cod died several days after a front of water released from the Menindee Lakes passed down the Lower Darling River. Examination of water quality information, observations made by landholders, and expert opinion from a workshop held in July 2004, suggests the deaths were a consequence of oxygen depletion (due most likely to a combination of extreme temperature, high algal respiration and organic loading), with the possible added stress of sulfide and/or ammonia toxicity released from anoxic sediments in hypolimnetic pools. The ecological significance of this fish kill is discussed in this report, and management guidelines aimed at avoiding similar events are presented.

[译文]：

对2004年2月达令河下游鱼类死亡的独立调查：将来达令河下游鱼类种群释放效果的指导

2004年2月，在达令河下游的下梅宁迪和蓬卡里之间墨瑞鳕鱼出现大量死亡。在延绵超过160km的流中出现死亡，主要是大的墨瑞鳕鱼。本地人和报纸估计墨瑞鳕鱼死亡数为2尾/km到30尾/km。一个调整的估计是整个受影响河段共3000尾墨瑞鳕鱼死亡（死亡数最低和最高估计分别为1000和5000尾）。就在死亡事件发生之前，达令河下游已经断流了，只剩下残留的水池。鱼群死亡的那一周，最高气温超过40℃，水温超过30℃。初步报告显示，大墨瑞鳕鱼是在梅宁迪湖前湖水流到达令河下游后的几天死亡的。本地人提供的水质检验报告和2004年7月举办的研讨会的专家意见表明，这些死亡是缺氧造成的（最可能结合极端温度，藻类耗氧量高和有机污染物负荷大），伴随着可能增加的来自池塘底层还原性沉积物中释放出来的有毒硫化物和/或氨。本报告讨论了这个鱼类死亡事件的生态意义，并提出了旨在避免类似事件发生的管理要点。

参考文献：Ellis, I. M. 2004. An independent review of the february 2004 lower darling river fish deaths: guidelines for future release effects on lower darling river fish populations [J]. Open Access This Report Has Been Reproduce with the Publishers Permission.

（罗钦、李冬梅、黄惠珍　译）

15. Diet and feeding habits of predatory fishes upstream and downstream of a low-level weir

[英文摘要]：A study investigating the influence of a low-level weir on the diets of three migratory percithyid species was undertaken on the Murrumbidgee River, Australia. A combination of fish community sampling and stomach flushing determined that Yanco Weir substantially impacted upon the abundance and feeding habits of golden perch Macquaria ambigua, Murray cod Maccullochella peelii peelii and trout cod Maccullochella macquariensis. The relative abundance of all three species was significantly greater downstream of the weir, where individuals attempting upstream migrations were obstructed. In areas of fish aggregations, some species displayed altered feeding strategies and exploited different prey taxa upstream and downstream of the weir. Diet overlap, and the proportion of individuals with empty stomachs, was also substantially greater from downstream zones. These results suggest that competition among species may be greater in areas of increased predator abundance and some species could be partitioning resources to minimize competitive interactions. Reducing accumulations of predatory species, by providing suitable fish passage facilities, would be an effective means to prevent such trophic interactions occurring at other low-level weirs.

[译文]：

低堰上游和下游的捕食鱼的饮食习惯

在澳大利亚 Murrumbidgee 河上进行了一项调查低级堰对三种迁徙性卵磷脂物种影响的研究。堰鱼类群落采样和胃部冲洗的组合确定了扬科堰显著影响金鲈，墨瑞鳕鱼和鳟鱼鳕鱼的丰度和摄食习性。所有三个物种的相对丰度在堰下游更为显著，因为在堰下游这个区域尝试上游迁移的个体受到阻碍。在鱼群聚集的地区，一些物种表现出改变后的喂养策略，并在堰的上下游开采不同的猎物类群。来自下游地区的饮食重叠和空腹个体的比例也大大增加。这些结果表明，物种之间的竞争可能会增加捕食量丰富的地区，有些物种可能会划分资源以减少竞争性相互作用。通过提供合适的鱼类通道设施来减少掠夺性物种的积聚，将是防止其他低水位堰层发生这种营养相互作用的有效手段。

参考文献：Baumgartner, L. J. 2007. Diet and feeding habits of predatory fishes upstream and downstream of a low-level weir [J]. Journal of Fish Biology, 70 (3), 879-894.

（罗钦、李冬梅、黄惠珍　译）

16. Ontogeny of critical and prolonged swimming performance for the larvae of six Australian freshwater fish species

[英文摘要]：Critical (<30min) and prolonged (>60min) swimming speeds in laboratory chambers were determined for larvae of six species of Australian freshwater fishes: trout cod Maccullochella macquariensis, Murray cod Maccullochella peelii, golden perch Macquaria ambigua, silver perch Bidyanus bidyanus, carp gudgeon Hypseleotris spp. and Murray River rainbowfish Melanotaenia fluviatilis. Developmental stage (preflexion, flexion, postflexion and metalarva) better explained swimming ability than did length, size or age (days after hatch). Critical speed increased with larval development, and metalarvae were the fastest swimmers for all species. Maccullochella macquariensis larvae had the highest critical [maximum absolute 46.4cm s^{-1} and 44.6 relative body lengths (LB) s^{-1}] and prolonged (maximum 15.4 cm s^{-1}, 15.6LB s^{-1}) swimming speeds and B. bidyanus larvae the lowest critical (minimum 0.1cm s^{-1}, 0.3LB s^{-1}) and prolonged swimming speeds (minimum 1.1cm s^{-1}, 1.0 LB s^{-1}).Prolonged swimming trials determined that the larvae of some species could not swim for 60min at any speed, whereas the larvae of the best swimming species, M. macquariensis, could swim for 60 min at 44% of the critical speed. The swimming performance of species with precocial life-history strategies, with well-developed larvae at hatch, was comparatively better and potentially had greater ability to influence their dispersal by actively swimming than species with altricial life-history strategies, with poorly developed larvae at hatch.

[译文]：

六种澳大利亚淡水鱼类幼体的临界游泳性能和长时间游泳性能的个体发育

测定在实验室六种澳大利亚淡水鱼（鳟鱼鳕鱼，墨瑞鳕鱼，金色圆尾麦氏鲈，澳洲银鲈，鲤鱼和墨累河彩虹鱼）幼体的临界（<30min）游泳速度和长时间（>60min）游泳速度。发育阶段（前屈，屈曲，后弯和超变体）比长度，大小或年龄（孵化后的天数）更好地解释了游泳能力。临界速度随着幼体发育而增加，并且超变体是所有物种中最快的游泳者。鳟鱼鳕鱼幼体具有最高的临界值［最大绝对值46.4cm/s和44.6相对体长（LB）/s］和延长值（最大值15.4cm/s，15.6LB/s）的游泳速度和澳洲银鲈幼体的最低临界值（最小值0.1cm/s，0.3LB/s）和长时间游泳速度（最小值1.1 cm/s，1.0LB/s））。长时间的游泳试验表明，某些物种的幼虫不能以任何速度游泳60min，而最佳游泳物种-鳟鱼鳕鱼幼体，可以以44%的临界速度游泳60min。相比性晚熟的鱼种幼体（即孵化期发育不良），性早熟的鱼种幼体（即孵化期发育良好）游泳性能相对较好，并且可能通过积极地游泳具备更大影响其分布的能力。

参考文献：Kopf, S. M., Humphries, P., Watts, R. J. 2014. Ontogeny of critical and prolonged swimming performance for the larvae of six australian freshwater fish species [J]. Journal of Fish Biology, 84 (6), 1820−1841.

（罗钦、李冬梅、黄惠珍　译）

17. The vertebrate fauna of Currawananna State Forest and adjacent agricultural and aquatic habitats in the New South Wales South Western slopes bioregion

[英文摘要]：The New South Wales South Western Slopes bioregion has been significantly altered by agricultural development and is likely to experience additional significant impacts as a result of anthropogenic climate change. A simple inventory survey of the vertebrate fauna of Currawananna State Forest, a small woodland remnant on the Murrumbidgee River in the south of the bioregion, over the period 2002—2010 identified 172 vertebrate fauna species. This included three species of national conservation concern (the Trout Cod Maccullochella macquariensis, Murray Cod Maccullochella peelii and Superb Parrot Polytelis swainsonii) and another 16 species (5 fishes, 8 birds and 3 mammals) of state-level concern, as well as diverse frog, reptile, woodland and wetland bird and microchiropteran bat assemblages. This study demonstrates that even small remnants play an important role in supporting biodiversity in agricultural landscapes.

［译文］：

新南威尔士州南部西部山坡生物区域 Currawananna 国家森林和毗邻的农业和水生栖息地的脊椎动物群

新南威尔士州南部西部山坡生物区域因农业发展而发生重大变化，并可能因人为气候变化而受到更多重大影响。2002—2010 年期间，在 Currawananna 国家森林的脊椎动物区系（生物区南部的马兰比季河上的一个小型林地残区）的简单调查确定了 172 种脊椎动物物种。这包括国家保护关注的三种物种（鳟鱼鳕鱼，墨瑞鳕鱼和超级鹦鹉）以及另外 16 种（5 种鱼，8 种鸟类和 3 种哺乳动物）的国家级动物，以及多样的青蛙，爬行动物，林地和湿地鸟类以及微型蝙蝠组合。这项研究表明，即使是小的残存物在支持农业景观的生物多样性方面也发挥着重要作用。

参考文献：Murphy, M. J. 2012. The vertebrate fauna of currawananna state forest and adjacent agricultural and aquatic habitats in the new south wales south western slopes bioregion [J]. Australian Zoologist, 36 (2), 209-228.

（罗钦、李冬梅、黄惠珍　译）

18. Determining Mark Success of 15 Combinations of Enriched Stable Isotopes for the Batch Marking of Larval Otoliths

［英文摘要］：Chemical marking of otoliths via immersion in solutions of enriched stable isotopesprovides a means of distinctively marking large batches of hatchery-produced fish. Four enriched stable isotopes (barium：137Ba and 138Ba; strontium：88Sr; magnesium：24Mg) were used individually and in combination to determine mark success and the ability to correctly classify 15 unique batch marks in the otoliths of larval Murray cod Maccullochella peelii. Marking with the enriched stable isotopes 137Ba, 138Ba, and 88Sr (individually or in combination) produced clear and distinctive marks (98% mark success) with 93% of fish correctly classified to their respective isotope mark. Despite exposure of the fish to an altered Mg isotope ratio in the water, a corresponding shift in the otoliths was not observed (8% mark success), and many 24Mg-enriched fish were misclassified. Due to the low cost and minimal effects on hatchery protocols, the use of Sr and Ba isotopes to mark hatchery-reared fish at the larval stage has the potential to be a powerful tool in the production and management of a wide range of fish species.

[译文]：

确定成批标记的幼鱼耳石中成功标记的 15 种富集稳定同位素幼鱼耳石

通过浸泡在富集稳定同位素溶液中，对耳石进行化学标记，为显著标记大批孵化场生产的鱼提供了一种方法。四种富含稳定同位素（钡：137Ba and 138Ba；锶：88Sr；镁：单独和组合使用以确定标记的成功和正确分类墨瑞鳕鱼幼体中 15 个独特批号的能力。利用富集的稳定同位素 137Ba，138Ba 和 88Sr（单独或组合）进行标记，产生清晰和独特的标记（标记成功率为 98%），其中鱼群 93% 正确地分类为它们各自的同位素标记。尽管鱼类在水中暴露于改变的 Mg 同位素比例，但没有观察到耳石的相应变化（标记成功率为 8%），许多富含 24Mg 的鱼类被错误归类。由于成本低，对孵化协议影响小，所以在幼鱼阶段使用 Sr 和 Ba 同位素标记人工孵化鱼有可能成为生产和管理各种鱼类的有力工具。

参 考 文 献：Skye H. Woodcock, Bronwyn M. Gillanders, Andrew R. Munro, *et al*. 2011. Determining mark success of 15 combinations of enriched stable isotopes for the batch marking of larval otoliths［J］. North American Journal of Fisheries Management, 31（5），843-851.

（罗钦、李冬梅、黄惠珍　译）

19. Phylogenetics and revised taxonomy of the Australian freshwater cod genus, Maccullochella

[英文摘要]：Determining the phylogenetic and taxonomic relationships among allopatric populations can be difficult, especially when divergence is recent and morphology is conserved. We used mitochondrial sequence data from the control region and three protein-coding genes（1253bp in total）and genotypes determined at 13 microsatellite loci to examine the evolutionary relationships among Australia's largest freshwater fish, the Murray cod, Maccullochella peelii peelii, from the inland Murray-Darling Basin, and its allopatric sister taxa from coastal drainages, the eastern freshwater cod, M. ikei, and Mary River cod, M. peelii mariensis. Phylogenetic analyses provided strong support for taxon-specific clades, with a clade containing both of the eastern taxa reciprocally monophyletic to M. peelii peelii, suggesting a more recent common ancestry between M. ikei and M. peelii mariensis than between the M. peelii subspecies. This finding conflicts with the existing taxonomy and suggests that ancestral Maccullochella crossed the Great Dividing Range in the Pleistocene and subsequently diverged in eastern coastal drainages. Evidence from the present study, in combination with previous morphological and allozymatic data, demonstrates that all extant taxa are genetically and morphologically distinct. The taxonomy of Maccullochella is revised, with Mary River cod now recognised as a species, Maccullochella mariensis, a sister species to eastern freshwater cod, M. ikei. As a result of the taxonomic revision, Murray cod is M. peelii.

[译文]：

澳大利亚淡水鳕鱼属鳕鲈属的系统学和修正的分类学

确定异种群体之间的系统发育关系和分类学关系可能是困难的，尤其是差别不明显并且形态学大致相同。我们使用来自对照区的线粒体序列数据和三个蛋白质编码基因（总共1253bp）以及在13个微卫星基因座确定的基因型，来检查内陆墨累-达令盆地中澳大利亚最大淡水鱼——墨瑞鳕鱼、来自沿岸排水区域的异种姐妹分类群的东部淡水鳕鱼：M. ikei，墨累河鳕鱼和鳟鱼鳕鱼之间的进化关系。系统发育分析为类群特异性进化提供了强有力的支持，有一个进化分支包含东部的类群和相反的单源的墨瑞鳕鱼，这表明 M. ikei 和 M. peelii mariensis 之间比 M. peelii 亚种之前有最接近的共同祖先。这一发现与现有分类学冲突，并且表明鳕鲈属祖先在更新世越过了大分水岭，并随后在东部沿海排水区域中分流。来自本研究的证据，结合之前的形态学和同种异体数据，表明所有现存的分类群在遗传和形态上都是不同的。鳕鲈属的分类标准被修订，玛丽河鳕鱼现在被认为是一种物种，叫作鳟鱼鳕鱼，是东部淡水鳕鱼 M. ikei 的姐妹物种。作为分类修订的结果，墨瑞鳕鱼是 M. peelii。

参考文献：Nock, C. J., Elphinstone, M. S., Rowland, S. J., *et al.* 2010. Phylogenetics and revised taxonomy of the Australian freshwater cod genus, Maccullochella [J]. Marine and Freshwater Research.

（罗钦、李冬梅、黄惠珍　译）

20. Biogeography and life history ameliorate the potentially negative genetic effects of stocking on Murray cod (Maccullochella peelii peelii)

[英文摘要]：Stocking wild fish populations with hatchery-bred fish has numerous genetic implications for fish species worldwide. In the present study, 16 microsatellite loci were used to determine the genetic effects of nearly three decades of Murray cod (Maccullochella peelii peelii) stocking in five river catchments in southern Australia. Genetic parameters taken from scale samples collected from 1949 to 1954 before the commencement of stocking were compared with samples collected 16 to 28 years after stocking commenced, and with samples from a local hatchery that supplements these catchments. Given that the five catchments are highly connected and adult Murray cod undertake moderate migrations, we predicted that there would be minimal population structuring of historical samples, whereas contemporary samples may have diverged slightly and lost genetic diversity as a result of stocking. A Bayesian Structure analysis indicated genetic homogeneity among the catchments both pre-and post-stocking, indicating that stocking has not measurably impacted genetic structure, although allele frequencies in one catchment changed slightly over this period. Current genetic diversity was moderately high (HE = 0.693) and had not changed over the period of stocking. Broodfish had a similar level of genetic diversity to the wild populations, and effective population size had not changed substantially between the two time periods. Our results may bode well for stocking programs of species that are undertaken without knowledge of natural genetic structure, when river connectivity is high, fish are moderately migratory and broodfish are sourced locally.

[译文]：

生物地理学和生活史改善了圈养墨瑞鳕鱼的潜在负面遗传影响

将孵化鱼与野生鱼类种群一同放养，对世界范围内的鱼类物种具有许多遗传影响。在本研究中，利用16个微卫星基因座确定了，在澳大利亚南部五条河流流域近30年的墨瑞鳕鱼放养的遗传效应。放养开始前，将来自从1949年至1954年收集的规模样本中的遗传参数，与放养16~28年后收集的样本，以及来自补充这些集水区的当地孵化场的样本分别进行比较。鉴于五个流域高度连通，成年墨瑞鳕鱼进行中度迁徙，我们预测历史样本的种群构成最少，而现代样本可能略有分歧，并且具有因放养造成的遗传多样性缺失。贝叶斯结构分析表明在放养之前和放养后的流域之间遗传均一性，表明放养对遗传结构没有显著影响，尽管在这一时期一个流域的等位基因频率略有变化。目前的遗传多样性中等偏高（HE=0.693），在放养期间没有变化。亲鱼的遗传多样性水平与野生种群相似，两个时期的有效种群规模没有大的变化。当河流连通性很高，鱼类适度迁移，而且亲鱼来源于当地时，我们的研究结果可能对没有自然遗传结构物种的养殖计划有利。

参考文献：Rourke, M. L., Mcpartlan, H. C., Ingram, B. A., *et al.* 2010. Biogeography and life history ameliorate the potentially negative genetic effects of stocking on murray cod (maccullochella peelii peelii) [J]. Marine & Freshwater Research, 61 (8), 918-927.

（罗钦、李冬梅、黄惠珍　译）

21. Calibration of a rapid non-lethal method to measure energetic status of a freshwater fish (Murray cod, Maccullochella peelii peelii).

[英文摘要]：The energetic status of freshwater fish provides a dynamic measure of their energy balance in response to the environment they occupy. Commercially available microwave technology (the 'energy meter') provides a rapid, non-lethal and inexpensive alternative to traditional laboratory methods for the determination of energy density. The energy meter requires species-specific confirmation of the water-lipid relationship, and comparison of energy meter readings with laboratory-determined estimates of the whole-body energy density. I explored the applicability of the energy meter to the threatened Murray cod (Maccullochella peelii peelii), using both hatchery and wild individuals. Although hatchery and wild fish varied in lipid content, water content and energy density, the parameter comparisons necessary to calibrate the energy meter were statistically consistent between both groups. Subsequently, a robust combined water-lipid relationship was identified for Murray cod, where energy density was strongly related to both water content and lipid content. Average energy meter readings were capable of providing a rapid, non-lethal and accurate assessment of Murray cod energy density. The successful calibration highlights the applicability of the energy meter to provide a dynamic measure of the energetic status of threatened freshwater fish throughout the world.

[译文]：

一种非致命的快速标定方法测量淡水鱼（墨瑞鳕鱼）的能量状态

淡水鱼的能量状态为它们所处的环境提供了其能量平衡的动态量度。商业上可用的微波技术（"能量计"）提供了一种快速，非致命和廉价的替代传统实验室测定能量密度的方法。能量计需要物种特异性的水-脂关系的确认，以及比较能量计读数与实验室测定的全身能量密度估计值。我探究了受威胁的养殖和野生的墨瑞鳕鱼对能量计的适用性。尽管养殖和野生鱼种的脂质含量，含水量和能量密度各不相同，但校准能量计所需的参数比较在两组之间都有统计学一致性。随后，确定了强大的水-脂联合关系，其中墨瑞鳕鱼的能量密度与水含量和脂质含量密切相关。平均能量计读数能够快速，非致命性和准确地评估墨瑞鳕鱼的能量密度。成功的校准强调了能量计的适用性，为全世界受威胁的淡水鱼的能量状态提供动态测量。

参考文献：Whiterod, N. S. 2010. Calibration of a rapid non-lethal method to measure energetic status of a freshwater fish (murray cod, maccullochella peelii peelii) [J]. Marine & Freshwater Research, 61 (5), 527-531.

(罗钦、李冬梅、黄惠珍　译)

22. Movements of Murray cod (Maccullochella peelii peelii) in a large Australian lowland river

[英文摘要]：This study of Murray cod (Maccullochella peelii peelii) movements in a large lowland river in south-eastern Australia indicated that the species was not sedentary, but undertook complex movements that followed a seasonal pattern. While there were sedentary periods with limited home ranges and high site fidelity, Murray cod also under took larger movements for considerable portions of the year coinciding with its spawning schedule. This generally comprised movements (up to 130km) from a home location in late winter and early spring to a new upstream position, followed by a rapid downstream migration typically back to the same river reach. Timing of movements was not synchronous amongst individuals and variation in the scale of movements was observed between individuals, fish size, original location and years.

[译文]：

墨瑞鳕鱼在澳大利亚的大型平原河流中的运动

在澳大利亚东南部大型低地河流中，对墨瑞鳕鱼的运动进行的这项研究表明，该物种并非不爱运动，而是采取了季节性模式的复杂运动。尽管有因家园范围受制和对家园依赖度高的久居时期，但是墨瑞鳕鱼在这一年内与其产卵时间表一致的很大一部分时间里也出现了较大的活动。这通常包括从晚冬和初春的家园到新的上游位置的移动（最多130km），其次是快速下游迁移，通常返回到同一河段。运动时间在个体之间不同步，并且从个体、鱼的大小，原始位置和年份之间观察到运动规模的变化。

参考文献：Koehn, J. D., Mckenzie, J. A., O'Mahony, D. J., et al. 2009. Movements of murray cod (maccullochella peelii peelii) in a large australian lowland river [J]. Ecology of Freshwater Fish, 18 (4), 594-602.

（罗钦、李冬梅、黄惠珍　译）

23. Traceability and discrimination among differently farmed fish: a case study on Australian Murray cod.

[英文摘要]：The development of traceability methods to distinguish between farmed and wild-caught fish and seafood is becoming increasingly important. However, very little is known about how to distinguish fish originating from different farms. The present study addresses this issue by attempting to discriminate among intensively farmed freshwater Murray cod originating from different farms (indoor recirculating, outdoor floating cage, and flow through systems) in different geographical areas, using a combination of morphological, chemical, and isotopic analyses. The results show that stable isotopes are the most informative variables. In particular, delta (13) C and/or delta (15) N clearly linked fish to a specific commercial diet, while delta (18) O linked fish to a specific water source. Thus, the combination of these isotopes can distinguish among fish originating from different farms. On the contrary, fatty acid and tissue proximate compositions and morphological parameters, which are useful in distinguishing between farmed and wild fish, are less informative in discriminating among fish originating from different farms.

[译文]：

不同养殖鱼类的溯源性和辨别性：澳大利亚墨累鳕鱼的案例研究

开发可追溯性方法来区分养殖和野生捕捞的鱼类和海鲜正变得越来越重要。然而，关于如何区分来自不同渔场的鱼，知之甚少。本研究试图通过形态学，化学和同位素分析的组合，在不同的地理区域中尝试区分源自不同养殖地区（室内循环，户外漂浮笼和通过系统的流动系统）集约化养殖的淡水墨瑞鳕鱼。结果显示稳定同位素是最具说服力的变量。特别是δ(13)C 和/或δ(15)N 将清楚地区分特定商业食用鱼，而δ(18)O 将鱼追溯到特定的水源。因此，这些同位素的组合可以区分来自不同养殖区域的鱼。相反，有利于区分养殖鱼和野生鱼的脂肪酸、组织的构成和形态参数，却不利于区分源自不同养殖地区的鱼。

参考文献：Turchini, G. M., Quinn, G. P., Jones, P. L., et al. 2009. Traceability and discrimination among differently farmed fish: a case study on australian murray cod [J]. Journal of Agricultural & Food Chemistry, 57 (1), 274-281.

（罗钦、李冬梅、黄惠珍 译）

24. Using radio telemetry to evaluate the depths inhabited by Murray cod (Maccullochella peelii peelii)

[英文摘要]：Radio telemetry is widely used in studies of freshwater fishes, but the vertical position of fish in riverine environments is rarely reported. The present study tested the application of radio transmitters fitted with depth sensors to determine the vertical position of Murray cod in the lower Ovens River in south-eastern Australia. As the scale of depths in rivers is usually limited (<10m in the present study), there is a greater need to assess measurement error. The study first involved trials to define depth measurement errors, and a mean relative bias of 9% (range 1.5~14.8%) towards greater depth was recorded. These data were then used to correct the depths recorded from tagged fish. Although data from this preliminary study are somewhat limited, results from the tagged fish showed that by day they all occupied the lower 15% of the water column, indicating that Murray cod exhibit demersal behaviour, using bottom rather than mid-water habitats. Although the present study highlights the importance of tag trials in determining errors, it also indicates the potential application of this technique to understanding the depth-integrated habitat preferences of Murray cod and other species.

[译文]:

使用无线电遥测来评估墨瑞鳕鱼（Maccullochella peelii peelii）栖息的深度

无线电遥测广泛用于淡水鱼的研究，但很少报道鱼类在河流环境中的垂直位置。目前的研究测试了配备深度传感器的无线电发射机，以确定澳大利亚东南部较低的 Ovens 河中墨瑞鳕鱼的垂直位置。由于河流深度的范围通常是有限的（本研究中<10m），因此更需要评估测量误差。该研究首先涉及确定深度测量误差的试验，记录了向更大深度的平均相对偏差 9%（范围 1.5%~14.8%）。这些数据然后用于校正从标记的鱼记录的深度。虽然这项初步研究的数据有限，但标记鱼的结果表明，白天它们都栖息在水位的 15% 以下，表明墨瑞鳕鱼的底部行为，是使用水层底部而不是中部水域栖息地。虽然本研究强调了标记试验在确定错误方面的重要性，但它也表明了该技术可用于探索墨累鳕鱼和其他物种的深度综合栖息地偏好。

参考文献：Koehn, J. D., 2009. Using radio telemetry to evaluate the depths inhabited by murray cod (maccullochella peelii peelii) [J]. Marine & Freshwater Research, 60 (4), 317-320.

（罗钦、李冬梅、黄惠珍　译）

25. 澳洲龙纹斑生物学特征及其繁养殖技术研究进展

摘要：澳洲龙纹斑 Maccullochella peelii peelii 是原产于澳洲墨瑞河的一种经济鱼类，是澳洲乃至世界上品质优良及种群类型较大的淡水鱼类之一。近年澳洲龙纹斑已经引入中国并开始驯化开发。其作为一项新兴的水产养殖产业，给人们带来新的开发途径，但开发者也将面临许多新的技术问题，如何有效地人工繁殖，保障健康养殖与有效防控病害等。本文就澳洲龙纹斑的生物学特性、人工繁殖养殖技术、经济价值和养殖技术以及疾病防控等方面进行综述介绍，同时结合先期的养殖实践，提出若干发展对策与建议。

Biological Characteristics, Propagation and Aquaculture of Murray Cod (Maccullochella peelii)

Abstract: Murray cod (Maccullochella peelii) is a famous economic fish originally grown in the Murray River in Australia. It is one of the largest freshwater fish in the world, and was recently introduced to China for studies. In domesticating the fish, detailedbreeding, propagating, cultivating and disease preventing/controlling operations must firstly be established. Thus, this article summarizes the relevant information regarding the biology, aquaculture, economics, marketing and disease management of Murray cod. In addition, some developmental strategies for the aquaculture are presented based upon previous experiences with similar fishes.

参考文献：翁伯琦，罗土炎，刘洋，等. 2016. 澳洲龙纹斑生物学特征及其繁养殖技术研究进展［J］. 福建农业学报，31（1）：89-94。

26. 澳洲鳕鲈的生物学特征及人工繁养技术

摘要：概述了澳洲鳕鲈的生物学特性以及人工繁育研究进展情况，为今后开展澳洲鳕鲈规模化人工养殖研究提供参考。

参考文献：郭松，王广军，方彰胜，等. 2012. 澳洲鳕鲈的生物学特征及人工繁养技术［J］. 江苏农业科学，40（12）：242-243。

注：无英文摘要。

27. 虫纹鳕鲈的生物学特性及人工养殖技术研究

摘要：虫纹鳕鲈（Maccullochella peelii peelii）是原产于澳洲墨累河流域的一种著名经济鱼类，是澳洲乃至世界上最大类型的淡水鱼类之一。对虫纹鳕鲈的生物学特性、人工养殖技术、经济价值和养殖前景进行了介绍。

Biology Characteristics and Artificial Breeding Techniques of Murray Cod（Maccullochella peelii peelii）

Abstract：Murray cod（Maccullochella peelii peelii）is a famous economic fish originally growing in Astralian Murray river. It is one of the biggest freshwater fish in Australia and even in world. This parer introduced the biology characteristics, artificial breeding techniques, economic value and market prospect of Murry cod.

参考文献：李娴，朱永安，钟君伟，等. 2013. 虫纹鳕鲈的生物学特性及人工养殖技术研究［J］. 湖北农业科学，52（9）：2114-2115。

28. 利用mtDNA COI基因序列分析引进的澳洲虫纹鳕鲈群体遗传多样性

摘要：根据线粒体COI基因序列分析了虫纹鳕鲈（Maccullochella peelii）引进群体的遗传多样性。用PCR技术扩增了线粒体C01的序列，PCR产物经纯化、克隆和测序后得到了652bp的核苷酸片段。运用CLUSTALX（1.83）软件比对了25个个体的序列，共检测到5个多态位点,其中4个为转换位点,1个为颠换位点;运用MEGA5.0软件构建了NJ系统树;用DNASP软件计算出的单倍型个数(H)为5,单倍型多样性(Hd)为0.300,核苷酸多样性(Pi)和平均核苷酸差异数(k)为分别为0.00073和0.473。结果表明,引进群体的虫纹鳕鲈mtDNAC01基因序列的变异程度较小,群体内遗传多样性较低。

Genetic Diversity of Introduced Murray Cod (Maccullochella peelii) Population Based on mtCOl Gene

Abstract:The genetic diversity of murray cod (Maccullochella peelii) introduced form Australia was studied by mtCOl gene se-quences to explore the genetic structure of the fish. The mtDNA COI gene was amplified in 25 individuals by PCR whose products Were purified, cloned and sequenced. The 652bp nucleotide sequences of partial COI gene were found, and the sequences of these sam-ples were compared by CLUSTAL X (1.83) each other. Five variation sites were observed including 4 transition sites and 1 transver-sion site. The NJ phylogenetic trees were constructed by MEGA 5.0 soft-ware. As estimated by DNASP, the haplotypes (H) was found to be 5 in number, the haplotype diversity (Hd) 0.300, the nucleotide diversity (Pi) 0.00073and the average number of nucleotide dif-ferences (k) 0.473. It can be concluded that the mtCOI gene of the introduced Maccullochella peelii population has relatively low level of variation and poor in genetic diversity.

参考文献:张龙岗,杨玲,李娴,等.2013.利用mtDNA COI基因序列分析引进的澳洲虫纹鳕鲈群体遗传多样性[J].水产学杂志,26(2):14-18。

29. 澳洲龙纹斑鱼种生长特性研究

摘要:在福建山区养殖条件下,进行澳洲龙纹斑Maccullochella peelii peeli鱼种的饲养试验。澳洲龙纹斑鱼种的平均体重和全长分别为(1.29±0.23)g和(4.56±0.30)cm;试验期水温22~28℃,投喂粗蛋白51.2%的商品饲料。经97d培育,澳洲龙纹斑鱼种的平均体重和全长分别达(12.84±1.30)g和(10.03±0.36)cm。澳洲龙纹斑鱼种体重和全长特定生长率分别为2.37%和0.81%;该鱼种体重与全长的最优回归方程为幂函数方程$W=0.016L^{2.884}$。以Gompertz非线性生长模型算得澳洲龙纹斑鱼种体重和全长生长拐点分别为11.8g和8.5cm,拐点日龄分别为170d和149d;全长生长拐点日龄的出现先于体重生长拐点。

Growth Performance of Juvenile Murray Cod, Maccullochella peelii peelii

Abstract:Feeding trials onjuvenile Murray cod, Maccullochella peelii peelii (Mitchell), were conducted in a mountainous area of northwest Fujian Province. The initial averagebody weight andthe total length of the juveniles were (1.29±0.23)g and (4.56±0.30)cm, respectively. The water temperature of the aquaculture pond ranged between 22℃ and 28℃. The fish were fed with acommercial feed (CP=51.2%). After 97 days of cultivation, the average body weight and total length of tjavascript：scrollTo (0, 0); he codincreased to (12.84±1.30)g and (10.03±0.36)cm, respectively. The daily specific growth rates on weight and length were 2.37% and 0.81%, respectively, with a mean condition factor of (1.35±0.26). The weight vs. length relationship was W=0.016L2.884. According to the Gompertz non-linear growth model, the growth inflection points on body weight and length were calculated to be 11.8g and 8.5cm, respectively; and, the day age of the growth inflection points onbody weight (170d) lagged behind that on the body length (149d) for the juvenile Murray cod.

参考文献：罗钦，罗土炎，林旋，等.2016.澳洲龙纹斑鱼种生长特性研究［J］.福建农业学报，31（1）：7-11。

30. 虫纹麦鳕鲈的形态和生物学性状

摘要：观察虫纹麦鳕鲈的外形，结合内脏解剖，描述其主要形态特征。它的身体延长，稍侧扁，呈纺锤形；口裂大，端位，下颌略长于上颌。背鳍2个，第一背鳍具11根棘，腹鳍前胸位，尾鳍圆形，体侧背部褐色，近腹部颜色渐淡，至白色。体表布有具黑缘的虫纹小斑块。胃发达，呈"Y"形，幽门盲囊2~4个，胆囊卵圆形，肠管短。

参考文献：王波，张艳华，韩茂森.2003.虫纹麦鳕鲈的形态和生物学性状［J］.水产科技情报，30（6）：266-267。

注：无英文摘要。

31. 虫纹鳕鲈线粒体COI基因片段的克隆与序列分析

［摘要］对虫纹鳕鲈（Maccullochella peelii）线粒体DNA细胞色素氧化酶Ⅰ亚基（cytochrome oxidase subunit I, COD基因进行了扩增、克隆和测序，并对COI序列进行了分析。结果显示，澳洲虫纹鳕鲈COI基因序列长度为652bp，A、C、T、G4种碱基平均比例分别为25.3%、16.6%、329%、252%，A+T比例（58.2%）明显高于G+C（41.8%）。与从GenBank中查到的其他6种真鲈科鱼类的同源序列比对，并采用最大简约法（MP）构建系统进化树，结果显示：7种真鲈科鱼类聚在一起，分为2个大的支系，麦鳕鲈属鱼类聚为一支，麦氏鲈属、鳜属和花鲈属的鱼类聚为一支。其中，花鲈属的花鲈与鳜属鱼类亲缘关系较近，与其他两属关系稍远。所得的聚类结果与传统的形态

分类基本一致。

参考文献：张龙岗，杨玲，张延华，等.2012.虫纹鳕鲈线粒体COI基因片段的克隆与序列分析.长江大学学报（自然科学版），9（8）：25-29。

注：无英文摘要。

第二章　澳洲龙纹斑繁育与养殖

第一节　专题研究概况

澳洲龙纹斑是原产于澳大利亚墨瑞河流域的一种著名淡水经济鱼类。该鱼为白肉鱼，在澳洲排名四大经济鱼类之首，是世界上最大的淡水鱼之一，与宝石鲈、黄金鲈、银鲈等原产于澳洲的淡水鱼并列为澳洲鱼。由于强度捕捞和环境污染及产卵场被破坏，澳洲龙纹斑资源量日益减少，而市场的需求量日益增加。进行澳洲龙纹斑的人工繁殖，并增加其鱼类资源已摆上人们的议事日程（陈永乐，1999）。

一、鱼苗繁殖

（一）自然繁殖

在澳大利亚，澳洲龙纹斑在自然水域可自行产卵孵化，繁殖季节通常是一尾雌鱼与多尾雄鱼在安全隐蔽处或有遮盖物的水域交配。水温21℃以上可以产卵，产卵后，雄鱼驱散雌鱼，担负护卵重任，时而游来游去，以使水流畅通，确保孵化顺利进行直至出苗（韩茂森，2003）。

在受监管的墨瑞河和附近的无管制的奥文斯河上，连续3年多对澳洲龙纹斑产卵的环境条件和时间进行了调查。从11月初开始在河流样品中收集幼鱼。澳洲龙纹斑幼鱼出现的时间长达十周。在墨瑞河上1995/6年的幼鱼明显比1994/5年的大，澳洲龙纹斑幼鱼为9.5~14.8mm。幼鱼大小与水温之间没有任何关系，但后来在墨瑞河上游河段的产卵时间与较低水温是一致的（Koehn，J. D.，2006）。

Humphries，P.（2005）调查了澳大利亚墨瑞-达令盆地南部布罗肯河中澳洲龙纹斑早期生活史的各部分。从1997到2001年间的10月中旬到12月中旬的每个繁殖季节，记录了2个河段的澳洲纹斑成鱼的丰度、长度、年龄和游离胚胎卵黄的数量和估计的产卵期。自由胚胎在 10 月底开始游离,而且一直持续到 12 月中旬至下旬。游离的自由胚胎的丰度在大多数年份没有明显的峰值,与排卵量无关。长度、卵黄量和年龄通常随着取样变化而变化,但只有年龄与温度呈强烈的负相关。因此,看上去好像是温度影响发展速度,而且那个发育阶段的决定性因素可能是自由胚胎离开雄鱼"巢穴"的时间而不是长度或年龄。大多数自由胚胎预计花费 5 到 7d 时间漂流。在布罗肯河,澳洲龙纹斑的产卵通常在 10 月中旬开始,一直持续到至少 12 月初。产卵需在 15℃ 及以上的温度,但排卵时高度变化,历年来仅此环境变量是一致的。产卵中期,通常在 11 月的第 1 周,被认为是一个更重要的时间,因为它可能与产卵高峰期相一致,而且对于存活的卵子和自由胚胎,在这段时间内和之后的条件预计是最佳的(Humphries, P., 2005)。

在澳大利亚的沃加沃,Greg Semple 于 1967 年在当地建立了首家澳洲龙纹斑孵化场,Located at Wagga Wagga。利用自然刺激成熟的亲鱼,育苗塘中的产卵盒和稚鱼天然浮游生物,使新南威尔士州的孵化场能够培育出高质量的澳洲龙纹斑鱼苗(Austasia Aquaculture, 2002)。

澳洲龙纹斑已经在野外放流计划中繁育了 30 多年。通过使用八个微卫星标记确定连续三个繁殖季节收集的 46 个单独产卵中的 1380 个后代亲本评估了这种野外放流计划种群的潜在遗传影响,并通过人口统计学和遗传方法估计其亲鱼年代的有效种群规模。结果显示多配偶制产卵(一雄多雌和一雌多雄)意想不到的情况,一个季节内两性的多次产卵,以及多个季节之间的鱼的重复交配。此外,在 3 年以上的研究期间,大约一半的亲鱼没产卵(Rourke, M. L., 2009)。

在 2012—2013 年,在马兰比季河 76km 的范围内,澳洲龙纹斑捕获率(鱼/每垂钓小时)分别为 0.228±0.047(平均值±SE;以船为基准)和 0.092±0.023(以河岸为基准),收获率(鱼/垂钓者小时)分别为 0.013±0.006(以船为基准)和 0.003±0.001(以海岸为基准)。在休渔期结束后的 5 个月,澳洲龙纹斑渔场捕获量和收获量都最大(Forbes, J. P., 2015)。

在墨瑞河(MR1 and MR2)和奥文斯河(OR)上的鱼苗采样位置见图 2-1。

(二)人工繁殖

在澳大利亚,经多年研究澳洲纹斑也可进行人工繁殖。近十年来人工繁殖生产苗种的技术更有把握(陈永乐,1999)。亲鱼可来自人工养殖而成或从野外捕回。捕回的野生性成熟亲鱼要经过 3 个月以上的培育。因为亲鱼经捕捞后,性腺发育要经过退化吸收的过程。除非即时催产注射。实验证明,繁殖前在培育池捕起 2 尾亲鱼作标记时,其卵

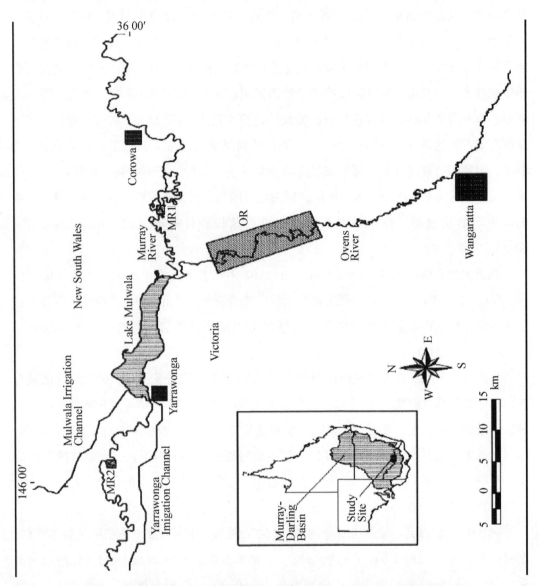

图 2-1 在墨瑞河（MR1 and MR2）和奥文斯河（OR）上的鱼苗采样位置
(JOHN D. KOEHN, 2005)

巢含有卵黄初级卵母细胞。在放回 14d 后重新捕起，其卵巢内有非常多的血、油和黏液及许多形状不一的半透明卵。而在同一池内未捕起处理的 4 尾雌鱼卵巢里没有萎缩卵母细胞的存在。通常情况下 4~5 龄体重为 2.5~4.0kg/尾可达性成熟。春季水温 20℃ 以上时注射 HCG 催产剂，保持水温 21℃，经 48~50h 即可产卵，卵为粘性，卵径平均 3.0~3.5mm。以特制的集卵箱收集受精卵。卵箱长 110mm，高 500mm 呈六角形。将卵箱移至保温、水流畅通的孵化槽中孵化，在 20~22℃ 水温条件下，受精卵经 5~7d 孵化出膜，孵化率可达90%。刚出膜的苗体长5.0~8.0mm，卵黄囊清晰可见，过8~10d后卵黄囊吸收完毕。体重2.5~3.0kg的雌鱼可产卵10000粒，体重5.0kg的雌体产卵14000~30000粒，体重23kg的雌鱼可产卵90000粒。随着外源激素（HCG、CPG）诱导鱼类产卵技术的应用和完善，在澳大利亚HCG和CPG已成功应用于澳洲龙纹斑的人工繁殖（图2-2至图2-5），孵化率可高达70%~90%（陈永乐，1999；韩茂森，2003）。

图2-2　澳洲龙纹斑（Jonathan Daly，2008）

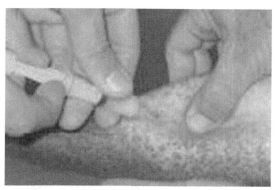

图2-3　人工取精（Jonathan Daly，2008）

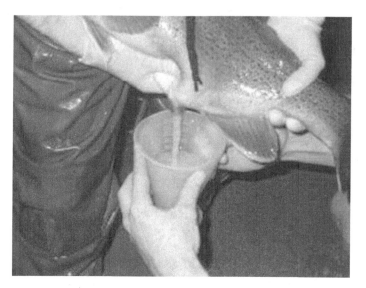

图2-4　人工取卵（Jonathan Daly，2008）

（三）人工繁殖关键技术

1. 亲鱼选择

Danem，N.（2008）用5MHz线性换能器，对组织学的性腺评估与超声图像进行比较。从所得图像中进行性别解释，并记录最大横截面性腺直径和面积，总的性别鉴别准

图 2-5 人工授精（Jonathan Daly，2008）

确率为96%。每个发育阶段卵巢的超声图像在视觉上与众不同，并且能够定性评估成熟度，从而补充定量 GI 评估（Danem，N.，2008）。

经 PHOTOTHERMAL 和 PHOTOPERIOD 处理的雌性直到 3 个月（六月份）和 4 个月（五月份）才分别达到成熟，超过对照组处理的雌鱼（十月份）。4 尾 PHOTOPERIOD 处理的雌鱼（13%）中也观察到一年成熟 2 次。成熟雌性的卵巢平均直径和相对繁殖力在试验组间是相似的（$p>0.05$），试验组中都排出了可用的鱼卵（100%排卵；14.02%~39.12%的平均孵化存活率）。相移的 E2 和 T 的卵巢直径和血浆水平保持基础水平和/或相对于对照组处理的贯穿成熟期的早期至中期的鱼显著降低（$p<0.05$）。然而，在相移雌性成熟开始前的 60 至 90d 内，血浆 T（0.54~4.39ng/ml 611）和卵巢直径（20.0~42.4mm）的快速增加表明了一个澳洲龙纹斑通过加速发育进程来补偿早期光敏响应延迟的能力。持续贯穿整个 PHOTOTHERMAL 和 PHOTOPERIOD 处理的雌鱼的成熟期的 E2 的低水平似乎没有显著地影响卵巢的生长。在恒定温度下对成熟雌性成功的光周期控制证明了这种方法可以提高亲体调节程序的通用性和成本效益（Danem，N.，2010）。

青春期鱼的卵巢发育在皮层小泡和脂滴产卵期之间有快速转变，与 2 龄以上雌鱼中血浆 17β-雌二醇显著性降低（p<0.05）相一致。来自 2 龄以上雌鱼的卵的平均成熟卵母细胞直径（2.44mm），受精后存活率（30.80%）和孵化率（0.99%）明显地比 3 龄以上成熟鱼的参数低（分别为 2.81mm，84.89% 和 23.58%）。青春期澳洲龙纹斑卵巢表现出卵黄化和排卵能力，然而，第 2 级卵母细胞生长功能异常可能导致卵子受精能力下降，因此，以高级卵巢阶段出现为基础的初始成熟期的评估可能高估了年轻的亲鱼种群的繁殖潜力（Newman，D. M.，2008）。

选择无病、无伤、体色正常而有光泽的健壮个体作为人工繁殖的亲鱼，雄鱼体重在 2~4kg，雌鱼体重为 4~5kg，雌雄比例为 2∶1。性腺发育良好的雌鱼腹部柔软，生殖孔突出有红肿，轻压溢出卵黄，呈黄色。（陈永乐 1999）用以人工繁殖的亲鱼全长 60~107cm，体重 3.4~31kg，年龄达 6 龄。当水温达 20℃时，从池塘捕起的成熟亲鱼腹部膨胀，生殖孔由微红色变为深红或紫红色，用直径 4mm 玻璃管吸出的卵子经镜检（×20），形状呈匀称的圆球状，直径约 3mm，琥珀色，半透明，单独的油球，均布在细胞质的卵黄粒清晰可见。而核未偏位的卵母细胞，不透明，白色，卵径 < 1.5mm。轻压性腺发育良好的雄鱼腹部，有乳白色精液流出。挤压雄鱼腹部流出来的精液，用生理盐水稀释后镜检（×200），检查其精子的相对密度和粘度。仅使用催产后具非常活力精子的雄鱼。考虑澳洲龙纹斑亲鱼个体较大，每次操作均用 25mg/L 奎钠丁（Quinaldine）溶液麻醉。操作后用放回池塘（陈永乐，1999；郭松，2012）。

2. 亲鱼培育

澳洲龙纹斑是肉食性鱼类，加之性成熟的澳洲龙纹斑亲鱼有区域性，培育密度以每公顷 50 尾为宜。性成熟的亲鱼须经过强化培育，通常在秋季和春季进行强化培育半年左右。强化培育过程中，一是要有充足的饵料鱼或者增加人工配合饲料的脂肪含量，二是要定期冲水以保证良好的水质。饲喂的活饵料有金鱼等小型鱼类（饲养方式类似于我国的鳜鱼）（陈永乐，1999；郭松，2012）。

3. 人工催产

Newman DM（2007）在模拟自然光热条件下，在澳洲龙纹斑 12 个月的卵巢成熟周期中，检查了卵母细胞生长，血浆性类固醇和体细胞能量储存的时间动态。除了卵黄卵泡无法进行最终的成熟，卵巢功能在被捕的鱼中相对不受影响，导致普遍的排卵前畸形。卵母细胞生长的季节性特征是在 3 月皮质肺泡积累，4 月份油脂沉积，5 月至 9 月卵黄生成。在澳洲龙纹斑卵巢中发现了两个不同批次的卵黄卵母细胞，表现出能多产卵的能力。17β 型-雌二醇和睾酮的血浆分布这 2 个在整个成熟期是高度可变的，表明这些激素在卵母细胞生长的不同阶段存在多重角色。在卵黄生长之前或生长鼎盛阶段发现肥满度、肝脏大小和内脏脂肪量都有增加。在繁殖周期之前，早于卵巢快速生长期，澳洲龙纹斑似乎有策略地提高摄食活动来积累能量储备（Newman DM，2007）。

澳洲龙纹斑卵母细胞和卵母细胞发育阶段的剖面图见图2-6。

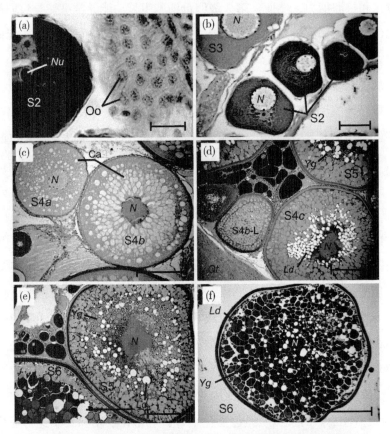

图2-6 澳洲龙纹斑卵母细胞和卵母细胞发育阶段的剖面图

（a）卵母细胞（Oo）和早期的卵母细胞核周围（S2）.（b）早期-（S2）和后期-（S3）卵母细胞核周围．（c）早期-（S4a）and中期（S4b）卵母细胞皮层小泡．（d）中期-（S4a）和后期-（S4b-L）卵母细胞皮层小泡，脂肪滴阶段卵母细胞（S4c），和早期卵黄的卵母细胞（S5）．（e）早期-（S5）和后期-（S6）卵黄的卵母细胞．（f）后期卵黄的卵母细胞（S6）. Nu，核仁；N，细胞核；Bb，卵黄核；Ld，脂肪滴；Yg，卵黄颗粒；Ot，卵巢膜．比例尺：10 μm（a），50μm（b），200 μm（c），300μm（d，e），1000μm（f）. 注：核仁染色质阶段的卵母细胞（S1）未显示．（Dane M. Newman，2007）

经历早期畸形的第6阶段卵母细胞的横切面见图2-7。

澳洲龙纹斑是澳洲维多利亚州东北部米特米塔河的本土鱼；然而，自1992年以后，这条河就没有找到常住的澳洲龙纹斑。研究表明没有卵和幼鱼能够存活在低13℃环境

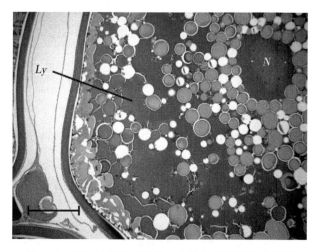

图 2-7　经历早期畸形的第 6 阶段卵母细胞的横切面

Ly，液化的卵黄；N，细胞核．比例尺：300μm.（Dane M. Newman，2007）

下，冷水排放过程对产卵后存活的影响是对澳洲龙纹斑种群生存能力的一个重大威胁（Todd，C. R.，2010）。

精子在 300mOsmol/kg 611 以上的离子和非离子溶液中保持失活。在淡水中离心和再悬浮后精子在 300 和 600mOsmol/kg611 溶液中的活力比在 900mOsmol/kg611 溶液中的好。在冷冻保护剂，二甲基亚砜，二甲基乙酰胺和甲醇的存在下，与 600mOsmol/kg611 相比，300mOsmol/kg611 的回收率更好。在含有甘油的所有介质中稀释的精子在水中离心和再悬浮后保持失活。由含有 10%甲醇的 300mOsmkg611d-山梨醇（DS）溶液组成的冷冻保存稀释液产生最好的解冻后活力（51.4%±3.4%），其次是含 10%甲醇的 Tris-蔗糖-钾（TSK）溶液（32.9%±4.7%）和含 10%甲醇的改性黑素培养基（27.5%±4.8%）。细胞膜全荧光染色（SYBR-14 和碘化丙啶）反映出解冻后活力的结果。使用在含 10%甲醇的 d-山梨醇溶液中冷冻精子的受精试验产生了 11.0%的孵化率（对照新鲜精子孵化率为 63.1±18.23%）和含 10%甲醇 Tris-蔗糖-钾溶液中产生了 6.4%的孵化率（对照新鲜精子孵化率为 58.5%±32.50%）。这标志着首次在澳洲龙纹斑精子中成功的冷冻保存（Jonathan，D.，2008）。

澳洲龙纹斑的平均渗透压分别为 221mOsmol/kg。精子活率在 300 和 600mOsmol/kg 下被抑制，在 900mOsmol/kg 下，恢复率显著降低。冷冻保护剂的加入降低了精子运动能力的恢复，但是澳洲龙纹斑在 300mOsmol/kg 培养基中的恢复更好。澳洲龙纹斑的解冻后活力最高为 51.4%。在受精试验中，澳洲龙纹斑使用冷冻保存精子的最佳孵化率（以对照的百分比表示）为 63.1±18.2%。使用 TSK+10%v/v 甲醇或 DS+10%v/v 甲醇实现了澳洲龙纹斑的相对高的受精率。这项工作代表了第一次在澳洲龙纹斑上成功使用冷冻保存的精子，并且是第一次成功研究出完全基于添加 D-山梨醇的冷冻保护剂的精液冷冻保存稀释剂的报告（Daly, J., 2008）。

人工催产季节通常在 5—8 月，当亲鱼成熟时，可进行人工催产，注射的药物有 HCG、LRH-A 和脑垂体，通常采用混合注射，药物用生理盐水配制，即配即用。用量依亲鱼体重而定，雄鱼剂量是雌鱼剂量的一半，最大水溶液 5mL/尾。均采用胸鳍基部分 2 次注射效果较好。在左腹鳍基部处作腹腔注射。使用 21G×38 针头。亲鱼注射通常是在池塘捕起的 1h 内，野外捕起的 2h 内完成。注射后亲鱼放在 2000L 圆池内，池上方用黑色聚乙烯布覆盖一半，水体保持充气和水温恒定。使用自然光。水温 26℃时，在 20h 左右的效应时间内即可发情产卵，产卵池流速不低于 20cm/s（陈永乐，1999；郭松，2012）。

使用 HCG100~750IU/kg 未能使澳洲龙纹斑产卵；800~900IU/kg 的，3 尾中有 2 尾产卵，这些亲鱼在注射后 49 或 49.5h 挤卵授精的孵化率是 90% 和 87%；而在 48.5 或 50h 挤卵授精的分别是 50% 和 70%。使用 1000 或 2000IU/kg 剂量都可使所有亲鱼排卵，这些亲鱼在注射前其卵巢样本均含有充满卵黄的初级卵母细胞，在 21±1℃ 水温下，剂量 HCG1000IU/kg 的，47h 开始排卵，48.5~49.5h 挤卵的平均孵化率为 79.3%，显示出在排卵后 2~3h 挤卵并授精的，孵化率高。剂量 2000IU/kg 的，注射后 46h 开始排卵，但孵化率最高只有 40%，表明了卵质可能是随着 HCG 剂量增加而衰退的。单独使用 CPG，每千克用 2、3、4 和 5mg 都有不同的反应。剂量 2mg/kg 和第 45h 挤卵的可使有 1 尾雌鱼的孵化率达 86%。而另 1 尾是 5%。在另外 1 尾亲鱼注射 CPG2mg/kg 后分别在 46、47 和 48h 挤卵授精，其孵化率从 80% 下降到 69% 和 42%。同样，剂量 3 和 5mg/kg，在第 45 和 46h 挤卵的 4 尾亲鱼的孵化率分别为 75%、3%、85% 和 3%。使用 CPG 时，应考虑粗制 CPG 制剂有效激素含量的不稳定性（陈永乐，1999）。

至于 HCG 和 CPG 混合使用的效果，如以 HCG200IU+CPG5mg/kg 计，注射后 45~46h 挤卵授精的孵化率高，达 70%~89%，但到 47h 才挤卵的只有 1%。诱导产卵成功与否，首先取决于卵巢发育的准确评估，只有卵黄积累完全的亲鱼对催产激素才有应答性反应。其次对催产药物的种类和剂量，有较强的种类属性。就澳洲龙纹斑而言，似使用 HCG900~1000IU/kg 或 CPG2~5mg/kg 为宜，加上适当的挤卵授精时间，就能得到理想的繁殖效果。因为不同鱼类卵子的成熟分裂也具有不同种的属性，此期间是卵子为精子进入作准备。在澳洲龙纹斑排卵后 1h 内或 4~6h 挤卵的，卵子可能未进入或已过成熟分裂阶段，从而降低卵子的孵化率，所以挤卵时间以澳洲龙纹斑排卵后 2~3h 挤卵为宜。雄鱼的催产剂量以500~2000mg/kg为宜。这样的剂量能使绝大多数雄鱼产生大量精液（陈永乐，1999）。

只有亲鱼开始排卵后2~3h才能挤卵，用麻醉液浸洗亲鱼。20kg以上的雄鱼侧放台上，台面铺设湿布，而较小的雌鱼和所有的雄鱼就1个人拿着。每次操作只挤5000粒卵左右。卵子用搪瓷碗盛载，碗内有4‰氯化钠和3‰脲混合受精液300mL，快速旋转碗的同时用力挤压雄鱼腹部滴几滴精液到碗内，旋转持续15分钟。1~2尾雄鱼的精液配5尾雌鱼的卵子。如果要用力地挤卵，则意味着那尾雌鱼的成熟不完全，应放回池里。每1碗的受精卵倒入置于Bow氏孵化槽（360×40×17cm）内带孔的小筐中。卵子到淡水后产生粘性，并缓慢地铺成单层（陈永乐，1999）。

4. 人工孵化

澳洲龙纹斑的受精卵为黏性卵，所以澳洲龙纹斑的人工孵化方式大多采用环道流水孵化或者是水流畅通的孵化槽中孵化。一是把附着在网片上的卵被放进直径为6m、深度为0.5m的100L水体的孵化池中孵化。自然产的卵运输箱运到孵化场后。慢慢地往运输箱中加孵化池中的水。使鱼卵慢慢地适应孵化池中水温及其他条件。然后将附着卵的网片剪成20~30cm的小方块，放入小型注满水的循环孵化池。孵化池放卵的密度一般为100m^{3}水体的小孵化池可放2.5万粒。二是用孵化筐在流水水槽中孵化。在人工授精结束后，估算出卵的数量。将卵放进孵化筐内，然后将孵化筐放进流水孵化槽中进行孵化。在进行孵化的过程中，要随时检查卵的孵化进展情况。一般发育的卵呈现出透明的珍珠状，并且呈黄灰至米黄色，或者呈奶油色。未发育的卵呈白的不透明状。通常每尾亲鱼产卵2万~7万粒，但成活率变化很大，从0到98%。孵化筐的规格为410mm×410mm×150mm。将装有受精卵的孵化筐放进水槽中。流水水槽的规格为2170mm×410mm×150mm。每个水槽可放4个孵化筐。筐底面有铝合金固定架。鱼卵孵化出膜时间与水温密切相关。孵化用水的流量为4~5L/min，水温保持在20~22℃。澳洲龙纹斑受精卵的孵化时间视孵化水温而定，在16.8~17.8℃水温条件下，10d才孵出；20~22℃下，受精卵经5~7d孵化出膜；水温为24~28℃时，孵化时间大大缩短，24.1~25.3℃下，4d孵出。通常水温20~22℃或19~24℃，开始孵出时间是5d，绝大多数是7d孵出，第8d全部孵出。孵化后24h内即可将仔鱼虹吸到干净的水槽。环道流水孵化时，孵化用水要求pH值为6.8~7.0、铵态氮含量小于0.5mg/L、溶氧量大于5.0mg/L。鱼苗出膜之前受精卵每天进行一次消毒处理，预防水霉和寄生虫感染，提高孵化率（陈永乐，1999；郭松，2012；翁伯琦，2016）。

二、养殖

Mellor, P. (2011) 研究了在咸水中养殖澳洲龙纹斑。在 15 和 20ng/L 试验组中，24h 内幼鱼的死亡率为 100%。急性暴露的 LC_{50} 值是 12.5ng/L。25d 后，养殖在 0 和 2.5ng/L 的幼鱼成活率明显（P<0.05）比养殖在 5，7.5 和 10ng/L 的高。在任何盐度的试验中，25d 后幼鱼的平均重量，平均长度和比生长率（SGR）差异不明显（p<0.05）。血液渗透压与试验盐渗透压浓度呈直接关系。结果表明，澳洲龙纹斑稚鱼等渗透点为 5.8ng/L。在转移到内陆盐水环境中 120h 后，幼鱼在 10ng/L 的鱼体水分含量明显（P<0.05）比淡水中稚鱼的高（Mellor, P., 2011）。

澳洲龙纹斑的特定生长率（SGRs）与放养规模和放养生物量均呈负相关。特定生长率和鱼类健康状况在放养几周后马上达到最高值，但后几周就下降了。存活率是变化的（0.2%~99%），但超过 13 个季节后普遍升高而且变化不明显。增加放养密度降低了鱼类的生长速度和健康状况，但不影响生存。生产指标（生长，存活和产量）与几种水的化学组成和生物参数（饵料丰度）之间是显著性相关的，但 3 种鱼中没有 1 个是始终如一的。一些水的化学组成参数达到了可能影响鱼类生长和存活的水平。在较高放养密度下鱼类的生长速度和健康状况在放养的后几周降低了，归因于过度放养导致首选的饵料供应减少了。澳洲龙纹斑对鱼苗池塘中浮游生物群落的影响在很大程度上取决于鱼类生物量和选择性捕食的规模（Ingram, B. A., 2009）。

Tonkin, Z. D. (2006) 研究了澳洲龙纹斑的形态学和摄食行为的遗传变化。在开始摄食时，澳洲龙纹斑游离胚胎体积大且发育良好，而另外两种开始外源性摄食的物种则更年幼，体型更小，发育较差。进行了饲养实验，向幼鱼提供了标准化的猎物混合物（包含相同数量的小型哲水蚤目桡足类，大型哲水蚤目桡足类，小型水蚤和大型水蚤）。经测试的大多数年龄和物种的幼鱼偏爱中型猎物（宽 300~500μm）。即使它们的嘴巴大大超过所提供的最大猎物，情况也是如此。水蚤比相似大小的桡足类消耗得更多。这项研究的结果表明，如果在幼年期没有宽度<500 的可捕食猎物，那么澳洲龙纹斑物种的存活将受到威胁。他们还认为可能会发生种间竞争猎物，尤其是当幼体非常小的时候（Tonkin, Z. D., 2006）。

澳洲龙纹斑稚鱼卵黄囊吸收完毕，开始摄食小型浮游动物，再过 4~6 周鱼苗发育至 15~20mm 时，转为摄食水生昆虫、线虫、小型甲壳类等。鱼苗的规格达 25~30mm 时，投喂人工配合饲料。在澳大利亚，集约化养殖是采用圆形玻璃钢鱼池，自动增氧充气、自动排污、封闭循环。养鱼用水经不锈钢自动滚筒过滤，经生化塔、紫外线消毒，同时注意控制光线，放养密度可达 10kg/m³。整个养鱼厂房内完全由电脑控制，由饲料加工厂专门生产颗粒料供货，硬颗粒饲料规格按鱼的大小生产，若放养密度合理，水温恒定，溶氧充足，投喂饲科学配方制成的硬颗粒饲料，饲料系数 1.3~1.5，产量一般可高达 50kg/m³。户外土池饲养也可取得好的效益，粗养 12 个月一般可长至 800g/尾（韩茂森，2003）。

（一）鱼苗培育

澳洲龙纹斑育苗培育常采用面积为 25~35m² 的水泥池或者流水环道方式，保持水质清新。一般采用三级鱼苗培育法。第 1 阶段从开口培育到体长为 0.8~1.2cm，培育时间为 5~7d，放养密度为 2.0 万~2.5 万尾/m³，该阶段以摄食小型浮游动物为主，国内主要用丰年虫饲喂，环道水流速度不低于 20cm/s。第 2 阶段从 0.8~1.2cm 培育到体长为 1.5~3.0cm，培育时间为 30~45d，放养密度为 1.5 万~1.8 万尾/m³，该阶段鱼摄食水生昆虫、线虫、小型甲壳类等，国内主要用水蚤饲喂，环道水流速度不低于 6cm/s。第 3 阶段从 1.5~3.0cm 培育到体长为 3.5~5.0cm，培育时间为 20~27d，放养密度为 1.0 万~1.2 万尾/m³，该阶段可以采用人工饲料进行驯化，人工配合硬颗粒饲料要求粗蛋白质含量大于 45、脂肪含量大于 15，育苗用水要求水温 22~28℃、pH 值 6.8~7.0、铵态氮含量小于 0.5mg/L、溶氧量大于 5.0mg/L。另外，Allen-Ankins（2012）等研究了在鱼苗养殖过程中水的浊度，饵料密度及养殖环境的复杂情况对澳洲龙纹斑鱼苗的影响，结果发现澳洲龙纹斑鱼苗能够适应比较浑浊的水环境，栖息环境的复杂性会增加饵料的消耗率（翁伯琦，2016；郭松，2012）。

通常用于养殖澳洲龙纹斑的水源有河水、湖水、水库水、地下水、家庭和城镇用水等。理想的水源应该是能保证正常供水量。没有病源、没有有机物、工业和城市污染。有关养殖澳洲龙纹斑用水质量的标准目前研究的还不太多。土池以及循环水中精养澳洲龙纹斑的水质要求如表 2-1。这些数据是正常生产所需要的环境，鱼在短时间内能够忍受的浓度远比这些数据高（曹凯德，2001）。

表 2-1　池塘及循环水精养墨累河鳕鱼水质要求（曹凯德，2001）

化学因子	土池养殖	循环水养殖（mg/L）
温度/℃	9~34	5~30
溶解氧	1.8~18.2	大于 5
pH	5.6~10.38	6.5~9
总氮量*	0.1~1.93	0.1~1.93

(续表)

化学因子	土池养殖	循环水养殖（mg/L）
亚硝酸盐	0~0.019	小于0.1
氮	0~0.58	20~400
硝基氮	0~1.5	
总无机氮	0.01~2.88	
磷酸根	0~2.8	
总碱	1.2~194.6	
透明度	0.12~1.3	
二氧化碳		小于15
氯		小于0.03
总硬度		50~400
硫化氢		小于0.002
硝		小于100
金属镉		小于0.003
钙		10~160
铜		小于0.006
铁		小于0.5
铅		小于0.03
镁		小于0.01
汞		小于0.002
锌		小于0.03

*毒性根据温度和pH值的变化而变化

（二）成鱼养殖

1. 池塘养殖

澳洲龙纹斑池塘养殖面积一般为2000~6660m²，池水深1.5m左右。鱼苗投放前，用112.5g/m²生石灰或20mg/L漂白粉消毒。1周后，确保水质无异常方可进行放养。根据养殖条件（水源、电源、气候）和养殖技术水平确定其放养密度，一般放养夏花鱼种1.3万~2.2万尾/hm²。澳洲龙纹斑是典型的肉食性鱼类，但可根据鱼体不同的生长阶段选择饵料，通常前期育苗在10g之前主要依次投喂红虫、鲜虾和野杂鱼，10g之后转为人工颗粒饲料，颗粒料主要包括鲟鱼颗粒料、大菱鲆颗粒料、海水鱼料，其中饲料营养蛋白质含量≥40%、粗脂肪含量≥8%、粗纤维含量≤5%、粗灰粉含量≤18%、食盐含量≤2%、钙含量≥2.5%、总磷含量≥1.2%、赖氨酸含量≥2.6%。每天早晚各投喂2次，投喂量宜为不同发育阶段体重的1.5%~3.5%。早晚巡塘，观察鱼的摄食、生长及水质变化情况。水中的溶氧量保持在 5.0mg/L 以上，pH 值为 6.8~7.0，铵态氮含量小于 0.5mg/L。每周换水 1 次，每次换水为 20cm。此外，还要注意防止鱼苗逃出池塘，合理使用增氧机（郭松，2012）。

2. 网箱养殖

网箱养殖既有利于大批量人工驯养，又可进行高密度集约化养殖，降低饵料成本。因此，网箱养殖澳洲龙纹斑是一条开发水域资源、发展优质渔业、向产业化迈进的有效途径。养殖澳洲龙纹斑的网箱面积通常约 20m^2、网目大小 2.0~2.5cm，放养密度常为 12~20 尾/m^2，以人工配合饲料为主。因养殖密度较大，应加强管理，注意溶氧量，经常洗网、查网，确保网箱水环境正常（郭松，2012）。

3. 集约化养殖

在澳大利亚集约化养殖是采用圆形玻璃钢鱼池，自动增氧充气、自动排污、封闭循环。养鱼用水经不锈钢自动滚筒过滤，经生化塔、紫外线消毒，同时注意控制光线，放养密度可达 10kg/m^3。整个养鱼厂房内完全由电脑控制，由饲料加工厂专门生产颗粒料供货。若放养密度合理，水温恒定，溶氧充足，投饲科学配方制成的硬颗粒饲料，饲料系数 1.3~1.5，产量一般可高达 50kg/m^3。户外土池饲养也可取得好的效益，粗养 12 个月一般可长至 800g/尾。人工配合硬颗粒饲料要求粗蛋白质>45%，脂肪>15%（韩茂森，2003；翁伯琦，2016）。

4. 循环水养殖

工厂化水产养殖是通过人工调控养殖环境各项理化因子，对适宜的养殖品种开展流水高密度养殖，实现高产高效的一种养殖方式，完美融合了工业化与信息化。在水产养殖中引入工业化理念，应平衡好发展与可持续的矛盾。工厂化循环水养殖与传统养殖方式相比，具有节水、节地、高密度集约化和排放可控的特点，符合可持续发展的要求，是水产养殖方式转变的必然趋势。福建省农业科学院已在福清渔溪养殖基地建立现代化的循环水育苗养殖中心，占地面积为 4500m^2。预计利用该基地饲养澳洲龙纹斑商品鱼，每 667m^2 产 16t，年产量 96t；该基地繁育澳洲龙纹斑种苗，年产鱼苗 200 万尾（翁伯琦，2016）。

5. 鱼菜共生养殖系统

澳洲龙纹斑和莴苣被用于在淡水鱼菜共生系统中测试 3 种底层水培系统（砾石床，浮筏和营养膜技术）之间的差异，鱼菜共生系统中植物的营养来自鱼的排泄物，在回养殖鱼池之前废水中的营养被植物吸收了。莴苣的产量很好，并且在生物量收益和产量方面，遵循着砾石床>浮筏>营养膜技术的关系，观测到所有试验之间的差异显著。营养膜技术处理比其他2种处理在去除硝酸盐方面效率更低（20%的较低效率），然而在所有试验之间，去除磷酸盐方面没有显著性差异。在溶解氧，水置换和电导率方面，在任何试验之间没有观察到显著性差异。总的来说，结果表明，在鱼菜共生系统中，营养膜技术的底层水培系统比砾石床或浮筏的底层水培系统在从鱼类养殖水中转移营养物质和生产植物生物量或产量这2个方面效率更低（Lennard，W. A.，2006）。

（三）饲养方式

澳洲龙纹斑作为一种密集化的合适的养殖品种正越来越受欢迎，特别是在封闭系统中。在五种饲养管理中对鱼的生长，存活，食物转化以及一系列其他相关参数包括鱼体组成成分进行了研究。实验中使用的喂养方式是每日用手投喂2次（SAT），按预测的体重1.2%的量每日用手投喂2次（HFR），和只在白天的机械带式投喂（B/D），仅在夜间的机械带式投喂（B/N）和24h的机械带式投喂（B/DN）。将5种投喂方式都随机分配了3个池塘（一式三份）。所有饲养方式都使用专门为澳洲龙纹斑制备的商业饲料，含有约50%的蛋白质和约16%的油脂。实验进行了84d。特定增长率为0.89 ± 0.01~1.07 ± 0.04%。饲料转化率（FCR）范围为1.09 ± 0.02~0.92 ± 0.03。在SAT投喂方式中观测到最快的生长速度和最大的最终体重；然而，在本投喂方式中也观测到最高的饲料转化率，内脏脂肪指数（VFI%）和肝脏体征指数（HSI%）。在B/D和SAT投喂方式中的鱼的生长速率和最终平均重量有显著的差异。B/N和B/DN投喂方式似乎对鱼最有利（Abery，N. W.，2015）。

考查了不同开口饲养措施对澳洲龙纹斑生物和经济的效果。对三组鱼进行了3种开口饲养措施测试：（1）先仅供应卤虫5d，然后供应卤虫加开口饲料7d，最后仅供应干饲料14d；（2）先供应卤虫加开口饲料12d，最后仅供应干饲料14d；（3）在整个实验期间直接供应干饲料。生长和饲料利用率无显着差异，而直接饲喂干饲料的鱼的死亡率（38.4 ± 0.35%）显著增高。在第1种措施中，鱼只在5d卤虫的时间段内减少了生长量，而在干饲料生长增强期得到了补偿。在不同处理条件下鱼类的组成没有显着差异。经济评估表明，同时供应卤虫和开口饲料的方式更为可取（Ryan，S. G.，2010）。

澳洲龙纹斑幼鱼对黑线虫的捕食具有Ⅱ型功能性反应，而Ⅱ型模型的参数在清澈（0NTU）和浑浊（150NTU）处理之间没有显着差异。进一步的实验表明，视觉可能不是澳洲龙纹斑幼鱼侦察和捕获猎物所必须的；光线与黑暗（$<1\times10^{-4}\mu Em^{-2}s^{-1}$）试验的无生命猎物消耗量无显著差异。这些结果表明，澳洲龙纹斑感官生理适应浑浊的视觉环境（Allen-Ankins，S.，2012）。

(四) 疾病防治

澳洲龙纹斑抗病能力强，常见的有侵袭性疾病和传染性疾病。侵袭性疾病通常有小瓜虫病、车轮虫病、斜管虫病、锚头鱼蚤病等，使用聚维酮碘、土霉素、福尔马林、硫酸铜、高锰酸钾等药物进行治疗，具有较好的疗效。传染性疾病主要是水霉病，此外还有少数的嗜水气单胞菌（Aeromonas hydrophila）引发的体表溃疡，全池使用2mg/L高锰酸钾的防治效果较好。澳大利亚的养殖专家认为传染性疾病多数与水质不良、管理不当或水温波动较大等因素直接相关。因此，在澳洲龙纹斑养殖的过程中保持良好的水质，保持养鱼设施的卫生条件，加强科学管理和防治，增强鱼自身的抗病能力，一般会减少疾病的发生与传染（郭松，2012）。

（罗钦、涂杰峰、陈红珊）

第二节　研究论文摘编

32. Recreational Fishing Effort, Catch, and Harvest for Murray Cod and Golden Perch in the Murrumbidgee River, Australia[①]

[英文摘要]：Recreational fishery management aims to prevent species decline and provide sustainable fisheries. Overfishing has been frequently suggested as a cause of historic fishery declines within the Murray-Darling Basin (MDB), Australia, but there have been few quantitative surveys for providing fishery-dependent data to gauge status. The Murray Cod Maccullochella peelii and the Golden Perch Macquaria ambigua are important species targeted by recreational fishers across the MDB. The fisheries are controlled by size and bag limits and gear restrictions (both species) as well as a closed season (Murray Cod only). A complemented fisher survey design was used to assess the recreational fishery for both species in a 76-km reach of the Murrumbidgee River in 2012-2013. Progressive counts were used to quantify boat-and shore-based fishing effort. Catch and harvest rate information was obtained from shore-based fishers via roving surveys and from boat-based fishers via bus route surveys. Murray Cod catch rates (fish/angler-hour) were 0.228 ± 0.047 (mean±SE; boat based) and 0.092 ± 0.023 (shore based), and harvest rates (fish/angler-hour) were 0.013 ± 0.006 (boat based) and 0.003 ± 0.001 (shore based). Golden Perch catch rates were 0.018 ± 0.009 (shore based)

---

① 研究论文摘编序号续上一章，全书同。

and 0.002±0.001 (boat based), and harvest rates were 0.006±0.002 (shore based) and 0.001±<0.001 (boat based). The Murray Cod fishery had maximal catch and harvest during the 5-month period after the closed season ended. The closed season aims to protect spawning Murray Cod, but this strategy's effectiveness may have been influenced by high fishing effort and deliberate bycatch during the closure period. To sustain and improve these MDB fisheries, we suggest quantification of catch-and-release impacts on spawning Murray Cod, provision of fish passage, re-stocking of Golden Perch, and education on fishing techniques that minimize Murray Cod bycatch during the closed season.

[译文]：

澳大利亚马兰比季河的休闲捕鱼力，捕捞和收获

休闲渔业管理旨在防止物种衰退并提供可持续渔业。过度捕捞经常被认为是造成澳大利亚墨累-达令盆地（MDB）历史性渔业下降的原因，但很少有提供以渔业为依据的数据来衡量状态的定量调查。圆尾麦氏鲈、澳洲龙纹斑和黄金鲈鱼是整个MDB地区休闲渔民的重要鱼种。渔业规模由鱼只大小和袋子（捕渔装备）限制（两个物种）以及封闭季节（仅限澳洲龙纹斑）决定。在2012—2013年，在马兰比季河76km的范围内，采用了补充渔民调查设计来评估这两个物种的休闲渔业。逐步计数用于量化基于船和河滨的捕捞力。渔获量和收获率信息是通过河滨渔夫巡游调查和船上渔民航线调查获得的。澳洲龙纹斑捕获率（鱼/每垂钓小时）分别为0.228±0.047（平均值±SE；以船为基准）和0.092±0.023（以河岸为基准），收获率（鱼/垂钓者小时）分别为0.013±0.006（以船为基准）和0.003±0.001（以海岸为基准）。黄金鲈鱼捕获率为0.018±0.009（以河岸为基础）和0.002±0.001（以船为基础），收获率为0.006±0.002（以河岸为基准）和0.001±<0.001（以船为基础）。在休渔期结束后的5个月，澳洲龙纹斑渔场捕获量和收获量都最大。封闭季节的目的是保护澳洲龙纹斑产卵，但这种策略的有效性可能受到高捕鱼力和关闭期间蓄意兼捕的影响。为了维持和改善这些MDB渔业，我们建议量化对产卵澳洲龙纹斑的捕捞和释放影响，提供鱼类通道，重新放养黄金鲈鱼，以及休渔期澳洲龙纹斑兼捕最小化的捕鱼技术训练。

参考文献：Forbes, J.P., Watts, R.J., Robinson, W.A., et al. 2015. Recreational fishing effort, catch, and harvest for murray cod and golden perch in the murrumbidgee river, australia [J]. North American Journal of Fisheries Management, 35 (4), 649-658.

(罗钦、李冬梅、黄惠珍 译)

33. Sexing accuracy and indicators of maturation status in captive Murray cod Maccullochella peelii peelii using non-invasive ultrasonic imagery.

[英文摘要]：Macroscopic-and histological-based assessments of gonad condition were compared with ultrasound images to determine the feasibility of this technology as a non-invasive diagnostic tool for identifying sex and assessing maturation status of Murray cod. Four age-classes (1+, 2+, 3+ and 6+ years), were sub-sampled at monthly intervals throughout their annual reproductive cycle and scanned with a 5 MHz linear transducer. An interpretation of sex was made from the resulting images and maximum cross-sectional gonad diameter and area were recorded. Fish were subsequently dissected to confirm gender, and the weights and maturation status of gonads determined and then compared with their respective image profile. Ovaries of females were usually a distinctive feature in ultrasound images, being particularly obvious in older and/or more developed fish. In contrast, the identification of male testis was more problematic. Nonetheless, identifying sex from ultrasound images was consistently achieved by recording the presence/absence of a female ovary (96% total sexing accuracy). Maximum cross-sectional ovary diameter and area were highly correlated with gonad weight ($r^2 = 0.90$ and 0.89, respectively) suggesting that indices of maturation status, comparable to the gonadosomatic index (GSI), can be obtained non-destructively from ultrasound scans of females. A less distinct relationship occurred between these dimensions and weight of testes ($r^2 = 0.41$). Significant increases ($p<0.05$) in mean gonad index (GI, calculated from gonad diameter) occurred for most gonad development stages. However, differences in mean GI between maturation stages were confounded by phenotypic variability, indicating that GI may be limited to population level studies. Nevertheless, ultrasound images of ovaries at each development stage were visually distinctive and enabled qualitative evaluations of maturity, thereby complementing quantitative GI assessments. Repeated serial-monitoring of the same population using ultrasound appears to have great potential for tracking maturation-induced changes in broodfish.

[译文]：

使用无创超声成像测定圈养的墨累-鳕鱼雌雄鉴别准确性和成熟指标

将基于宏观和组织学的性腺评估与超声图像进行比较，以确定该技术作为识别性别和评估澳洲龙纹斑成熟状态的无创性诊断工具的可行性。四个年龄阶段（1+、2+、3+和6+年）在整个年度生殖周期内每月间隔抽样，并用5MHz线性换能器进行扫描。从所得图像中进行性别解释，并记录最大横截面性腺直径和面积。随后解剖鱼以确认性别,确定性腺的重量和成熟状态,然后与它们各自的图像轮廓进行比较。雌性的卵巢通常是超声图像中的一个显著特征,在成年和/或发育更完全的鱼类中尤其明显。相比之下,雄性性腺的鉴定是不确定的。尽管如此,通过记录女性卵巢的存在/不存在,足以确定超声图像中的性别(总的性别鉴别准确率为96%)。最大横截面卵巢直径和面积与生殖腺重量高度相关(分别为$r^2=0.90$和0.89),表明成熟状态指数与生殖腺指数(GSI)可以从无损的雌性超声扫描图像中获得。这些因素和睾丸重量之间发生较不明显的关系($r^2=0.41$)。大部分性腺发育阶段的平均性腺指数(GI,由性腺直径计算)显著增加($p<0.05$)。然而,成熟期之间平均GI的差异被表型变异混淆,表明GI可能仅限于种群水平研究。尽管如此,每个发育阶段卵巢的超声图像在视觉上与众不同,并且能够定性评估成熟度,从而补充定量GI评估。反复使用超声对同一群体进行连续监测,对跟踪成熟诱发的亲鱼变化具有很大的潜力。

参考文献:Danem, N., Paull, J., Bretta, I. 2008. Sexing accuracy and indicators of maturation status in captive murray cod maccullochella peelii peelii using non-invasive ultrasonic imagery [J]. Aquaculture, 279 (1-4), 113-119.

(罗钦、李冬梅、黄惠珍 译)

34. Ontogeny of feeding in two native and one alien fish species from the Murray-Darling Basin, Australia.

[英文摘要]:Investigations into the feeding of the early stages of fishes can provide insights into processes influencing recruitment. In this study, we examined ontogenetic changes in morphology and feeding behaviour of two native Australian freshwater species, Murray cod, Maccullochella peelii peelii, and golden perch, Macquaria ambigua, and the alien species, common carp, Cyprinus carpio. Murray cod free embryos are large and well developed at the onset of feeding, whereas the other two species begin exogenous feeding much younger and are smaller and less-developed. Carp commence exogenous feeding 302 days earlier than golden perch, and show more advanced development of the eyes and ingestive apparatus. We conducted feeding experiments, presenting larvae of the three species with a standardised prey mix (comprising equal numbers of small calanoid copepods, large calanoid copepods, small Daphnia, and large Daphnia). Larvae of most tested ages and species showed a preference for mid-sized prey (300-500μm wide). This was true even when their gapes substantially exceeded the largest prey offered. Daphnia were consumed more than similar-sized copepods. The results of this study suggest that survival through their larval period will be threatened in all three species if catchable prey<5000208m in width are not available throughout such time. They also suggest that interspecific competition for prey may occur, especially when larvae are very young. The precocious development of structures involved in feeding and the extended transition from endogenous to exogenous feeding of early carp larvae are likely to have contributed to the success of this species since its introduction to Australia.

[译文]：

澳大利亚墨累-达令盆地两种本地鱼类和一种外来鱼类的个体发育

对鱼类早期摄食的调查可以提供对增员过程的了解。在这项研究中，我们研究了两种澳大利亚原生淡水物种，澳洲龙纹斑和黄金鲈鱼以及外来物种鲤鱼的形态学和摄食行为的遗传变化。在开始摄食时，澳洲龙纹斑游离胚胎体积大且发育良好，而另外两种开始外源性摄食的物种则更年幼，体型更小，发育较差。鲤鱼开始外源性摄食比黄金鲈鱼早302d，并表现出更加发达的眼睛和摄食器官。我们进行了饲养实验，向三种物种的幼体提供了标准化的猎物混合物（包含相同数量的小型ca状桡足类，大型ca状桡足类，小型水蚤和大型水蚤）。经测试的大多数年龄和物种的幼体偏爱中型猎物（宽300~500μm）。即使它们的嘴巴大大超过所提供的最大猎物，情况也是如此。水蚤比相似大小的桡足类消耗得更多。这项研究的结果表明，如果在幼年期没有宽度<500的可捕食猎物，那么所有三种物种的存活将受到威胁。他们还认为可能会发生种间竞争猎物，尤其是当幼体非常小的时候。早期鲤鱼仔鱼摄食结构涉及的早熟发育，从内源性向外源性摄食的延伸转变，都可能是该物种自澳大利亚引进成功的原因。

参考文献：Tonkin, Z. D., Humphries, P., Pridmore, P. A. 2006. Ontogeny of feeding in two native and one alien fish species from the murray-darling basin, australia [J]. Environmental Biology of Fishes, 76 (2-4), 303-315.

（罗钦、李冬梅、黄惠珍 译）

35. A visit to Murray Cod Hatcheries

[英文摘要]：Using natural stimuli to mature broodfish, spawning boxes in the brood ponds and natural plankton for the young fry enable a New South Wales hatchery to produce high quality Murray cod fingerlings. The oldest privately owned native finfish hatcheryin% Australia, the Murray Cod Hatcheries, has been operating since 1967. Located at Wagga Wagga. 273km south-west of Sydney, the operation has been owned and operated by Greg Semple over six seasons. It operates as a hatchery facility for Murray cod (Maccullochella peelii peelii), siver perch (Bidyanus bidyanus), golden perch (Macquaria ambigua), freshwater catfish (Tandanus tandanus) and yabbies (Cherax destructor), these are mainly used a food source for brood fish. Water on the property is fromthreesources--catchment runoff, town supply or 6 bores. Two underground rivers run through the property, says Greg, Ones 10-14m down and the other is at 36m. I have a bore capacity to pump at a rate of 7ML per day if required. The standing water table is at 10m and it is bedrock granite at 46m below the surface.

[译文]：

参观澳洲龙纹斑孵化场

利用自然刺激成熟的亲鱼，育苗塘中的产卵盒和稚鱼天然浮游生物，使新南威尔士州的孵化场能够培育出高质量的澳洲龙纹斑鱼苗。在澳大利亚距离悉尼西南侧373km的沃加沃有一家自1967年以来一直经营着当地最早的原生澳洲龙纹斑私人孵化场。Located at Wagga Wagga。该厂由 Greg Semple 拥有并已经营了六个季节。它主要用于澳洲龙纹斑，澳洲银鲈，黄金鲈鱼，澳洲鳗鲶和鳌虾等幼体孵化。该厂有三方水源：集水区径流，城镇供水或6孔。格雷格说，两条地下河流穿过该厂，其中一条在10~14m以下，另一条在36m以下。如果需要，我有一个能够每天7mL的泵送能力。该设施的地下水位在10m处，而在地表以下46m处是基岩花岗岩。

参考文献：Austasia Aquaculture, 28, April/May 2002.

（罗钦、李冬梅、黄惠珍　译）

36. Advanced ovarian development of Murray cod Maccullochella peelii peelii via phase-shifted photoperiod and two temperature regimes

[英文摘要]：The timing and characteristics of reproductive development in adult female Murray cod exposed to a simulated seasonal photothermal cycle (14：45 to 09：45 daylight h; 12~26℃) (CONTROL) were compared to the development of females exposed to a three month phase-shifted (advanced) seasonal photothermal cycle (PHOTOTHERMAL) and to females exposed to a three month phase-shifted (advanced) photoperiod cycle in combination with constant temperature (19.5℃) (PHOTOPERIOD). Females in PHOTOTHERMAL and PHOTOPERIOD treatments reached maturity up to three (June) and four (May) months in advance of CONTROL fish (October), respectively. Biannual maturation was also observed in four PHOTOPERIOD females (13%). Mean ovary diameter and relative fecundity of mature females were similar between treatments ($p > 0.05$), and viable eggs were produced in all groups (100% ovulated; 14.02%~39.12% mean survival to hatching). Ovary diameters and plasma levels of E_2 and T in phase-shifted females remained at basal levels and/or were significantly reduced (p<0.05) relative to CONTROL fish throughout the early to mid phases of the maturation period. However, rapid increases in plasma T (0.54~4.39ng ml 61 1) and ovary diameter (20.0~42.4mm) in the 60 to 90 days preceding the onset of maturity in phase-shifted females revealed a capacity of Murray cod to accelerate development processes to compensate for earlier delays in photo-responsiveness. Low levels of E 2 that persisted throughout the maturation period of PHOTOTHERMAL and PHOTOPERIOD females did not appear to greatly affect ovarian growth. The successful maturation of photoperiodically-manipulated females under constant temperature demonstrates an alternative approach for influencing maturation patterns in Murray cod that may improve the versatility and cost-effectiveness of broodstock conditioning procedures.

[译文]：

经过相移光周期和 2 个温度状态澳洲龙纹斑的晚期卵巢发育

暴露在一个模拟季节性的光热循环（14：45~09：45 日光 H；12~26℃）（CONTROL）的成年雌性澳洲龙纹斑生殖发育的时间和的特征与一个 3 个月的相移（晚期）季节性的光热循环（PHOTOTHERMAL）和暴露于一个 3 个月的相移（晚期）光周期结合恒温（19.5℃）（PHOTOPERIOD）的雌鱼的发育相比较。PHOTOTHERMAL 和 PHOTOPERIOD 处理的雌性直到 3 个月（六月份）和 4 个月（五月份）才分别达到成熟，超过对照组处理的雌鱼（十月份）。4 尾 PHOTOPERIOD 处理的雌鱼（13%）中也观察到一年成熟 2 次。成熟雌性的卵巢平均直径和相对繁殖力在试验组间是相似的（p>0.05），试验组中都排出了可用的鱼卵（100%排卵；14.02%~39.12%的平均孵化存活率）。相移的 E2 和 T 的卵巢直径和血浆水平保持基础水平和/或相对于对照组处理的贯穿成熟期的早期至中期的鱼显着降低（p<0.05）。然而，在相移雌性成熟开始前的 60~90d 内，血浆 T（0.54~4.39ng/ml 61 1）和卵巢直径（20.0~42.4mm）的快速增加表明了一个澳洲龙纹斑通过加速发育进程来补偿早期光敏响应延迟的能力。持续贯穿整个 PHOTOTHERMAL 和 PHOTOPERIOD 处理的雌鱼的成熟期的 E2 的低水平似乎没有显著地影响卵巢的生长。在恒定温度下对成熟雌性成功的光周期控制证明了这种方法可以提高亲体调节程序的通用性和成本效益。

参考文献：Danem, N., Paull, J., Bretta, I. 2010. Advanced ovarian development of murray cod maccullochella peelii peelii via phase-shifted photoperiod and two temperature regimes [J]. Aquaculture, 310 (1), 206-212.

（罗钦、李冬梅、黄惠珍　译）

37. Age-related changes in ovarian characteristics, plasma sex steroids and fertility during pubertal development in captive female Murray cod Maccullochella peelii peelii

[英文摘要]: Age-related changes in ovarian development characteristics and plasma sex steroids in female Murray cod were examined throughout their second, third and fourth years of life to better understand the physiological and endocrine processes associated with puberty in this species in captivity. Spawning performance of 2+ and 3+ year old females was also assessed to identify ontogenetic differences in egg fertility. Puberty was acquired in 38% of 1+ year old females and 100% of age 2+ females. By age 3+, all females had developed full (adult) reproductive function. Ovarian development in pubertal fish was characterised by a rapid transition between cortical alveoli and lipid droplet oogenic phases, coinciding with significantly lower plasma 17β-oestradiol in age 2+ females ($p<0.05$). Mean mature oocyte diameter (2.44 mm), post-fertilisation viability (30.80%) and hatchability (0.99%) of eggs from age 2+ females were significantly reduced relative to age 3+ adults (2.81 mm, 84.89% and 23.58%, respectively). Ovaries of pubertal Murray cod exhibited both vitellogenic and ovulatory capacities, yet functional abnormalities during secondary oocyte growth are likely to have contributed to poor egg fertility and consequently, evaluations of age-at-first maturity based on the presence of advanced ovarian stages may overestimate the reproductive potential of younger broodstock populations.

[译文]:

被捕的墨瑞鳕雌鱼在青春期发育时，卵巢特征、血浆性类固醇激素和生育率的增龄变化

检查了墨瑞鳕雌鱼生命中的第2，第3和第4年中的卵巢发育特征和血浆性类固醇的增龄变化，为了更好地了解人工饲养的该鱼种中的生理和内分泌过程与青春期的关系。还评估了2龄和3龄以上雌鱼的产卵性能，以确定受精卵个体发育的差异。达到青春期阶段的1龄以上雌鱼占38%，2龄以上雌鱼占100%。3岁以上的所有雌鱼生育功能已经发育完全（成熟的）。青春期鱼的卵巢发育在皮层小泡和脂滴产卵期之间有快速转变，与2龄以上雌鱼中血浆17β-雌二醇显著性降低（$p<0.05$）相一致。来自2龄以上雌鱼的卵的平均成熟卵母细胞直径（2.44mm），受精后存活率（30.80%）和孵化率（0.99%）明显地比3龄以上成熟鱼的参数低（分别为2.81mm，84.89%和23.58%）。青春期澳洲龙纹斑卵巢表现出卵黄化和排卵能力，然而，第2级卵母细胞生长功能异常可能导致卵子受精能力下降，因此，以高级卵巢阶段出现为基础的初始成熟期的评估可

能高估了年轻的亲鱼种群的繁殖潜力。

参考文献：Newman, D. M., Jones, P. L., Ingram, B. A. 2008. Age-related changes in ovarian characteristics, plasma sex steroids and fertility during pubertal development in captive female murray cod maccullochella peelii peelii [J]. Comparative Biochemistry & Physiology Part A, 150 (4), 444.

（罗钦、李冬梅、黄惠珍 译）

38. Cryopreservation of sperm from Murray cod, Maccullochella peelii peelii

[英文摘要]：Freshwater fish sperm are inactive in the male reproductive tract and seminal plasma. They are activated under hypotonic conditions, with a brief period of motility at the time of fertilization. The aims were to find a dilution medium that would maintain sperm inactivity, to assess the cytotoxicity of a range of cryoprotectants, and test the effectiveness of selected diluents for cryopreservation of Murray cod sperm. Sperm remained immotile in ionic and non-ionic solutions above 300mOsm kg 611. Sperm motility after centrifugation and re-suspension in fresh water was better for solutions of 300 and 600mOsm kg 611 than 900mOsm kg 611. In the presence of cryoprotectants, dimethylsulfoxide, dimethylacetamide, and methanol, recovery was better at 300mOsm kg 611 compared to 600mOsm kg 611. Sperm diluted in all media containing glycerol remained inactive after centrifugation and re-suspension in water. A cryopreservation diluent composed of 300mOsm kg 611 d-Sorbitol (DS) solution with 10% methanol produced the best post-thaw motility (51.4±3.4%), followed by Tris-Sucrose-Potassium (TSK) solution with 10% methanol (32.9±4.7%), and Modified Kurokuras medium with 10% methanol (27.5 ± 4.8%). Fluorescent staining for membrane integrity (SYBR-14 and Propidium Iodide) reflected the post-thaw motility results. Fertilization trials using sperm frozen in DS solution with 10% methanol produced a hatch rate of 11.0% (63.1±18.23% of a control fresh sperm hatch rate) and 6.4% (58.5±32.50% of fresh sperm hatch rate) using sperm frozen with TSK with 10% methanol. This represents the first successful cryopreservation of sperm from Murray cod. Sperm cryopreservation will facilitate both conservation and aquaculture of the Murray cod and related endangered Maccullochella species.

[译文]：

澳洲龙纹斑精子的冷冻保存

淡水鱼精子在雄性生殖道和精浆中无活性。它们在低渗条件下被激活，在受精时具有短暂的动力。目的是找到一种可保持精子失活的稀释介质，以评估一系列冷冻保护剂的细胞毒性,并测试了所选的用于低温保存澳洲龙纹斑精子的稀释剂的效果。精子在300mOsmol/kg611以上的离子和非离子溶液中保持失活。在淡水中离心和再悬浮后精子在300和600mOsmol/kg611溶液中的活力比在900mOsmol/kg611溶液中的好。在冷冻保护剂,二甲基亚砜,二甲基乙酰胺和甲醇的存在下,与600mOsmol/kg611相比,300mOsmol/kg611的回收率更好。在含有甘油的所有介质中稀释的精子在水中离心和再悬浮后保持失活。由含有10%甲醇的300mOsm kg611d-山梨醇(DS)溶液组成的冷冻保存稀释液产生最好的解冻后活力(51.4±3.4%),其次是含10%甲醇的Tris-蔗糖-钾(TSK)溶液(32.9±4.7%)和含10%甲醇的改性黑素培养基(27.5±4.8%)。细胞膜全荧光染色(SYBR-14和碘化丙啶)反映出解冻后活力的结果。使用在含10%甲醇的d-山梨醇溶液中冷冻精子的受精试验产生了11.0%的孵化率(对照新鲜精子孵化率为63.1±18.23%)和含10%甲醇Tris-蔗糖-钾溶液中产生了6.4%的孵化率(对照新鲜精子孵化率为58.5±32.50%)。这标志着首次在澳洲龙纹斑精子中成功的冷冻保存。精子冷冻保存将促进澳洲龙纹斑及相关濒危鳕鲈属鱼类的保护和水产养殖。

参考文献:Jonathan, D., David, G., William, B., et al. 2008. Cryopreservation of sperm from murray cod, maccullochella peelii peelii [J]. Aquaculture, 285(1-4), 117-122.

(罗钦、李冬梅、黄惠珍 译)

39. Cryopreservation of sperm from Murray cod (Maccullochella peelii peelii) and silver perch (Bidyanus bidyanus)

[英文摘要]:Murray cod and silver perch are Australian freshwater species of interest for both conservation and aquaculture. Murray cod is listed as Vulnerable under the EPBC Act (1999), while the Australian Society for fish Biology and the IUCN list silver perch as Vulnerable. Captive breeding techniques, involving natural spawning (Murray cod only) and hormone induction, have been established. Breeding programs produce juveniles for stock enhancement to support recreational fisheries, and there is a developing commercial aquaculture industry for both species in Australia and overseas for human consumption. Sperm cryopreservation has numerous practical applications in teleost breeding, including optimal utilisation of gametes, transport of genetic material between facilities, selective breeding activities, genetic banking, and biodiversity conservation. The aim of the present research was to develop a sperm cryopreservation protocol for Murray cod and silver perch to assist in the management of breeding programs. Mature broodstock were maintained in a recirculating aquaculture system at the Marine and Freshwater Fisheries Institute, Snobs Creek. Experiments were carried out on freshly stripped milt to determine osmolality effects on sperm inactivation and cryoprotectant toxicity. These experiments guided the design of cryopreservation media, and treatments that produced the best post-thaw motility were then used in fertilisation trials. Broodstock were induced to spawn with hCG and the hand-stripped eggs were fertilised with either fresh or thawed sperm. The mean osmolality of Murray cod and silver perch milt was 221 and 320 mOsmol/kg, respectively. In Murray cod, sperm motility was inhibited at 300 and 600 mOsmol/kg, but recovery was significantly reduced at 900mOsmol/kg. In silver perch, motility was inhibited at 600 and 900 mOsmol/kg, but not at 300mOsmol/kg. The addition of cryoprotectants reduced the recovery of sperm motility, but recovery was better in 300mOsmol/kg media for Murray cod and in 600mOsmol/kg for silver perch. Post-thaw motility of sperm frozen in D-Sorbitol (DS) solution (300mOsmol/kg for Murray cod and 600mOsmol/kg for silver perch) with 10% v/v methanol were significantly better than for either Tris-Sucrose-Potassium (TSK) solution or Modified Kurokuras (MK) medium. The highest post-thaw activity for Murray cod was 51.4% and for silver perch 50.0%. In fertilisation trials, the best hatch rates (expressed as a percentage of controls) using cryopreserved sperm, were 63.1±18.2% and 16% for Murray cod and silver perch, respectively. Relatively high fertilisation rates in Murray cod were achieved with TSK + 10% v/v methanol or DS + 10% v/v methanol, while in silver perch the highest was obtained with MK + 10% v/v methanol. This work represents the first successful use of cryopreserved sperm on Murray cod and silver perch, and is the first report of a successful sperm cryopreservation diluent based solely on D-Sorbitol with the addition of a cryoprotectant. Sperm cryopreservation methods developed as part of this study will provide a basis for improved management of breeding programs for aquaculture, stock enhancement and conservation of these and other valuable freshwater fish species.

[译文]:

澳洲龙纹斑和银鲈精液的冷冻保存

澳洲龙纹斑和银鲈鱼是既受保护又在养殖的澳洲淡水鱼类。根据EPBC法案（1999年）澳洲龙纹斑被列为濒危物种，同时澳洲鱼类生物学协会和自然保护联盟将银鲈列为濒危物种。包括自然产卵（仅澳洲龙纹斑）和激素引产的捕获育种技术已经建立。育种计划生产的幼鱼用于加强放流从而支持休闲渔业，并且这2种鱼类在澳洲和海外都有一个发展中的商业水产养殖业供人类消费。精子冷冻保存在硬骨鱼类育种有许多实际的应用,包括配子的最佳利用,设施之间的遗传物质的运输,选择性育种活动,基因银行,和生物多样性保护。本研究的目的是为澳洲龙纹斑和银鲈制定精子冷冻保存方案,以协助育种计划的管理。成熟的亲体养殖在海水淡水渔业研究所Snobs Creek的循环水产养殖系统中。对新鲜剥离的乳清进行实验以确定渗透压浓度对精子活性的影响和冷冻保护剂毒性。这些实验指导了冻存培养基的设计,并在受精试验中使用了能够在解冻后得到最佳活力的处理方法。亲鱼采用绒毛膜促性腺激素进行诱导产卵和用新鲜或解冻的精子受精的手工剥离产卵。澳洲龙纹斑和银鲈的平均渗透压分别为221和320mOsmol/kg。在澳洲龙纹斑,精子活率在300和600mOsmol/kg下被抑制,在900mOsmol/kg下,恢复率显著降低。在银鲈鱼中,活性在600和900mOsmol/kg下,而不是300mOsmol/kg时被抑制。冷冻保护剂的加入降低了精子运动能力的恢复,但是澳洲龙纹斑在300mOsmol/kg培养基和银鲈鱼的600mOsmol/kg培养基中的恢复更好。用10%v/v甲醇在D-Sorbitol(DS)溶液(300mOsmol/kg Murray cod和600mOsmol/kg用于银鲈鱼)中冷冻的精液的解冻动力明显优于Tris-蔗糖-钾(TSK)溶液或改良黑素(MK)培养基。澳洲龙纹斑的解冻后活力最高为51.4%,银鲈鱼为50.0%。在受精试验中,澳洲龙纹斑和银鲈鱼使用冷冻保存精子的最佳孵化率(以对照的百分比表示)分别为的63.1±18.2%和16%。使用TSK+10%v/v甲醇或DS+10%v/v甲醇实现了澳洲龙纹斑的相对高的受精率,而在银鲈鱼中,用MK+10%v/v甲醇获得最高的受精率。这项工作代表了第一次在澳洲龙纹斑和银鲈上成功使用冷冻保存的精子,并且是第一次成功研究出完全基于添加D-山梨醇的冷冻保护剂的精液冷冻保存稀释剂的报告。作为本研究的一部分,精子冷冻保存方法将为改进水产养殖育种计划、加强和养护这些珍贵淡水鱼类提供依。

参考文献:Daly, J., Galloway, D., Bravington, W., et al. 2008. Cryopreservation of sperm from Murray cod (Maccullochella peelii peelii) and silver perch (Bidyanus bidyanus) [J]. Skretting Australasian Aquaculture. Innovation in a Global Market (3-6 August 2008, Brisbane).

(罗钦、李冬梅、黄惠珍 译)

40. The impact of cold water releases on the critical period of post-spawning survival and its implications for Murray cod (Maccullochella peelii peelii): a case study of the Mitta Mitta River, southeastern Australia

[英文摘要]: The effects of cold water releases, as a by-product of storing irrigation water in large dams, has been a source of great concern for its impact on native freshwater fish for some time. The Mitta Mitta River, northeast Victoria, is impacted by altered thermal regimes downstream of the fourth largest dam in Australia, Dartmouth dam, with some daily temperatures 10~12C below normal. Murray cod (Maccullochella peelii peelii) were endemic to the Mitta Mitta River; however, resident Murray cod have not been found in this river since 1992. The response of eggs and hatched larvae from Murray cod to different temperature gradients of water were measured and the post-spawning survival recorded. As a case study, post-spawning survival was then inferred from flow data for each year of operation of Dartmouth Dam, recorded since first operation in 1978, and included in a stochastic population model to explore the impact of the altered (historical) thermal regime on population viability. Experimental results revealed no egg and larval survival below 13C and predicted historical temperature regimes point to more than 15 years of low temperatures in the Mitta Mitta River. Population modelling indicates that the impact of cold water releases on post-spawning survival is a significant threatening process to the viability of a Murray cod population. Additionally, we consider changes to the thermal regime to explore how the thermal impact of large dams may be minimized on downstream fish populations through incrementally increasing the temperature of the releases. The modelled Murray cod population responds to minor increases in the thermal regime; however, threats are not completely removed until an increase of at least 5~6℃.

[译文]：

冷水排放对产卵后存活关键期的影响及其对澳洲龙纹斑的影响：以澳洲东南部的米塔米塔河为例

冷水排放作为一个在大型水坝中储存灌溉水的副产品，其结果有时是影响本地淡水鱼的一个重要因素。澳洲维多利亚州东北部米塔米塔河受到澳洲第4大的达特茅斯大坝下游热状况改变的影响，有些日均温度才10~12℃，低于常温。澳洲龙纹斑是特米特米塔河的本土鱼；然而，自1992年以后，这条河就没有找到常住的澳洲龙纹斑。测量了澳洲龙纹斑卵和孵化仔鱼对不同温度梯度的水的反应，并记录了产卵后的存活率。作为一个案例研究，从1978年首次运行以来记录的达特茅斯大坝每年运行的流量数据来推算出产卵存活率，并且被列入一个随机的种群模型中，探索改变（历史的）热状况对种群生存力的影响。实验结果显示没有卵和幼鱼能够存活在低13℃环境下，并且预测的历史的温度模式表明米塔米塔河的低温时间超过15年。种群模拟表明，冷水排放过程对产卵后存活的影响是对澳洲龙纹斑种群生存能力的一个重大威胁。此外，我们考虑改变热状况，探索如何通过梯度地提高排放温度使得大型水坝对下游鱼类种群的热影响最小化。模拟显示，澳洲龙纹斑种群在热状况里轻微的增加；然而，威胁没有完全去除，直到至少增加5~6℃。

参考文献：Todd, C. R., Ryan, T., Nicol, S. J., *et al.* 2010. The impact of cold water releases on the critical period of post-spawning survival and its implications for murray cod (maccullochella peelii peelii): a case study of the mitta mitta river, southeastern australia [J]. River Research & Applications, 21 (9), 1035-1052.

（罗钦、李冬梅、黄惠珍 译）

41. Temporal dynamics of oocyte development, plasma sex steroids and somatic energy reserves during seasonal ovarian maturation in captive Murray cod Maccullochella peelii peelii.

［英文摘要］：The temporal dynamics of oocyte growth, plasma sex steroids and somatic energy stores were examined during a 12 month ovarian maturation cycle in captive Murray cod Maccullochella peelii peelii under simulated natural photothermal conditions. Ovarian function was found to be relatively uninhibited in captivity, with the exception that post-vitellogenic follicles failed to undergo final maturation, resulting in widespread pre-ovulatory atresia. Seasonal patterns of oocyte growth were characterised by cortical alveoli accumulation in March, deposition of lipids in April, and vitellogenesis between May and September. Two distinct batches of vitellogenic oocytes were found in Murray cod ovaries, indicating a capacity for multiple spawns. Plasma profiles of 17beta-oestradiol and testosterone were both highly variable during the maturation period suggesting that multiple roles exist for these steroids during different stages of oocyte growth. Condition factor, liver size and visceral fat stores were all found to increase prior to, or during the peak phase of vitellogenic growth. Murray cod appear to strategically utilise episodes of high feeding activity to accrue energy reserves early in the reproductive cycle prior to its deployment during periods of rapid ovarian growth.

［译文］：

被捕的澳洲龙纹斑在季节性卵巢成熟过程中卵母细胞发育、血浆性类固醇激素和体细胞能量储备的时间动态

在模拟自然光热条件下，被捕的澳洲龙纹斑在12个月的卵巢成熟周期中，检查了卵母细胞生长，血浆性类固醇和体细胞能量储存的时间动态。除了卵黄卵泡无法进行最终的成熟，卵巢功能在被捕的鱼中相对不受影响，导致普遍的排卵前畸形。卵母细胞生长的季节性特征是在3月皮质肺泡积累，4月份油脂沉积，5月至9月卵黄生成。在澳

洲龙纹斑卵巢中发现了两个不同批次的卵黄卵母细胞，表现出能多产卵的能力。17β型-雌二醇和睾酮的血浆分布这2个在整个成熟期是高度可变的，表明这些激素在卵母细胞生长的不同阶段存在多重角色。在卵黄生长之前或生长鼎盛阶段发现肥满度、肝脏大小和内脏脂肪量都有增加。在繁殖周期之前，早于卵巢快速生长期，澳洲龙纹斑似乎有策略地提高摄食活动来积累能量储备。

参考文献：Newman D M, Jones P L, & Ingram B A. 2007. Temporal dynamics of oocyte development, plasma sex steroids and somatic energy reserves during seasonal ovarian maturation in captive murray cod maccullochella peelii peelii [J]. Comparative Biochemistry and Physiology, Part A, 148（4），876-887.

（罗钦、李冬梅、黄惠珍　译）

42. Environmental conditions and timing for the spawning of Murray cod (Maccullochella peelii peelii) and the endangered trout cod (M. macquariensis) in southeastern Australian rivers

[英文摘要]：The environmental conditions and timing of spawning of Murray cod and trout cod were investigated over three successive years in the regulated Murray River and in the nearby, unregulated Ovens River. Larvae were collected in drift samples from early November. Murray cod larvae were present for up to ten weeks, but trout cod larvae were present for only about two weeks. Cod larvae were collected in both rivers in each year sampled, despite a range of flow conditions. Spawning periods, estimated by back-calculating larval ages and egg incubation times, were in part, concurrent for the two species, beginning in October when water temperatures had exceeded 15℃, allowing the occasional hybridization that has been noted between these two species. Trout cod larvae (10.0~18.2mm) were significantly larger than Murray cod larvae (9.5~14.8mm) in both years and the larvae of both species were significantly larger in 1995/6 than in 1994/5 in the Murray River. There was no relationship between larval size and water temperature, but later spawning times at the upper Murray River site coincided with lower water temperatures. Larval abundance varied significantly between sites, samples and years, with peak larval abundances occurring in November. Murray cod larval abundance was best explained by the explanatory variables of year, day length and change in flow over the previous 7d. Environmental conditions for the spawning of Murray cod and trout cod are similar, and both species exhibit a similar larval dispersal strategy by emergence into the drift. Spawning occurred regularly under a range of flow conditions and it is likely that recruitment of these species in these rivers is driven by the subsequent survival of larvae and juveniles.

[译文]：

澳洲东南部的澳洲龙纹斑和濒危吻麦鳕鲈的产卵环境条件和时间

在受监管的墨瑞河和附近的无管制的奥文斯河上，连续3年多对澳洲龙纹斑和吻麦鳕鲈的产卵的环境条件和时间进行了调查。从11月初开始在河流样品中收集幼鱼。墨瑞鳕幼鱼出现的时间长达十周，但吻麦鳕鲈幼鱼出现的时间只有约2周。尽管有一系列的流动条件，每年采集了2条河流中的鳕鱼幼鱼。产卵期，通过反推计算幼鱼年龄和鱼卵孵化时间来估计，是部分的，两种鱼类时间相同，在10月当水温超过15℃时开始，这2种鱼类之间偶尔出现杂交。这2年吻麦鳕鲈幼鱼（10.0~18.2mm）明显比墨瑞鳕幼虫（9.5~14.8mm）大，并且在墨瑞河上1995/6年的这2种幼鱼明显比1994/5年的大。幼鱼大小与水温之间没有任何关系，但后来在墨瑞河上游河段的产卵时间与较低水温是一致的。仔鱼资源量在河段，采样和年份之间有显著性差异，仔鱼资源量高峰期出现在11月份。通过年份，日长和7d前流量变化的解释变量能最好地解释澳洲龙纹斑的仔鱼丰富。澳洲龙纹斑和吻麦鳕鲈的产卵环境条件是相似的，而且这2种鱼类都采用相似的通过涌现在漂流中的策略进行幼鱼扩散。在一系列流动条件下产卵有规律地发生，随后存活的仔稚鱼可能是这些河流中这些鱼类的补充。

参考文献：Koehn, J. D., Harrington, D. J. 2006. Environmental conditions and timing for the spawning of murray cod (maccullochella peelii peelii) and the endangered trout cod (m. macquariensis) in southeastern australian rivers [J]. River Research & Applications, 22 (3), 327-342.

（罗钦、李冬梅、黄惠珍　译）

43. Spawning time and early life history of Murray cod, Maccullochella peelii (Mitchell) in an Australian river

[英文摘要]：I investigated aspects of the early life history of Murray cod in the Broken River, southern Murray-Darling Basin, Australia. I documented patterns in abundance, length, age and the amount of yolk of drifting free embryos and estimated spawning periods for adult Murray cod in two reaches throughout each breeding season between mid-October and mid-December from 1997 to 2001. Free embryos began drifting in late-October and continued until mid-to late-December. Abundances of drifting free embryos showed no obvious peak in most years and were unrelated to discharge. Length, amount of yolk and age typically varied with sampling date, but only age showed strong and negative correlations with temperature. Thus, it appears that temperature affected rates of development, and that developmental stage, not length or age, was likely to be the determinant for when free embryos left the paternal 'nest'. Most free embryos were estimated to have spent between 5 and 7 days drifting. Spawning of Murray cod in the Broken River usually commenced in mid-October and continued until at least early December. Initiation of spawning was associated with temperatures of 15°C and above, but discharge was highly variable, and no other environmental variables were consistent across years. The mid-point of spawning-usually in the first week of November-is considered a more significant time, because it likely coincides with peak spawning, and conditions during and immediately after this time are expected to be optimal for the survival of eggs and free embryos.

[译文]：

澳洲河流中澳洲龙纹斑的产卵期和早期生命史

我调查了澳洲墨瑞-达令盆地南部布罗肯河中澳洲龙纹斑早期生活史的各部分。从1997到2001年间的10月中旬到12月中旬的每个繁殖季节，我都记录了2个河段的成年澳洲龙纹斑的丰度、长度、年龄和游离胚胎卵黄的数量和估计的产卵期。自由胚胎在10月底开始游离，而且一直持续到12月中旬至下旬。游离的自由胚胎的丰度在大多数年份没有明显的峰值，与排卵量无关。长度、卵黄量和年龄通常随着取样变化而变化，但只有年龄与温度呈强烈的负相关。因此，看上去好像是温度影响发展速度，而且那个发育阶段的决定性因素可能是自由胚胎离开雄鱼"巢穴"的时间而不是长度或年龄。大多数自由胚胎预计花费5到7d时间漂流。在布罗肯河，澳洲龙纹斑的产卵通常在10月中旬开始，一直持续到至少12月初。产卵需在15℃及以上的温度，但排卵时高度变化，历年来仅此环境变量是一致的。产卵中期，通常在11月的第1周，被认为是一个更重要的时间，因为它可能与产卵高峰期相一致，而且对于存活的卵子和自由胚胎，在这段时间内和之后的条件预计是最佳的。

参考文献：Humphries, P. 2005. Spawning time and early life history of murray cod, maccullochella peelii, (mitchell) in an australian river [J]. Environmental Biology of Fishes, 72 (4), 393-407.

（罗钦、李冬梅、黄惠珍 译）

44. Culture of juvenile Murray cod, trout cod and Macquarie perch (Percichthyidae) in fertilised earthen ponds.

[英文摘要]：Juvenile Murray cod (Maccullochella peelii peelii), trout cod (Maccullochella macquariensis), and Macquarie perch (Macquaria australasica) (family Percichthyidae) are reared in fertilised earthen ponds to produce fingerlings for aquaculture and stock enhancement purposes to support recreational fisheries and conservation efforts in south eastern Australia. A total of 102 pond stocking trials were conducted over 13 production seasons to quantify growth, survival, condition and yield of juvenile percichthyids and to explore interactions between these variables and a range of biotic and abiotic factors. Water chemistry and zooplankton within ponds fluctuated widely from 1 week to the next and varied considerably between ponds. Zooplankton was composed of rotifers, cladocerans (mainly Moina and Daphnia), calanoids and cyclopoids. Fish Length-weight, length-age and weight-age relationships are described. Specific growth rates (SGRs) were negatively correlated with both size-at-stocking and stocking biomass for all species. SGRs and fish condition were highest in the weeks immediately following stocking, but declined in latter weeks. Survival rates were variable (0.2% ~ 99%), but generally increased and became less variable over the 13 seasons. Increasing stocking density reduced the growth and condition of fish, but did not affect survival. Significant correlations were detected between various fish production indicators (growth, survival and yield) and several water chemistry and biota parameters (prey abundance), but none were consistent for all three species of fish. Some water chemistry parameters reached levels that may have stressed fish and reduced growth and survival. Declines in growth and condition at higher densities, and in latter weeks following stocking, were attributed to a reduction in the availability of preferred prey due to overgrazing by fish. The influence of percichthyid fish on plankton communities in nursery ponds depends largely on fish biomass and size-selective predation. This study has contributed significantly to the knowledge of nursery pond ecosystems, and led to improved pond management and increased reliability of the production of fingerling percichthyids in Australia.

[译文]：

在受精的池塘中养殖澳洲龙纹斑、吻麦鳕鲈和澳洲麦氏鲈鱼苗

澳洲龙纹斑、吻麦鳕鲈和澳洲麦氏鲈鱼苗（人工饲养真鲈科鱼类）在受精土池中养殖，生产出用于水产养殖和加强放流目的的鱼苗，用于支持澳洲东南部的休闲渔业和鱼类保护工作。在13个生产季节进行了102次池塘放养试验，以确定真鲈科鱼苗的生长率，存活率，健康状况和产量，并探讨这些变量与一系列生物和非生物因素之间的相互作用。池塘中水的化学组成和浮游动物从第1周到下1周的波动很大，而且在池塘之

间变化也很大。浮游动物的组成是轮虫，枝角类（主要是墨西哥和水蚤），镖水蚤和剑蚤幼体。对鱼的长度-重量，长度-年龄和体重-年龄关系进行了说明。所有鱼类的特定生长率（SGRs）与放养规模和放养生物量均呈负相关。特定生长率和鱼类健康状况在放养几周后马上达到最高值，但后几周就下降了。存活率是变化的（0.2%～99%），但超过13个季节后普遍升高而且变化不明显。增加放养密度降低了鱼类的生长速度和健康状况，但不影响生存。各种鱼类生产指标（生长，存活和产量）与几种水的化学组成和生物参数（饵料丰度）之间是显著性相关的，但3种鱼中没有1个是始终如一的。一些水的化学组成参数达到了可能影响鱼类生长和存活的水平。在较高放养密度下鱼类的生长速度和健康状况在放养的后几周降低了，归因于过度放养导致首选的饵料供应减少了。真鲈科鱼对鱼苗池塘中浮游生物群落的影响在很大程度上取决于鱼类生物量和选择性捕食的规模。本研究对育苗池生态系统的了解作出了重大贡献，并且使得池塘管理水平的提高并且使得澳洲鲈鱼苗的产量增加可靠性。

参考文献：Ingram, B. A. 2009. Culture of juvenile murray cod, trout cod and macquarie perch (percichthyidae) in fertilised earthen ponds [J]. Aquaculture, 287 (1), 98-106.

（罗钦、李冬梅、黄惠珍　译）

45. Physiological responses of Murray cod (Maccullochella peelii peelii) (Mitchell 1839) larvae and juveniles when cultured in inland salinewater

[英文摘要]：Two trials were conducted to investigate the (i) survival, acute salinity tolerance levels (LC sub (50) -96 hour) and growth of Murray cod (Maccullochella peelii peelii) larvae and (ii) osmoregulatory capacity and body moisture levels of juvenile Murray cod when cultured in inland saline water. During the first trial, survival, length, wet and dry weights were measured over a two-week period when 12 day-post-hatch (dph) larvae were exposed to inland salinity levels of 2.5, 5, 7.5, 10, 15 and 20 ppt. Probit method analysis was used to calculate LC sub (50) values for acute inland salinity exposure to cod larvae. During the second trial, juvenile cod (2.1~5.5g; $n=48$) were subjected to 0, 2.5, 5, 7.5 and 10 ppt of inland salinity. The blood osmolality and body tissue moisture levels were measured at 2, 24 and 120 hours after transfer to the test salinities. In both trials freshwater (0 ppt) was used as a control. Hundred percent larval mortality occurred within 24 hours at 15 and 20 ppt. The LC sub (50) values for acute exposure was 12.5 ppt. Twenty five-dph larvae reared at 0 and 2.5 ppt showed significantly ($p<0.05$) higher survival than larvae reared at 5, 7.5 and 10 ppt. The mean weights, mean lengths and specific growth rates (SGR) of 25-dph larvae were not significantly (p<0.05) different at any of the salinity levels tested. During the second trial, direct transfer to any inland salinity did not influence the survival of juvenile cod for 120 hours. Blood osmolality showed a direct relationship with the osmolality of test salinities. The results showed that juveniles are osmoregulators with an isosmotic point of 5.8 ppt. Body moisture level of juveniles at 10 ppt following 120 hours of transfer to inland saline conditions was significantly higher (p<0.05) than body water of juveniles in freshwater.

[译文]：

内陆咸水养殖墨瑞鳕仔稚鱼的生理反应

进行2次试验，为了研究在内陆盐水中培养时，（i）墨瑞鳕幼鱼的存活率，墨急性盐分耐受度（LC_{50}~96h））和生长速度，和（ii）墨瑞鳕幼鱼的渗透调节能力和鱼体水分含量。在第1次试验中，将孵化12d后（dph）的幼鱼暴露在2.5，5，7.5，10，15和20ppt的内陆盐度中两周以上，测量存活率，长度，湿重和干重。用概率分析法计算急性内陆盐度对鳕鱼幼鱼的LC_{50}值。在第2次试验中，鳕鱼稚鱼（2.1~5.5g；n=48）暴露在0，2.5，5，5.5和10ppt的内陆盐度中。当鱼转移到盐度试验2，24和120h后检测血液渗透压和鱼体组织水分。在2次试验中，淡水（0ppt）作为对照组。在15和20ppt试验组中，24h内幼鱼的死亡率为100%。急性暴露的LC_{50}值是12.5ppt。25d后，养殖在0和2.5ppt的幼鱼成活率明显（p<0.05）比养殖在5，7.5和10ppt的高。在任何盐度的试验中，25d后幼鱼的平均重量，平均长度和比生长率（SGR）差异不明显（p<0.05）。在第2次试验中，直接转移到任何内陆盐度的试验组中，120h的鳕鱼存活不受影响。血液渗透压与试验盐渗透压浓度呈直接关系。结果表明，鳕鱼稚鱼等渗透点为5.8ppt。在转移到内陆盐水环境中120h后，幼鱼在10ppt的鱼体水分含量明显（p<0.05）比淡水中稚鱼的高。

参考文献：Mellor, P., Fotedar, R. 2011. Physiological responses of murray cod (maccullochella peelii peelii) (mitchell 1839) larvae and juveniles when cultured in inland saline water [J]. Indian Journal of Fisheries.

（罗钦、李冬梅、黄惠珍 译）

46. Polygamy and low effective population size in a captive Murray cod (Maccullochella peelii peelii) population: genetic implications for wild restocking programs

[英文摘要]：Stocking of freshwater fish species with hatchery-bred fish is a common response to depleted wild stocks, but may have numerous genetic implications. Murray cod, Maccullochella peelii peelii (Mitchell), have been produced in captivity for wild stocking programs for more than 30 years. The potential genetic impacts of this stocking program on wild populations was investigated by using eight microsatellite markers to determine the parentage of 1380 offspring from 46 separate spawnings collected over three consecutive breeding seasons, and by estimating the effective population size of the broodfish generation through demographic and genetic methods. Results revealed unexpected incidences of polygamous spawnings (both polygyny and polyandry), multiple spawnings by both sexes within a season and repeated matings between pairs of fish across multiple seasons. Furthermore, approximately half of the broodfish failed to spawn at all over the 3-year study period. This likely contributed to the estimated effective population size of around half of the census size, moderate but significant reductions in allelic richness in all three cohorts investigated and a small but significant reduction in heterozygosity in two cohorts. These results allowed us to make recommendations regarding captive husbandry that will maximise genetic diversity of fish intended for stocking.

[译文]：

在捕获的澳洲龙纹斑种群中多配偶制和低效群体规模：野外再放流计划的遗传影响

放流人工孵化的淡水鱼种是对濒危野生种群普遍的育种措施，但可能有许多遗传问题。澳洲龙纹斑已经在野外放流计划中繁育了30多年。通过使用八个微卫星标记确定连续三个繁殖季节收集的46个单独产卵中的1380个后代亲本评估了这种野外放流计划种群的潜在遗传影响，并通过人口统计学和遗传方法估计其亲鱼年代的有效种群规模。结果显示多配偶制产卵（一雄多雌和一雌多雄）意想不到的情况，一个季节内两性的多次产卵，以及多个季节之间的鱼的重复交配。此外，在3年以上的研究期间，大约一半的亲鱼没产卵。这可能有助于估计有效体规模大约为现有群体规模的一半，调查的所有3个种群中的基因丰富度都逐渐地显著性下降，2个种群的杂合性也小幅度的显著性下降。这些结果使我们能够就用于放流鱼的遗传多样性的最大化的人工饲养方面提出建议。

参考文献：Rourke, M. L., Mcpartlan, H. C., Ingram, B. A., et al. 2009. Polygamy and low effective population size in a captive murray cod (maccullochella peelii peelii) population: genetic implications for wild restocking programs [J]. Marine & Freshwater Research, 60 (60), 873-883.

（罗钦、李冬梅、黄惠珍　译）

47. A comparison of three different hydroponic sub-systems (gravel bed, floating and nutrient film technique) in an Aquaponic test system.

[英文摘要]：Murray Cod, Maccullochella peelii peelii (Mitchell), and Green Oak lettuce, Lactuca sativa, were used to test for differences between three hydroponic subsystems, Gravel Bed, Floating Raft and Nutrient Film Technique (NFT), in a freshwater Aquaponic test system, where plant nutrients were supplied from fish wastes while plants stripped nutrients from the waste water before it was returned to the fish. The Murray Cod had FCR's and biomass gains that were statistically identical in all systems. Lettuce yields were good, and in terms of biomass gain and yield, followed the relationship Gravel bed>Floating>NFT, with significant differences seen between all treatments. The NFT treatment was significantly less efficient than the other two treatments in terms of nitrate removal (20% less efficient), whilst no significant difference was seen between any test treatments in terms of phosphate removal. In terms of dissolved oxygen, water replacement and conductivity, no significant differences were observed between any test treatments. Overall, results suggest that NFT hydroponic sub-systems are less efficient at both removing nutrients from fish culture water and producing plant biomass or yield than Gravel bed or Floating hydroponic sub-systems in an Aquaponic context. Aquaponic system designers need to take these differences into account when designing hydroponic components within aquaponic systems.

[译文]：

鱼菜共生系统中3种不同底层水培系统的比较（砾石床，浮筏和营养膜技术）

澳洲龙纹斑和莴苣被用于在淡水鱼菜共生系统中测试3种底层水培系统（砾石床，浮筏和营养膜技术）之间的差异，鱼菜共生系统中植物的营养来自鱼的排泄物，在回养殖鱼池之前废水中的营养被植物吸收了。澳洲龙纹斑在所有系统中具有统计学上相同的饲料转化率和生物量收益。莴苣的产量很好，并且在生物量收益和产量方面，遵循着砾石床>浮筏>营养膜技术的关系，观测到所有试验之间的差异显著。营养膜技术处理比其他2种处理在去除硝酸盐方面效率更低（20%的较低效率），然而在所有试验之间，去除磷酸盐方面没有显著性差异。在溶解氧，水置换和电导率方面，在任何试验之间没有观察到显著性差异。总的来说，结果表明，在鱼菜共生系统中，营养膜技术的底层水培系统比砾石床或浮筏的底层水培系统在从鱼类养殖水中转移营养物质和生产植物生物量或产量这2个方面效率更低。鱼菜共生系统设计人员在设计鱼菜共生系统中的水培组分时，需要考虑到这些差异。

参考文献：Lennard, W. A., Leonard, B. V. 2006. A comparison of three different hydroponic sub-systems (gravel bed, floating and nutrient film technique) in an aquaponic test system [J]. Aquaculture International, 14 (6), 539–550.

(罗钦、李冬梅、黄惠珍 译)

48. Performance of murray cod, Maccullochella peelii peelii (Mitchell) in response to different feeding schedules

[英文摘要]: The Australian freshwater fish Murray cod, Maccullochella peelii peelii (Mitchell) is gaining popularity as a suitable species for intensive culture, particularly in closed systems. The aim of this study was to evaluate the performance of Murray cod in response to different feeding schedules.

Growth, survival, food conversion and a range of other related parameters including carcass proximate composition were evaluated for fish in five feed management regimes. The feeding regimes used in the experiment were hand fed to satiation twice daily (SAT), a pre-determined ration of 1.2% of the body weight day which was hand fed twice daily (HFR), and belt fed through the day only (B/D), belt fed through the night only (B/N) and belt fed for 240h (B/DN). Each of the five feeding regimes was randomly allocated to three tanks (triplicates). All of the feeding regimes used a commercially prepared diet formulated specifically for Murray cod, containing ≈ 50% protein and ≈ 16% lipid. The experiment was conducted for 84 days. Specific growth rate ranged from 0.89 ± 0.01 to $1.07 \pm 0.04\%$ day611. Food conversion ratio (FCR) ranged from 1.09 ± 0.02 to 0.92 ± 0.03. The fastest growth and greatest final body weight were observed in the SAT treatment; however, the highest FCR, visceral fat index (VFI %) and hepatosomatic index (HSI %) were also observed in this treatment. Significant differences were found in specific growth rate and final mean weight between fish in the B/D and SAT treatments. B/N and B/DN feeding regimes appeared to result in the most favourable fish performance.

[译文]:

澳洲龙纹斑对不同饲喂方式的反应

澳洲龙纹斑作为一种密集化的合适的养殖品种正越来越受欢迎,特别是在封闭系统中。本研究目的是评估澳洲龙纹斑对不同喂养计划的反应。在5种饲养管理中对鱼的生长,存活,食物转化以及一系列其他相关参数包括鱼体组成成分进行了研究。实验中使用的喂养方式是每日用手投喂两次(SAT),按预测的体重1.2%的量每日用手投喂2次(HFR),和只在白天的机械带式投喂(B/D),仅在夜间的机械带式投喂(B/N)和24h的机械带式投喂(B/DN)。将五种投喂方式都随机分配了3个池塘(一式三份)。所有饲养方式都使用专门为澳洲龙纹斑制备的商业饲料,含有约50%的蛋白质和约16%

的油脂。实验进行了84d。特定增长率为0.89±0.01~1.07±0.04%。饲料转化率（FCR）范围为1.09±0.02~0.92±0.03。在SAT投喂方式中观测到最快的生长速度和最大的最终体重；然而，在本投喂方式中也观测到最高的饲料转化率，内脏脂肪指数（VFI%）和肝脏体征指数（HSI%）。在B/D和SAT投喂方式中的鱼的生长速率和最终平均重量有显著的差异。B/N和B/DN投喂方式似乎对鱼最有利。

参考文献：Abery, N. W., De Silva, S. S. 2015. Performance of murray cod, maccullochella peelii peelii (mitchell) in response to different feeding schedules [J]. Aquaculture Research, 36 (5), 472-478.

（罗钦、李冬梅、黄惠珍 译）

49. Evaluation of Weaning Strategies for Intensively Reared Australian Freshwater Fish, Murray Cod, Maccullochella peelii peelii

[英文摘要]：The Australian Murray cod supports a growing national industry. However, with regard to the process of weaning fry, there is a lack of information and optimal procedures need to be developed. The aim of the present investigation was to test the biological and economic efficacy of different weaning strategies for Murray cod. Three weaning strategies were tested on triplicate groups of fish：(1) only Artemia for 5d, 7d on Artemia plus starter diet, and 14d on dry diet only; (2) 12d on Artemia plus starter diet and 14 d on dry diet only; and (3) directly to dry diet for the entire experimental period. No significant differences were recorded in the growth and feed efficiency, while significantly higher mortality (38.4±0.35%) was recorded in fish weaned directly onto dry diet. Fish subjected to the first 5 d on Artemia only showed a growth reduction during this period, which was compensated by a phase of enhanced growth during the dry-diet phase. No significant differences were noted in the proximate composition of fish under the different treatments. The economic evaluation suggested that the treatment with the simultaneous supply of Artemia and starter diet is preferable.

[译文]：

对规模化饲养的澳洲淡水澳洲龙纹斑开口饲养措施的评价

澳洲澳洲龙纹斑的养殖是受支持的日益壮大的国民产业。然而，关于其开口饲养的过程缺乏信息，需要研究最佳的开口饲养措施。本研究的目的是考查不同开口饲养措施对澳洲龙纹斑生物和经济的效果。对三组鱼进行了3种开口饲养措施测试：（1）先仅供应卤虫5d，然后供应卤虫加开口饲料7d，最后仅供应干饲料14d；（2）先供应卤虫加开口饲料12d，最后仅供应干饲料14d；（3）在整个实验期间直接供应干饲料。生长和饲料利用率无显着差异，而直接饲喂干饲料的鱼的死亡率（38.4±0.35%）显著增高。在第1种措施中，鱼只在5d卤虫的时间段内减少了生长量，而在干饲料生长增强期得到了补偿。在不同处理条件下鱼类的组成没有显着差异。经济评估表明，同时供应卤虫和开口饲料的方式更为可取。

参考文献：Ryan, S. G., Smith, B. K., Collins, R. O., et al. 2010. Evaluation of weaning strategies for intensively reared australian freshwater fish, murray cod, maccullochella peelii peelii [J]. Journal of the World Aquaculture Society, 38 (4), 527-535.

（罗钦、李冬梅、黄惠珍　译）

50. The effects of turbidity, prey density and environmental complexity on the feeding of juvenile Murray cod Maccullochella peelii

[英文摘要]：Juvenile Murray cod Maccullochella peelii exhibited a type II functional response while preying on blackworms Lumbriculus variegatus, and the parameters of the type II model did not differ significantly between clear (0 NTU) and turbid (150 NTU) treatments. Further experiments showed that vision may not be necessary for prey detection and capture by juvenile M. peelii; consumption of inanimate prey was not significantly different between light and dark ($<1 \times 10^{-4} \mu E\ m^{-2} s^{-1}$) trials. These results imply that the sensory physiology of M. peelii is well adapted to a turbid visual environment. In addition, habitat complexity increased the food consumption rate of juvenile M. peelii, perhaps by relaxing innate predator avoidance behaviours that depress foraging in more open environments.

[译文]：

浊度、养殖密度和环境复杂度对墨瑞鳕幼鱼的饲养影响

墨瑞鳕幼鱼对黑线虫的捕食具有Ⅱ型功能性反应，而Ⅱ型模型的参数在清澈（0NTU）和浑浊（150NTU）处理之间没有显着差异。进一步的实验表明，视觉可能不是墨瑞鳕幼鱼侦察和捕获猎物所必须的；光线与黑暗（$<1 \times 10^{-4} \mu Em^{-2}s^{-1}$）试验的无生命猎物消耗量无显著差异。这些结果表明，澳洲龙纹斑感官生理适应浑浊的视觉环境。此外，栖息地复杂性增加了墨瑞鳕幼鱼的食物消耗率，也许放松固有的捕食者逃避行为，在更开放的环境降低觅食。

参考文献：Allen-Ankins, S., Stoffels, R. J., Pridmore, P. A., et al. 2012. The effects of turbidity, prey density and environmental complexity on the feeding of juvenile murray cod maccullochella peelii [J]. Journal of Fish Biology, 80 (1), 195-206.

（罗钦、李冬梅、黄惠珍　译）

51. Who's your mama? Riverine hybridisation of threatened freshwater Trout Cod and Murray Cod

[英文摘要]: Rates of hybridization and introgression are increasing worldwide because of translocations, restocking of organisms and habitat modifications; thus, determining whether hybridization is occuring after reintroducing extirpated congeneric species is commensurately important for conservation. Restocking programs are sometimes criticized because of the genetic consequences of hatchery-bred fish breeding with wild populations. These concerns are important to conservation restocking programs, including those from the Australian freshwater fish family, Percichthyidae. Two of the better known Australian Percichthyidae are the Murray Cod, Maccullochella peelii and Trout Cod, Maccullochella macquariensis which were formerly widespread over the Murray Darling Basin. In much of the Murrumbidgee River, Trout Cod and Murray Cod were sympatric until the late 1970s when Trout Cod were extirpated. Here we use genetic single nucleotide polymorphism (SNP) data together with mitochondrial sequences to examine hybridization and introgression between Murray Cod and Trout Cod in the upper Murrumbidgee River and consider implications for restocking programs. We have confirmed restocked riverine Trout Cod reproducing, but only as inter-specific matings, in the wild. We detected hybrid Trout Cod-Murray Cod in the Upper Murrumbidgee, recording the first hybrid larvae in the wild. Although hybrid larvae, juveniles and adults have been recorded in hatcheries and impoundments, and hybrid adults have been recorded in rivers previously, this is the first time fertile F1 have been recorded in a wild riverine population. The F1 backcrosses with Murray cod have also been found to be fertile. All backcrosses noted were with pure Murray Cod. Such introgression has not been recorded previously in these two species, and the imbalance in hybridization direction may have important implications for restocking programs.

[译文]：

母鱼是谁？河流杂交威胁淡水吻麦鳕鲈和墨瑞鳕鱼

因为易位，鱼类再放流和栖息地变异造成杂交率和基因渗入率在世界范围内急剧增加；因此，在重新放流已灭绝的同属的鱼类后是否会发生杂交的确定对环境保护来说相当重要。有时批评重新放流的计划是因为人工培育的鱼类与野生鱼类繁殖的遗传后果。这些担忧对于保护性重新放流的计划非常重要，包括来自澳洲淡水鱼家族的真鲈科鱼类。澳洲两个已知最好的真鲈科是墨瑞鳕鱼和吻麦鳕鲈，其曾在墨瑞-达令盆地广泛分布。在穆伦比基河的大部分地区，吻麦鳕鲈和墨瑞鳕鱼都是在同一地区生存的，直到

20世纪70年代末,吻麦鳕鲈被灭绝。在这里,我们使用遗传单核苷酸多态性(SNP)数据和线粒体序列检查了在穆伦比基河上游的墨瑞鳕鱼和吻麦鳕鲈之间的杂交和基因渗入,并且考虑对重新放流计划的影响。我们已经确认了吻麦鳕鲈可以重新繁殖,但在野外只有种间才交配。我们检测到在穆伦比基河上杂交的吻麦鳕鲈-墨瑞鳕鱼,记录了野生的第一个杂交鱼苗。虽然杂交鱼苗,杂交幼鱼和成年杂交鱼已经在孵化场和湖泊中有记录,而且成年杂交鱼之前在河流中也有记录,但是这次是在河流野生鱼类中首次发现能生育的F1。F1与墨瑞鳕鱼的回交也发现是能生育的。所有回交都是使用纯种的墨瑞鳕鱼。在这2种鱼类中这样的基因渗入之前没有记录,杂交方向的不平衡可能对重新放流方案有重要的影响。

参考文献:Couch, A. J., Unmack, P. J., Dyer, F. J., et al. 2016. Who's your mama? riverine hybridisation of threatened freshwater trout cod and murray cod [J]. Peerj, 4 (8), e2593.

(罗钦、李冬梅、黄惠珍 译)

52. The influence of saline concentration on the success of Calcein marking techniques for hatchery-produced Murray cod ('Maccullochella peelii')

[英文摘要]:Chemical marking is a useful technique to determine natal origin of fish and is increasingly used to determine the success of fish stocking programs. This study sought to optimise an osmotic-induction batch marking technique, using the calcium-binding chemical, Calcein, to enable future identification of hatchery-marked Murray cod (Maccullochella peelii). It was hypothesised that higher saline concentrations would create a more reliable bone mark but it was unknown whether saline exposure would influence fish survival. A laboratory trial was undertaken to determine the optimum saline concentration required to maximise survival of Murray cod and marking of bony body parts. Fish were exposed to a no salt control, a no Calcein control or one of three different saline concentration treatments then housed in either 60 L aquarium tanks or hatchery ponds and monitored for 43 days post marking. There was no significant difference in mortality rates among the three treatments under controlled aquarium conditions or among marked fish released into hatchery ponds. Whilst saline concentration did not influence fish survival, marking using concentrations less than seawater produced a detectable mark and reduced stress on Murray cod fingerlings. Mark intensity, however, was greater when fish were exposed to higher saline concentrations.

[译文]:

盐度对用于墨瑞鳕鱼孵化的成熟的钙黄绿素标记技术的影响

化学标记是一种有用的技术来确定鱼的出生来源，并越来越多地用来确定成功的鱼类放流计划。本研究旨在优化一批次的渗透诱导标记技术，使用钙结合化学品，钙黄绿素，以使今后能鉴定出孵化场标记的墨瑞鳕鱼。假设较高的盐度会产生更可靠的骨标记，但不知盐水暴露是否会影响鱼的存活。通过实验室试验确定了在墨瑞鳕鱼存活率并且骨骼部分标记物最大化的条件下的最佳盐度。将鱼暴露于无盐，无钙黄绿素的对照组或3种不同盐度的试验组中，置于60L的水族箱或孵化池中，并在标记后监测43d。在水族箱或孵化池条件下3种试验组标记鱼的死亡率无显著差异。虽然盐度不影响鱼的生存，但是使用低于海水浓度的标记产生了一个可检测的标记并且减少对墨瑞鳕幼鱼的压力。然而，当鱼暴露于更高的盐度时标记强度更高。

参考文献：Baumgartner, L. J., Mclellan, M., Robinson, W., et al. 2012. The influence of saline concentration on the success of calcein marking techniques for hatchery-produced murray cod (maccullochella peelii) [J]. Journal and Proceedings-Royal Society of New South Wales, 145 (445), 168-180.

（罗钦、李冬梅、黄惠珍 译）

53. 澳洲"国宝鱼"墨瑞鳕养殖关键技术

摘要：墨瑞鳕鱼，又称澳洲鳕鲈、澳洲龙纹斑、虫纹鳕鲈、虫纹石斑鱼。是澳洲驰名的淡水养殖良种，有"国宝鱼"之称，属鲈形目鮨鲈科鳕鲈属。墨瑞鳕刺少、肉细滑无腥味、且有一种淡淡的特殊香味，外型理想，市场价值较高。2014年，广东东莞市安业水族养殖有限公司宣布全人工繁育养殖墨瑞鳕获得成功，并且掌握一套养殖关键技术。

参考文献：欧小华，罗汝安. 2017. 澳洲"国宝鱼"墨瑞鳕养殖关键技术 [J]. 海洋与渔业（5）：56-57。

注：无英文摘要。

54. 澳洲虫纹鳕鲈工厂化养殖技术

摘要：虫纹鳕鲈（MaccuHocbel Iapeelil）俗称为墨瑞鳕、河鳕、东洋鳕、淡水鲈鱼。在我国一般称之为虫纹鳕鲈、虫纹石斑鱼。主要分布于澳洲东南部墨瑞达令河流域，体长一般55~65cm，最大个体可达1.8m。虫纹鳕鲈的温度适应范围在4~40℃，最适合生长温度为23~32℃。在温度适宜条件下，养殖1周年个体重可达0.5~0.6kg，之后生长速度逐渐加快。鱼体呈纺锤形，头后部稍隆起口端位，口裂大。

参考文献：张志山，朱树人，朱永安. 2016. 澳洲虫纹鳕鲈工厂化养殖技术 [J]. 齐鲁渔业（2）：15-16。

注：无英文摘要。

55. 澳洲龙纹斑工厂化养殖技术

摘要：澳洲龙纹斑（Murray cod），又称墨瑞鳕、河鳕、东洋鳕、鳕鲈、澳洲淡水鳕鲈等，中国称为澳洲龙纹斑、虫纹鳕鲈或虫纹石斑。

参考文献：杨小玉，郭正龙.2013.澳洲龙纹斑工厂化养殖技术［J］.水产养殖，34（2）：26-27。

注：无英文摘要。

56. 澳洲龙纹斑工厂化养殖技术

摘要：澳洲龙纹斑又称墨瑞鳕、河鳕、东洋鳕、鳕鲈、澳洲淡水鳕鲈等，中国称澳洲龙纹斑、虫纹鳕鲈或虫纹石斑。澳洲龙纹斑是澳洲的原生鱼种，为最顶级的白肉鱼，在澳洲排名四大经济鱼类之首，是世界上最大的淡水鱼之一，与宝石鲈、黄金鲈、银鲈等原产于澳洲的淡水鱼并列为澳洲鱼。

参考文献：郭正龙，杨小玉，孟庆宇.2012.澳洲龙纹斑工厂化养殖技术［J］.科学养鱼，34（12）：26-27。

注：无英文摘要。

57. 澳洲墨累河鳕鱼养殖技术

摘要：墨累河鳕鱼为一大型热、温带鱼。能生活在海水、咸淡水和淡水水域中。介绍了生物学特性、对水质要求、繁殖技术以及鱼种、成鱼的养殖技术。

参考文献：曹凯德.2001.澳洲墨累河鳕鱼养殖技术［J］.水生态学杂志，21（1）：16-18。

58. 虫纹鳕鲈苗种培育影响因素初步分析

摘要：在室内培育条件下研究了不同温度（16℃、21℃、26℃）、不同饵料（枝角类、水丝蚓）对虫纹鳕鲈鱼苗生长的影响，结果表明：鱼苗的摄食率在16~26℃范围内随温度的升高及个体的成长而升高．培育前期投喂枝角类较好，后期以水丝蚓较佳。

参考文献：左瑞华，汪学军.2001.虫纹鳕鲈苗种培育影响因素初步分析［J］.安徽农学通报，7（3）：57-59。

注：无英文摘要。

59. 澳大利亚墨累河鳕鱼养殖技术

摘要：墨累河鳕鱼为一大型热、温带鱼。能生活在海水、咸淡水和淡水水域中。介绍了生物学特性、对水质要求、繁殖技术以及鱼种、成鱼的养殖技术。

参考文献：曹凯德.2001.澳大利亚墨累河鳕鱼养殖技术［J］.水利渔业，21（1）：

16-118。

注：无英文摘要。

60. 墨累鳕的养殖技术

主要内容：墨累鳕是澳大利亚迄今发现的最大的淡水鱼类。它的自然分布范围贯穿墨累-达令盆地。西部划分范围从昆士兰东南，经过新南威尔士州到维多利亚和澳大利亚南部。

参考文献：陈杰，杨兴丽，申秀英．2002．墨累鳕的养殖技术［J］．河南水产，(2)：9。

注：无中文和英文摘要。

61. 澳洲龙纹斑的繁殖方法（专利）

摘要：本发明提供澳洲龙纹斑的繁殖方法，采用人工繁殖与自然繁殖相结合的；包括以下步骤：1）准备亲鱼培育池；2）选择亲鱼；3）亲鱼产前培育；4）设置产卵器；5）产卵及受精；6）受精卵的孵化；采用本发明提供的澳洲龙纹斑的繁殖方法，各步骤操作简单方便，有效提高了澳洲龙纹斑鱼卵的受精率（受精卵可达80%~90%）以及受精卵的孵化成功率（孵化成功率可达90%以上），此外，繁殖过程中不会损伤亲鱼、有效避免了受精卵孵化过程被水霉等病害所威胁、降低了养殖成本和提高了养殖密度。

参考文献：罗钦，罗土炎，饶秋华．2016．澳洲龙纹斑的繁殖方法．CN105660479A。

注：无英文摘要。

62. 澳洲龙纹斑鱼苗的培养方法（专利）

摘要：本发明提供澳洲龙纹斑鱼苗的培养方法，包括以下步骤：1）鱼苗暂养；2）鱼苗开口培育；3）鱼苗转口驯化；4）准备鱼苗培育池；5）鱼苗培育；6）鱼苗培育池的管理；7）病害的预防和治疗。本发明澳洲龙纹斑鱼苗的培养方法中各步骤操作简单方便，能对病害进行及时预防和治疗；在投喂配合饲料前，有对鱼苗进行40~60d的开口培育以及5~7d的转口驯化；鱼苗培育池管理规范，定时排污，筛选出过小或过大的鱼苗另行培育，有效地提高了鱼苗较高的成活率、提高了经济效益。

参考文献：罗钦，罗土炎，饶秋华．2016．澳洲龙纹斑鱼苗的培养方法．CN105706971A。

注：无英文摘要。

63. 澳洲龙纹斑受精卵孵化器（专利）

摘要：本实用新型提供澳洲龙纹斑受精卵孵化器，包括缸体、水循环系统、曝气系统、筛网和支撑架，所述的水循环系统和曝气系统设置在缸体上使缸体内的液体循环和通气溶氧，所述的支撑架包括至少一对设置在缸体底部的竖杆，所述的筛网的两个相对的侧边上设置有联接件，筛网由联接件可拆卸的设置在一对竖杆之间，筛网上用于固定澳洲龙纹斑受精卵。本实用新型的澳洲龙纹斑受精卵孵化器具有如下优点：操作简单方便，成本低，有效避免受精卵感染水霉菌，提高澳洲龙纹斑受精卵孵化成功率。

参考文献：罗钦，罗土炎，饶秋华. 2015. 澳洲龙纹斑受精卵孵化器. CN205093374U。

注：无英文摘要。

64. 一种新品种龙纹斑温室规模化养殖方法（专利）

摘要：本发明涉及一种新品种龙纹斑温室规模化养殖方法，具体包括建造温室、建造养殖池、设置微孔增氧系统、设置鱼巢、配制颗粒饵料、建立养殖操作规程、添加微生态制剂七部分。通过本发明的工艺养殖出的澳洲龙纹斑，生长快、营养均衡、免疫力强、含肉率高、成活率保持在95%以上，养殖密度高达50kg/m^{3}水体，并且保留有龙纹斑的鲜味，具有独特的品质、风味和丰富的营养，经济效益和生态效益得到显著提高。

参考文献：蔡星，秦巍仑，周忠良. 2014. 一种新品种龙纹斑温室规模化养殖方法. CN103947585A。

注：无英文摘要。

65. 一种澳洲龙纹斑受精卵的孵化方法（专利）

摘要：本发明提供澳洲龙纹斑受精卵的孵化方法，采用两端开口、外表面覆盖有遮布、内表面的底面上设置有筛网的中空筒体收卵设备；采用包括有水循环系统、曝气系统、第一支撑架的孵化设备；采用包括有曝气系统、第二支撑架的消毒设备；孵化方法包括以下步骤：1）受精卵的收集；2）受精卵的孵化：a 受精卵的消毒处理、b. 死卵的剔除、c. 受精卵的分散及另行孵化；d. 鱼苗的破膜；3）鱼苗的收腹；采用本发明提供的澳洲龙纹斑受精卵的孵化方法，各步骤操作简单方便，有效提高了澳洲龙纹斑鱼卵的孵化率，解决了现有澳洲龙纹斑的孵化率低、死卵易感染霉菌并感染周围健康受精卵、孵化池中水质不良及缺氧等问题。

参考文献：罗土炎，张志灯，饶秋华，等. 2017. 一种澳洲龙纹斑受精卵的孵化方法. CN201710667649.2。

注：无英文摘要。

66. 一种龙纹斑幼鱼的仿自然生态投喂模式（专利）

摘要：本发明公开了一种龙纹斑幼鱼的仿自然生态投喂模式，即改变全池撒喂的传统投喂方式，在适合集约化高密度养殖环境下，利用幼鱼喜贴壁、惧强光趋弱光、代谢消化快、喜动饵的生活习性，采用弱光引诱、贴壁连续的投喂方式，使龙纹斑幼鱼集群摄食。

本发明采用贴壁连续投喂和灯光引诱的投喂模式，日投喂量占鱼体重的2%左右，每天投喂24h；且采用四种粒径依次不同的颗粒饵料，根据幼鱼的口径大小，增强幼鱼的适口性，大大提高了幼鱼的摄食率，显著提高了幼鱼的成活率、饵料利用率以及生长速度，减少了因抢食导致的个体生长差异。

参考文献：钱晓明，朱永祥，刘大勇，等．2015．一种龙纹斑幼鱼的仿自然生态投喂模式．CN201510125321.9。

注：无英文摘要。

67 一种新品种龙纹斑池塘生态养殖技术（专利）

摘要：本发明涉及一种龙纹斑池塘生态养殖技术，具体包括养殖池塘准备、培育水生植物、培育鲜活饵料生物链、微生态制剂调节水质、投喂人工配合饲料五方面。通过本发明的工艺养殖出的龙纹斑，生长快、营养均衡、免疫力强、含肉率高、成活率保持在95%以上，并且保留有野生龙纹斑的鲜味，具有独特的品质、风味和丰富的营养，经济效益和生态效益得到显著提高。

参考文献：蔡星，秦巍仑，周忠良．2014．一种新品种龙纹斑池塘生态养殖技术．CN201410146331.6。

注：无英文摘要。

68. 一种新品种龙纹斑的自然繁殖方法（专利）

摘要：本发明涉及一种新品种龙纹斑的自然繁殖方法，龙纹斑从澳洲引进中国的新品种，澳洲在南半球气候、季节与中国不同，本发明通过：①建造保温亲鱼培育池，②培育亲鱼生物饵料，③设置亲鱼鱼巢，④培育亲鱼，⑤增加水流刺激性腺发育，⑥产卵及收集受精卵，⑦受精卵孵化，⑧仔鱼培育，创造与澳洲相似自然生态气候、环境的繁殖条件，实现了龙纹斑在中国自然繁殖。

参考文献：蔡星，周忠良，秦巍仑．2014．一种新品种龙纹斑的自然繁殖方法．CN201410146971.7。

注：无英文摘要。

69. 一种新品种龙纹斑的人工繁殖方法（专利）

摘要：本发明涉及一种新品种龙纹斑的人工繁殖方法，模仿该新品种在原产地澳洲繁殖条件，包括如下几个步骤：建造控温亲鱼培育池，亲鱼鱼巢设置，亲鱼培育，催产，人工受精，受精卵人工孵化，仔鱼培育，通过这种方法，龙纹斑的人工繁殖取得成功，填补国内空白。

参考文献：蔡星，周忠良，郭正龙．2015．一种新品种龙纹斑的人工繁殖方法. CN201410146334. X。

注：无英文摘要。

70. 一种通用型集约化循环养殖池（专利）

摘要：本发明公开了1种通用型集约化循环养殖池，包括水体循环系统、环绕微孔增氧系统、终端检测系统3个部分，其中水体循环系统包括排水管筒、卸放管路、抬升水箱、进水管路、锥形储水进水器，环绕微孔增氧系统包括PVC增氧管、进气管路、空气压缩机。该养殖池不仅满足多元化养殖要求，而且能够持续稳定增氧和高效的水利用率。

参考文献：钱晓明，金加余，温松来. 2017. 一种通用型集约化循环养殖池. CN201710612275. 4。

注：无英文摘要。

71. 一种用于养殖池水过滤回用处理方法（专利）

摘要：本发明公开了1种用于养殖池水过滤回用处理方法，包括养殖池、弧形网筛槽、泡沫分馏器、筛绢网槽、生物膜净化池、填充式流化床及相应的管路和泵压管路，养殖池中的水依次经过弧形网筛槽过滤、泡沫分馏器分离、筛绢网槽2次过滤、生物膜净化池分解、填充式流化床吸附。本发明有效的过滤、分解了养殖水体中的残饵、排泄物等。同时通过网筛对于未食用的鱼饵起到了回收作用，大大节约了养殖过程中水资源，且能耗较低。

参考文献：钱晓明，温松来，凌诚，等. 2016. 一种用于养殖池水过滤回用处理方法. CN201610345392. 4。

注：无英文摘要。

72. METHOD FOR THE PRODUCTION OF FISH PROGENY（专利）

Abstract：The invention relates to a method for reducing the variation of genetic characteristics of fish progeny in a fish population containing groups varying in relation to the degree of relatedness. The invention also relates to a fish fry and fish progeny produced by this method. In the method of the invention, at least two fish groups (PA, PB) are formed of the fish population (P) so that the average genetic degree of relatedness (r) between the fish groups (RA, RB) is lower than in the fish population (P) on average; such members (RAK×RAN, RNK x RBN) of the fish group (RA, RB) are paired with each other, the average degree of relatedness (r) between which is higher than in the fish population (P) on average to form at least two separate parent generations (PA, PB), and members (PAN×PBK, PAK×PBN) of at least two separate parent generations (PA, PB) are cross-paired to form fish progeny (F1).

[译文]：

一种鱼类繁殖方法

摘要：本发明涉及一种减少鱼类种群中鱼类后代的遗传特征变化的方法，该方法包含不同于亲缘程度的群体。本发明还涉及该方法产生的鱼苗和鱼后代。在本发明的方法中，至少有两个鱼类种群（PA，PB）是由鱼类种群（P）组成的，因此鱼类种群（RA，RB）之间的亲缘关系的平均遗传程度（r）低于鱼类种群（P）的平均值；这样的成员（RAK×RAN，RNK×RBN）鱼的集团（RA，RB）相互搭配，平均程度的关联性（r）是高于鱼人口之间（P）平均形成至少两个独立的父代（PA、PB）和成员（PAN×PBK，PAK×PBN）至少两个独立的父代（PA、PB）交叉配对形成鱼后代（F1）。

参考文献：KAUSE ANTTI, MAENTYSAARI ESA, JAERVISALO OTSO. 2009. METHOD FOR THE PRODUCTION OF FISH PROGENY. FI：2007000301：W。

（罗钦、李冬梅、黄惠珍　译）

73. THERMAL CONTROLLED AQUACULTURE TANKS FOR FISH PRODUCTION（专利）

Abstract：A fish farming system is provided with a cooling tank (20) and a heating tank (22). The water in which the fish are bred is passed through heat exchangers (26, 36) in the heating and cooling tanks (20, 22) as required to maintain the water at a substantially constant temperature.

[译文]：

用于鱼类生产的热控制水产养殖罐

文摘：养鱼系统提供冷却罐（20）和加热槽（22）。养殖鱼类的水通过热交换器（26，36）在加热和冷却水箱（20，22）中，以保持水在相当恒定的温度下维持。

参考文献：MERRYFULL ALBERT. 2016. THERMAL CONTROLLED AQUACULTURE TANKS FOR FISH PRODUCTION. AU：2004000152：W。

（罗钦、李冬梅、黄惠珍　译）

第三章 澳洲龙纹斑疾病与防控

第一节 专题研究概况

澳洲龙纹斑抗病能力强，在其原产地主要是侵袭性鱼病，此外是传染性鱼病。危害较大的是小瓜虫病（*Ichthyophthirius multifllils*），发病时水温在 12~27℃，有时在孵化场的池塘内，由于放苗太密，一旦饵料供不应求，连续五天溶解氧低于 3.5mg/L 时，会导致澳洲龙纹斑因小瓜虫的侵袭而大批死亡。此外车轮虫（*Trichodina sp*）病于每年的 10 月至来年的 3 月是侵袭苗种的高峰期，常致其大量死亡。斜管虫（*Chilodonella hexasticha*）也不容忽视，在澳大利亚内陆的一个渔业站曾发生过澳洲龙纹斑斜管虫病，死亡率高达 24%。其他如杯体虫（*Apiosoma sp*），寄生甲壳类的锚头鱼蚤病（*Lernaea sp*）亦时有发生，其传染源来自野生鲤鱼。不容忽视的是澳洲龙纹斑在养殖过程中以及亲鱼鱼卵于冬季极易被水霉菌（*Saprolegnia sp*）传染。病毒性鱼病主要是 EHNV 病毒引起澳洲龙纹斑患出血性坏疽病（韩茂森，2003）。

澳大利亚鱼病防治药物与我国相似，福尔马林是常用药，养殖中经常用 NaCl 溶液浸浴鱼体。澳大利亚新威尔士州在防治澳洲龙纹斑苗种的车轮虫病时，曾采用 40mg/L 的福尔马林，但效果不佳。有研究者认为，水霉病是由真菌科的水霉菌（*Saprolegnia*）和绵霉菌（*Achlya*）引起，诱发原因是拉网、运输等过程中操作不细心，致使掉鳞或擦伤后而引起的。为避免因操作不慎而致病，运鱼时采用麻醉法，用 Benzocaine，剂量为 20mg/L，如需重麻醉，则用量为 50mg/L，同时所运亲鱼在入塘前再用 1.0mg/L 的甲基蓝浸浴，隔离检疫不少于 48h，然后再进行一次 1% 的 NaCl 溶液浸浴。对严重感染水霉的亲鱼可用饱和的高锰酸钾液，涂抹患处。在澳洲龙纹斑养殖史上，还发现过嗜水气单胞杆菌（*Aeromonas hydrophila*）引发的体表溃疡，澳大利亚的专家认为多数溃疡病，均与水质不良、管理不当等因素直接相关。一般认为，在澳洲龙纹斑养殖中保持良好的水质，保持养鱼设施的环卫条件，加强科学管理，增强鱼自身的抗病能力，必然会减少疾病的传染（韩茂森，2003）。

近年来，澳洲龙纹斑人工繁育已经成功实现，大规模人工养殖已悄然兴起。澳洲龙纹斑抗病能力较强，但其在养殖的过程中，也会受到各种鱼病的威胁。随着养殖规模不断扩大，澳洲龙纹斑各种病害问题也日益突出，澳洲龙纹斑对养殖条件要求严格，特别是在其养殖过程中发生的各种鱼病要注重防治。在卵化阶段主要威胁是由水霉引起的真菌性病害；鱼苗和成鱼阶段的主要威胁为侵袭性鱼病，危害较大的有车轮虫病，小瓜虫病，斜管虫病和盾纤毛虫病，另外聚缩虫病也不容忽视；在澳洲龙纹斑的生长过程中，细菌主要引起继发性感染，从而产生烂鳃病、烂尾病等细菌性疾病；另外由鱼病毒造血坏死病毒（epizootic haematopoietic necrosis virus（EHNV））引起的病毒性疾病也应受到高度重视，该病曾在澳洲本地养殖过程中造成过重大的经济损失。从1988—1993年，在南部春季中后期和夏季初期，通过收集捕获的养在池塘的亲鱼在产卵箱中的受精卵进行自然孵化，进行澳洲龙纹斑人工繁育。被称为"卵黄囊肿胀综合症"（SYSS）的疾病导致了澳洲龙纹斑卵和新孵出幼鱼的死亡率越来越高，卵和幼鱼的死亡通常发生在孵化和苗种培育阶段（Gunasekera，R. M.，1998；罗土炎，2015）。

通过对国内外文献检索、整理，关于澳洲龙纹斑病害研究的有刊载的文献最早出现于1998年，1998—2017年（20年）国外期刊上发表的澳洲龙纹斑病害研究文献共19条，年平均0.95篇。从不同年代发文期刊数和载文量来看是呈上升趋势，1998—2002年（5年）年平均发文量是0.2篇；2003—2007年（5年）年平均发文量是0.8篇；2008—2012年（5年）年平均发文量是1.0篇；2013—2017年（5年）年平均发文量是1.8篇；2013—2017年（5年）刊载的发文数是1998—2002年的9倍，表明澳洲龙纹斑病害的研究在不断加强，研究深度和研究领域在不断加深，2014年后澳洲龙纹斑病害的相关文献报道量激增，这可能与我国科研人员开始参与澳洲龙纹斑病害研究，并且得到相关部门的支持力度有关。

表3-1 1998—2017年澳洲龙纹斑病害研究文献量与发表刊数分布

发表年代	发表刊数/种	载文量/篇	平均年发表量/篇
1998—2002	1	1	0.2
2003—2007	4	4	0.8
2008—2012	3	5	1.0
2013—2017	6	9	1.8

核心作者是指发表该类文献最多的第一作者,通过对这 19 篇文献的署名分析,由表 3-2 可知,发表 3 篇的即为核心作者,核心作者共 2 人,分别是罗土炎和 Aaron G. Schultz。

表 3-2 文献核心作者群分析

作者名字	发表文章/篇
罗土炎	3
Aaron G. Schultz	3
Jeffrey Go	2
ANDREW J. HARFORD	2
罗钦	1
刘洋	1
Joy A. Becker	1
Ceaig A. Boys	1
Rona Barugahare	1
J E Baily	1
A E Rimmer	1
Rasanthi M. Gunasekera	1
Giana Bastos Gomes	1
合计	19

合作度是指每篇论文的平均作者数,合著率是合著论文与发表总论文的比例。由表 3-3 可知,文献的合著率为 100%,其中作者数为 5 人的最多,占 26.31%;其次为 3 人的,占 21.05%;而 9 人以上的合作论文无刊载。

表 3-3 作者合作情况分析

作者人数/人	篇数/篇	比例/%
2	1	5.26
3	4	21.05
4	2	10.05
5	5	26.31
6	3	15.79

(续表)

作者人数/人	篇数/篇	比例/%
7	3	15.79
8	1	5.26
合计	19	100

对澳洲龙纹斑病害19篇论文的产出单位进行统计,结果表明,主要产出单位集中在澳大利亚、中国和英国的学校和研究所,其中高校共计12篇、研究所共7篇。澳大利亚共13篇,占总论文数量的68.42%;中国共6篇,占总论文数量的31.58%;英国共1篇,占总文献数量的5.26%。以第一单位发表论文最多的是中国的福建省农业科学院农业质量标准与检测技术研究所和澳大利亚的The University of Sydney,都是5篇。详见表3-4。

表3-4 澳洲龙纹斑病害文献主要产出单位和区域分析

第一单位名称	发表文章/篇	国籍	单位性质
福建省农业科学院农业质量标准与检测技术研究所	5	中国	研究所
The University of Sydney	5	澳大利亚	高校
Deakin University	4	澳大利亚	高校
RMIT University	2	澳大利亚	高校
Marine and Aquaculture Sciences, College of Science and Engineering and Centre for Sustainable Tropical Fisheries and Aquaculture	1	澳大利亚	研究所
Port Stephens Fisheries Institute	1	澳大利亚	研究所
University of Stirling	1	英国	高校
合计	19		

澳洲龙纹斑病害研究文献主要期刊分布见表3-5,发表此类论文最多的期刊是《Aquaculture》,共5篇,其次是《福建农业学报》和《Journal of Fish Diseases》,各3篇。由此可见,发表此类论文较集中,刊载的期刊专业性较强,综合性期刊较少。

表3-5 澳洲龙纹斑病害研究文献主要期刊分布

期刊名称	刊载文量/篇
Aquaculture	5
福建农业学报	3

（续表）

期刊名称	刊载文量/篇
Journal of Fish Diseases	3
中国预防兽医学报	1
Aquatic Toxicology	1
Genome Announcements	1
Veterinary Parasitology	1
Environmental Toxicology & Chemistry	1
Journal of Aquatic Animal Health	1
PloS ONE	1
Molecular & Cellular Probes	1
合计	19

按照研究内容划分，人们对澳洲龙纹斑病害研究的范围越来越深和越来越广，涵盖了寄生虫病、病毒病、细菌病、真菌病和有毒有害物质中毒这五大病害。其中，澳洲龙纹斑病害研究最多的是寄生虫病，共9篇，其次是病毒病，共5篇，未明确病因的论文1篇。详见表6。由于国内外关于澳洲龙纹斑病害方面的研究开始时间较迟，研究内容较少，研究程度不深，整体现状表现为各国对澳洲龙纹斑病害不够重视。因此，我国引进澳洲龙纹斑时，相关检验检疫部门，应尽量对澳洲龙纹斑原水域的病原体和寄生虫进行检验检疫，尽可能从无病虫害的水域引进澳洲龙纹斑，同时做好消毒、隔离、防逃等工作，防止原产地的异病通过澳洲龙纹斑传播到我国，危害我国的水产养殖业。同时，我国应更加重视对澳洲龙纹斑病害方面的研究，给予更多的科研人员、科研经费和科学技术的投入，特别是在病毒病和细菌病方面的发现、诊断、鉴定、防治方案及其疫苗和特效药的研究，为培育澳洲龙纹斑这一优质高值的新兴产业提供重要的科技支持。

随着澳洲龙纹斑人工繁育养产业化发展，人们对澳洲龙纹斑病害的研究越来越深入，通过对国内外20篇文献的研究内容进行分析，澳洲龙纹斑病害的研究最多的是寄生虫病，占总文献数量的45%；占总文献数量的25%；未明确病因的1篇，占总文献数量的5%。详见表3-6。

表3-6 澳洲龙纹斑病害论文研究内容分析

病害分类	研究内容	篇数/篇
寄生虫	斜管虫病、隐孢子虫病、小瓜虫病、孢子寄生虫病	8（+1）

(续表)

病害分类	研究内容	篇数/篇
病毒	虹彩病毒病、造血坏死病毒病	4（+1）
有毒有害物质	重金属离子与消毒药物中毒、有机锡中毒、农药中毒	3
细菌	链球菌病	1（+1）
真菌	卵菌亚纲真菌病	1（+1）
综述	车轮虫病、小瓜虫病、斜管虫病、盾纤毛虫病、聚缩虫病、造血坏死病毒、虹彩病毒、RAN病毒、细菌性烂鳃病、细菌性烂尾病、真菌性水霉病。	1
未知病因	卵黄囊肿胀综合病症	1
合计		19

备注：（+1）表示把综述的这篇文章加入

一、真菌性水霉病的危害及其特征

体外寄生溃疡综合症（EUS）是一种全球性的鱼类疾病并且可向国际兽疫局报告。在2010年6月，从澳洲新南威尔士州的布鲁克和布雷沃里纳之间的墨瑞-达令河流域（MDRS）中的有严重溃疡的骨鲱鱼，黄金鲈鱼，澳洲龙纹斑和金色鲻鲈鱼中取样。组织病理学和聚合酶链反应识别出了真菌媒介丝囊霉卵菌，即体外寄生溃疡综合症的病原体。除了之前在骨鲱鱼上的一个记录之外，体外寄生溃疡综合症在野外仅在澳洲沿海流域被记录。本研究首次报道了真菌媒介丝囊霉在墨瑞-达令河流域的野生鱼群中，并且首次确认在黄金鲈鱼，澳洲龙纹斑和金色鲻鲈鱼中体外寄生溃疡综合症的记录。在同一流行期间收集的溃疡鲤鱼没有发现感染体外寄生溃疡综合症，支持先前关于该鱼类对这种疾病免疫的说法。由于缺乏之前的临床证据，在200km范围内这种流行病的大量新宿主（n=3）严重的溃疡和明显的高致病性表明相对近期被真菌媒介丝囊霉入侵。本流行病和相关的环境因素在其可能进入墨瑞-达令河流域的环境中进行了记录和讨论，并制定关于持续监测、研究和生物安全的建议（Boys，C.A.，2012）。

该病由卵菌纲中 *Saprolegnia* 属，*Achlya* 属和 *Aphanomyces* 属等多种霉菌感染引起，此类霉菌最适生长温度为13~18℃，28℃以上受到抑制，故水霉病25℃以下都会发生，一年四季都能感染。病鱼鱼体体表受伤组织和死卵容易受到感染，病鱼发病早期肉眼不易察觉，随着水霉菌在伤口处大量繁殖，向内、外生长并扩散，向内入浸上皮和真皮组织产生内菌丝，向外生长出外菌丝，外菌丝形似灰白色棉絮（图3-1），严重时患处肌肉腐烂，体表黏液增多，行动迟钝，食欲减退，最后瘦弱死亡（罗土炎，2015）。

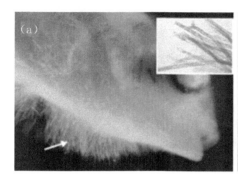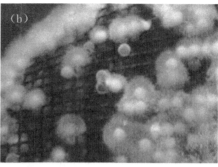

图3-1 水霉菌感染澳洲龙纹斑

（a）水霉菌感染死去的鱼，其中斜箭头指示的为霉菌菌丝；（b）澳洲龙纹斑鱼卵感染水霉菌的情况。参考图片来至于Ingram et al（罗土炎，2015）

二、侵袭性鱼病及其特征

（一）车轮虫病的危害及其特征

车轮虫病病原为车轮虫或小车轮虫，二者的差别是小车轮虫无向中心的齿棘，体表寄生的车轮虫形体较大，鳃上寄生的则略小。车轮虫主要寄生在澳洲龙纹斑体表和鳃上，大量寄生时，使鱼苗、鱼种成群绕池狂游，呈"跑马"症状；造成鱼体衰弱，运动迟缓，不摄食；刺激鳃和鱼体体表产生大量黏液，呼吸困难；常造成鱼苗大量死亡。严重感染时，鱼昂头浮游，皮肤出现蓝灰色斑块，幼鱼体表常因虫体刺激而炎性水肿变得破烂不堪。诊断时很难通过外观来判断，必须通过镜检进行确诊（罗土炎，2015）。

（二）小瓜虫病的危害及其特征

小瓜虫病病原为 *Ichthyophthirius multifiliis*，其生长周期分为营养体和游离体两种状态：营养体寄生到宿主上皮细胞下，肉眼下可见白色斑点，营养体可以在宿主之间传播；营养体成熟后脱离宿主，形成被囊膜包裹的分裂体，分裂体在囊膜内以二分裂的形式产生幼虫。此病的危害性较大，幼鱼、成鱼均易发生。在病鱼的鳃、皮肤、鳍条上，肉眼可见0.5~1.0mm的白色针尖状小点胞囊，严重时体表覆盖一层白色薄膜。严重感染时因病灶处细菌继发感染，使体表发炎，或局部鳞片脱落，鳍条烂裂。病鱼食欲下降，鱼体消瘦，反应迟钝，或漂浮水面。最终因细菌继发感染引发败血症而死亡（罗土炎，2015）。

(三) 斜管虫病的危害及其特征

斜管虫寄生在澳洲龙纹斑口腔、鳃部、体表，刺激寄主分泌大量黏液，使寄主皮肤表面形成苍白色或淡蓝色的黏液层。病鱼食欲差，鱼体消瘦发黑，在水中急躁不安，病鱼离群独游，头昂到水面以上或侧卧水中，呼吸困难，不久即可导致死亡（罗土炎，2015）。

斜管虫在全球对淡水鱼养殖造成了相当大的经济损失。一些斜管虫的遗传研究已经表明许多种斜管虫在属内可能形成隐蔽种复合物。从热带北昆士兰（QLD），温带维多利亚州（Vic）和新南威尔士州（NSW）鱼场的受感染的尖吻鲈鱼和澳洲龙纹斑中分离出来斜管虫，并进行了遗传和形态学分析。形态分析显示有4种不同形态的斜管虫感染了人工养殖的尖吻鲈鱼和澳洲龙纹斑。从尖吻鲈鱼中分离出3种虫（斜管

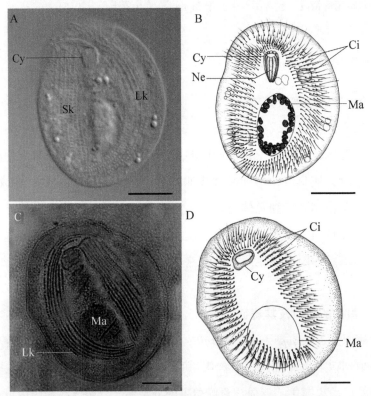

图3-2 推定的斜管虫（A，B）来自人工养殖的澳洲尖吻鲈的鳃部和皮肤，尖吻鲈在热带北部的昆士兰和已鉴定的斜管虫（C，D）来自人工养殖的澳洲龙纹斑的鳃部和皮肤，澳洲龙纹斑来自澳大利亚新南威尔士州。（A）新鲜的细胞样本显示短的（Sk）和长的（Lk）动胞器列圈和细胞口（Cy）。（B）手工画中的细胞纤毛（Ci）；细胞口（Cy）；口腔（Ne）和大细胞核（Ma）。（C）银色细胞（图片由Brett Ingram提供）；（D）细胞的手工画。其他的缩写词至于（A）和（B）。比例尺：10um。(Gomes, G. B. 2017)

虫，斜管蚓和斜管线虫，见图3-2、图3-3、图3-4）和在澳洲龙纹斑中分离出1种（斜管虫鱼蛭）。然而，系统发育分析仅检测到3种不同的基因型，其中斜管虫和斜管虫鱼蛭共享100%序列一致性，表明分离出的斜管虫和斜管虫鱼蛭是同一属（Gomes，G. B.，2017）。

（四）盾纤毛虫病的危害及其特征

盾纤毛虫病的病原为盾纤毛虫，一年四季均有发生，冬春两季为发病高峰。病鱼食欲减退，常聚群浮于水面，呼吸困难，体表黏液增多，体色变暗。病鱼发病初期出现口

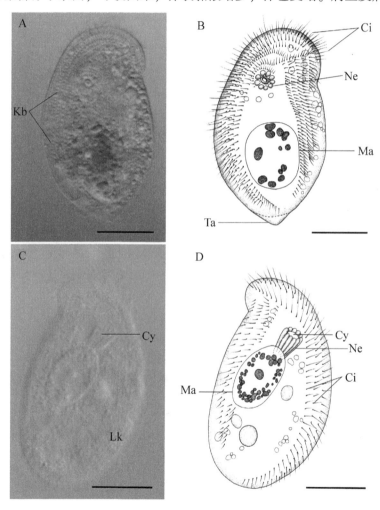

图3-3 推定的斜管虫属种（A，B）和钩刺斜管虫（C，D）最初是从人工养殖的尖吻鲈的鳃部和皮肤上分离得到，来自澳大利亚热带北部的昆士兰，并且保存在试管中。（A）新鲜的细胞样本用苏木精轻微染色以提高对比度：Kb=动胞器列圈；（B）手工画中的细胞尾部附器（Ta）。（C）新鲜的细胞样本；（D）细胞的手工画．其他的缩写词至于（A）和（B）。比例尺：10m。（Gomes，G. B.，2017）

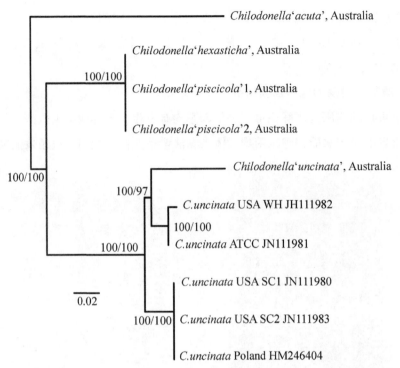

图3-4 不同种斜管虫属的关系是以贝叶斯推理和最大的线粒体 SSU-rDNA 可能性分析为基础的。节点（BI/ML）证明了后验概率和支持率在。系统发生图是正中央生根的。斜管虫属鱼蛭属1和2再次分别从维多利亚和新南威尔士中分离到。本研究中顺序的生成显示在加粗字体中（Gomes, G. B., 2017）。

唇部、鳃盖、鳍边缘发白的症状，严重者出现体表溃烂和充血，眼睛充血和皮下肌肉组织溃疡（图3-5）。镜检病灶部位发现有大量纤毛虫寄生；剖检发现腹腔积水，肝脏充血，脾脏花状，肾脏暗红，肠中有大量黄色黏液。而在成鱼的内脏组织器官极少发现纤毛虫。另外病灶处常与细菌复合感染，造成组织溃烂（罗土炎，2015）。

（五）聚缩虫病的危害及其特征

聚缩虫病的病原菌为聚缩虫，一年四季均有发生，养殖水质环境较差时尤为严重。聚缩虫主要寄生于鱼的鳍和受伤的体表，患病鱼苗漂浮水面，多于塘边或下风口处集群。病鱼食欲减退，甚至拒食，行动迟钝，病鱼常聚集到池壁或固定物摩擦体表，体表黏液增多。发病初期体表，尾鳍或背鳍边缘出现发白的斑点，有絮状物附着，严重者出现背鳍和尾鳍隆起，充血和溃烂（图3-6），并发细菌感染，引起死亡（罗土炎，2015）。

图 3-5 澳洲龙纹斑亲鱼感染盾纤毛虫的症状，白色箭头所指的位置为感染病灶。（a）澳洲龙纹斑感染盾纤毛虫后眼睛充血、突出；（b）盾纤毛虫感染皮下组织，尾部病灶处的溃疡组织在显微镜下可观察到病原体；（c）和（d）感染后期，澳洲龙纹斑出现多处溃疡（罗土炎，2015）。

图 3-6 感染聚缩虫的澳洲龙纹斑

（a）病灶位于背侧；（b）病灶在背鳍附近（罗土炎，2015）。

（六）隐孢子虫病的危害及其特征

Rona Barugahare（2011）首次报道了隐孢子虫感染澳洲龙纹斑（图 3-7 至图 3-13）。来自一个澳洲水产设施养殖场中的 3 种澳洲龙纹斑胃中的原生动物形态与隐孢子虫形态在组织结构上明显一致。感染隐孢子虫的澳洲龙纹斑死亡率低，食欲降低，生长慢，胃和小肠肿胀充液，鱼体发育不良并且游动异常。在显微镜下，传染病与轻度至中度的胃黏膜单核白细胞渗透及偶尔的胃黏膜上皮细胞坏死和脱落有关。使用分子生物学鉴定的系统发育重建与在鱼中所有可利用的隐孢子虫基因型证实了隐孢子虫属于同一种群（Rona Barugahare，2011）。

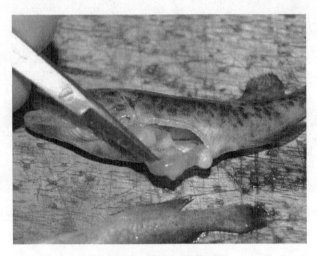

图 3-7 澳洲龙纹斑

与其他年份的鱼相比，这条鱼的尺寸非常小，而肠道内被液体膨胀（Rona Barugahare，2011）。

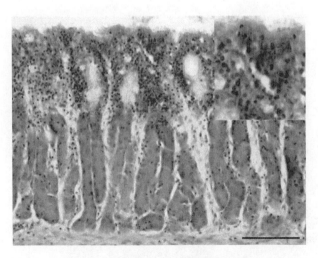

图 3-8 澳洲龙纹斑的胃隐孢子虫病

在超级胃黏膜内的单核白细胞（插图）。HE. Bar=80m（Rona Barugahare，2011）。

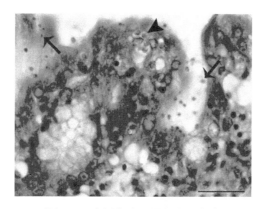

图 3-9 澳洲龙纹斑胃隐孢子虫病

几种胃隐孢子虫与表层（箭）有关系而且是在胃黏膜上皮细胞的细胞质（箭头）内。HE. Bar=25m（Rona Barugahare，2011）。

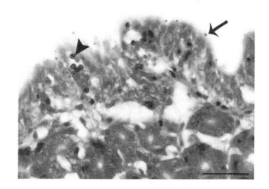

图 3-10 澳洲龙纹斑胃隐孢子虫病

胃隐孢子虫表层阶段染成深蓝色（箭）并且细胞质内阶段染成紫色（箭头）。Geimsa. Bar=25m（Rona Barugahare，2011）。

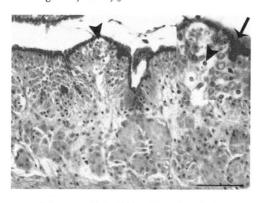

图 3-11 澳洲龙纹斑胃隐孢子虫病

胃隐孢子虫表层阶段染成淡蓝色（箭，插图）并且细胞质内阶段染成亮粉色（箭头）。Periodic acid-Schiff. Bar=50m（Rona Barugahare，2011）。

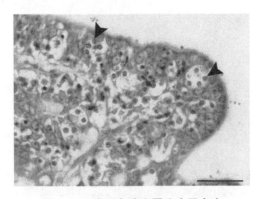

图 3-12 澳洲龙纹斑胃隐孢子虫病

胞浆内的卵原细胞阶段（箭头）染成淡紫色。Ziehl-Neelsen acid fast. Bar=25m（Rona Barugahare，2011）。

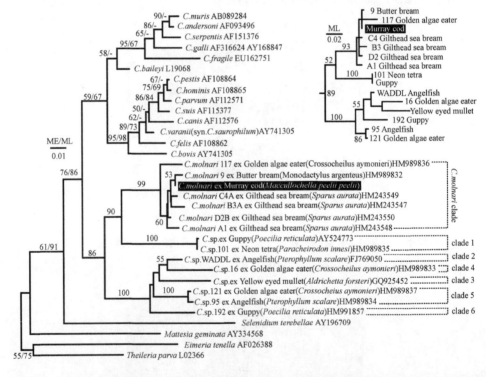

图 3-13 澳洲龙纹斑隐孢子虫病的系统发育关系是以分子生物学鉴定为基础的。系统发生图是用最小进化法（ME）（Tamura et al.，2007）和最大似然估计法（ML）（Guindon and Gascuel 2003）推测的。鱼的序列进化枝名称显示在右边。列队由693个位置组成并且执行了成对的删除选项以便重新构建ME树。这些线上面的数字代表了自引支持度是从使用ME（1000复制）和ML（100复制）的1000个复制中推算出来的。有不到50%的自引支持度的分支被收缩。树的根部是血虫（泰勒虫）和球虫（艾美球虫）。隐孢子虫的进化枝是在簇虫（Mattesia，Selenidium）序列中嵌入。插图，ML树对鱼类基因型进化枝和C. molnari分支有支持（Rona Barugahare，2011）。

三、细菌性疾病的危害及其特征

(一) 细菌性烂鳃病

细菌性烂鳃病的病原菌主要是柱状黄杆菌。细菌性烂鳃病主要危害鱼苗和成鱼,对幼苗的危害尤其严重,在养殖密度高、养殖水质恶化的养殖池塘易发生。病鱼游动缓慢,食欲不振,反应迟钝,呼吸困难,喜欢单独在池边或池角游动,体色加深,胸鳍基部充血;病鱼鳃丝腐烂带有污泥,不易洗去,鳃盖骨的内表皮往往充血,中间部分的表皮常腐蚀成一个圆形不规耻的透明小窗(俗称开天窗)。在显微镜下观察,澳洲龙纹斑鳃瓣感染细菌以后,引起的组织病变不是发炎和充血,而是病变区域的细胞组织呈现不同程度的腐烂、溃烂和"侵蚀性"出血等症状(罗土炎,2015)。

(二) 细菌性烂尾病

烂尾病病原菌主要为嗜水气单胞菌,杀鲑气单胞菌和柱状黄杆菌,发病季节大多集中在春夏季。烂尾病发病早期鱼尾柄处苍白,黏液流失,尾鳍慢慢被蛀蚀,然后尾柄尾鳍及尾鳍尾柄处充血,尾鳍鳍条开始溃烂。严重时鱼尾尾鳍大部分或全部断裂,尾柄肌肉出血、溃烂,骨骼外露,造成病鱼死亡(罗土炎,2015)。

慢性腐皮病(CED)是一种集约化养殖澳洲龙纹斑的疾病,其与使用地下水(从浅水井机械提取)养殖有关。慢性腐皮病导致鱼头部和两侧的皮肤局部溃疡。感染慢性腐皮病的澳洲龙纹斑,2~3周后病变出现,伴随着感觉气管毛孔周围的组织退化。这些毛孔随着气管的轴线扩张是由气管上的组织缺失造成。在组织病理学上,井水养殖3周后观察到管道上皮层增生,并且这个组织随后坏死和脱落导致通道顶层损毁(Baily,J. E.,2005)(图3-14至图3-25)。

慢性溃疡性皮肤病(CUD)是一种对在未处理的地下水中培养的澳洲龙纹斑造成腐皮影响的病。电解质富集或紫外线照射预处理可以延迟慢性溃疡性皮肤病的发生。经过覆盖着植被的土池或当代大片的人工造林预处理的地下水大大降低了慢性溃疡性皮肤病的发病率和严重程度,超过90%的鱼没有出现看得见的病症。血液学和血液参数被用来进行对未来诊断潜力的评估,但是即使在晚期受慢性溃疡性皮肤病感染的鱼中也没有观察到血液参数变化。流经土池和当代大片的人工造林的地下水处理方法是两种预防在澳洲龙纹斑幼鱼中发生慢性溃疡性皮肤病的高效方法(Aarong,S.,2011)。

Schultz(2008)调查了在集约化水产养殖中受慢性溃疡性皮肤病(CUD)影响的澳洲龙纹斑中的渗透压调节能力。这种情况似乎只出现在利用地下水的设施养殖中,

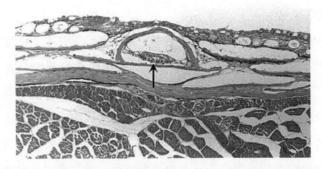

图 3-14 侧线管与神经末梢的横断面（↑），健康的澳洲龙纹斑（H&E，bar ¼ 100μm）（Baily，J.E.，2005）。

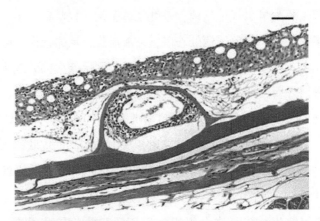

图 3-15 一条健康金鱼的侧线管横截面（H&E，bar ¼ 100 μm）（Baily，J.E.，2005）。

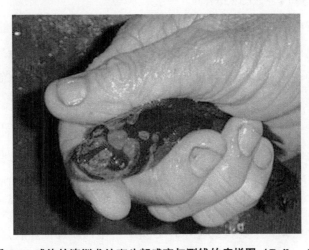

图 3-16 受 CED 感染的澳洲龙纹斑头部感官与测线的病样图（Baily，J.E.，2005）

图 3-17 受 CED 感染的澳洲龙纹斑鳍部（Baily，J. E.，2005）

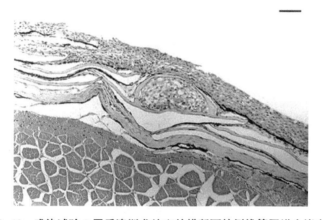

图 3-18 感染试验 3 周后澳洲龙纹斑的横断面的侧线管因增生堵塞了
（H&E，bar $^{1/4}$ 100μm）（Baily，J. E.，2005）。

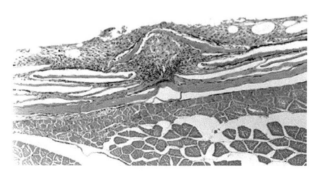

图 3-19 感染试验 4 周后澳洲龙纹斑的侧线管的横断面与增生组织堵塞
并且这个组织显示坏死而且延伸到隐伏鳞片（Baily，J. E.，2005）。

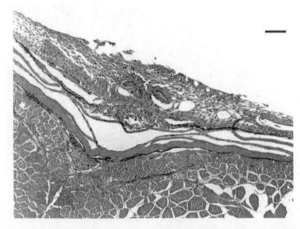

图 3-20 感染试验 5 周后澳洲龙纹斑的横断面增生和侧线管的坏死与表皮组织的脱落（H&E，bar $^{1/4}$ 100μm）（Baily, J. E., 2005）。

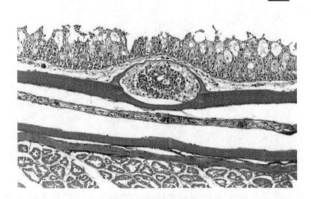

图 3-21 感染试验 2 周后金鱼的侧线管的横断面含有增生和坏死上皮细胞。（H&E，bar $^{1/4}$ 50μm）（Baily, J. E., 2005）。

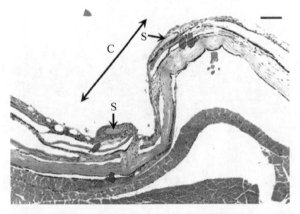

图 3-22 感染试验 4 周后澳洲龙纹斑的侧线管的横断面（C）显示缺乏正常管道织构和大量碎片的出现（S）（H&E，bar $^{1/4}$ 50μm）（Baily, J. E., 2005）。

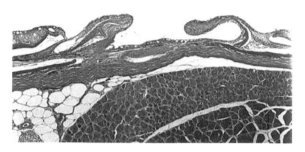

图3-23 运行系统2个月后澳洲龙纹斑的再生侧线管截面（H&E，bar ¼ 100μm）（Baily，J. E.，2005）。

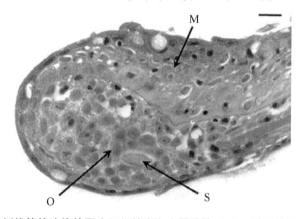

图3-24 在侧线管的边缘的再生组织的脊和大量碎片（S），活跃的骨母细胞（O）和马氏细胞（M）（H&E，bar ¼ 10μm）（Baily，J. E.，2005）。

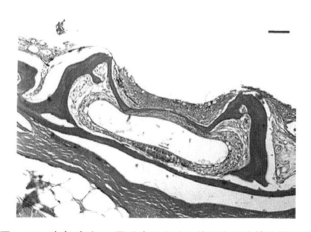

图3-25 康复试验10周后澳洲龙纹斑的再生侧线管的横断面（H&E，bar ¼ 25μm）（Baily，J. E.，2005）。

病原体是通过水传播的。将健康的澳洲龙纹斑（约700g）转移到受慢性溃疡性皮肤病感染的养殖场中，监测该症状的进展，并且约5个月后开始出现慢性溃疡性皮肤病的症状。受慢性溃疡性皮肤病感染的澳洲龙纹斑的血浆电解质浓度和渗透压被确定与非受慢性溃疡性皮肤病感染的鱼的参考值一致。与非受慢性溃疡性皮肤病感染的鱼相比，在受慢性溃疡性皮肤病感染的养殖场中培养的澳洲龙纹斑观察到更多数量的鳃黏液细胞。我们还发现一种未确定的细胞类型，仅存在于受慢性溃疡性皮肤病感染的澳洲龙纹斑的鳃中。与转移到受慢性溃疡性皮肤病感染的养殖场的个体相比，受严重的慢性溃疡性皮肤病感染的澳洲龙纹斑的鳃中 Na^+，K^+-三磷酸核苷酸活性显着升高（Schultz，2008）。

A G Schultz（2014）在受慢性溃疡性皮肤病（CUD）感染的澳洲龙纹斑鳃中鉴定出一种未知的细胞种类（图3-26至图3-29），慢性溃疡性皮肤病是一种导致头部周围的表皮和侧感线严重侵蚀的情况。这种情况出现在利用地下水的水产设施养殖中，原因可能是一种未知污染物。光镜和透射电镜被用于受慢性溃疡性皮肤病感染的澳洲龙纹斑中

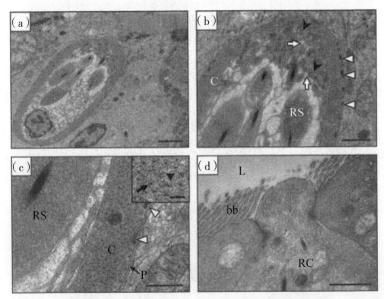

图3-26　(a) 一个澳洲龙纹斑肾小管成熟的杆状细胞的电子显微照片。(b) 线粒体（▶）和囊泡（⇨）在胶囊中的杆状细胞和外围斑块（▷）的尖端区。(c) 胶囊（C）的高倍放大电子显微照片在质膜（P）的下面围绕这杆状细胞。注，高电子密度的外围斑块（▷）在质膜的下面并且插图，厚的（→）和薄的（▶）丝状体在胶囊里（比例尺＝1lm）。(d) 杆状细胞（RC）的顶端区释放它的细胞内容物到一个肾近球小管的管腔（L）里。注，近端小管的刷毛缘（bb）。RS，小棒囊。比例尺：a＝2μm；b and d＝1μm；and c＝0.5μm（A G Schultz，2014）。

未知细胞的形态描述和量化。细胞被鉴定为罗氏细胞,其特征是椭圆形或圆形,细胞核位于基部,厚的纤维囊包围着细胞,以及在细胞内含有一个中央电子密集核的多个小棒状囊。在非受慢性溃疡性皮肤病感染和受慢性溃疡性皮肤病感染的澳洲龙纹斑鳃,肾和肠中存在罗氏细胞,然而,在鱼群之间罗氏细胞被观察到的数量存在差异。在鱼鳃和受慢性溃疡性皮肤病感染的鱼导管中明显地观测到更多的细胞数。本文首次报道了在澳洲龙纹斑中的罗氏细胞,并且我们认为在受慢性溃疡性皮肤病感染的澳洲龙纹斑中增加的细菌数量可能对使慢性溃疡性皮肤病增加的存在于地下水中的知污染物有反应(A G Schultz,2014)。

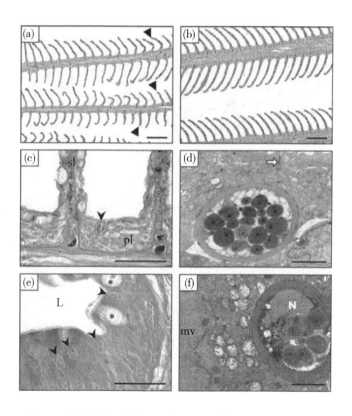

图3-27 (a)受CUD感染的澳洲龙纹斑的鳃和(b)未受CUD感染的澳洲龙纹斑的鳃被染成亮绿色和曙红色。所有受CUD感染的澳洲龙纹斑鳃部有不规则的鳃瓣(▶)。(c)杆状细胞(➤)是位于主要鳃瓣(pl)的上皮层,在中等鳃瓣(sl)之间。(d)澳洲龙纹斑鳃的杆状细胞的电子显微照片。一个紧紧的结点(⇨)能在上皮细胞间看见。(e)未受CUD感染的澳洲龙纹斑的肠道被染成亮绿色和曙红色。杆状细胞是位于肠道的上皮层。(f)澳洲龙纹斑肠道杆状细胞的电子显微照片。L,管腔;(*)黏液细胞;N,核;和mv,微绒毛。比例尺,a,b=200μm;c,e=50 lm;and d,f=2μm(A G Schultz,2014)。

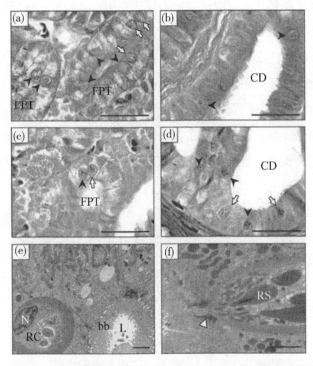

图3-28 （a，b）受CUD感染的和（c，d）未受CUD感染的澳洲龙纹斑肾脏被染成亮绿色和曙红色。杆状细胞（▶）是位于澳洲龙纹斑第一个近端小管（FPT）和（b，d）收集管（CD）的上皮层（a，c）。一些杆状细胞（⇨）明显释放它们的细胞内容物到近端小管和收集管腔中。（e）澳洲龙纹斑肾脏中典型的杆状细胞（RC）的电子显微照片。（f）一个杆状细胞在释放它的细胞内容物进入近端小管腔前的纵断面。注，细胞质的扩大（▷）在细胞的顶端区。L，管腔；bb，刷毛缘；N，核；RS，小棒囊。比例尺：a–d = 50μm；e = 5μm；and f = 2μm（A G Schultz，2014）。

（三）链球菌病

Liu，Y.（2016）从淡水澳洲龙纹斑中分离出来的X13SY08可能呈现出一种新型的链球菌。并提出了一个注释该物种的基因组序列草图，这将改善对其生理和发病机制的理解（Liu，Y.，2016）。

四、病毒性疾病的危害及特征

流行性造血坏死病毒（EHNV）是澳洲南部特有的疫情，自然暴发导致野生红鳍鲈鱼（也称为欧亚鲈鱼）大规模死亡。澳洲龙纹斑接触病毒后没有感染病毒，说明澳洲龙纹斑不是EHNV的宿主（Becker，J. A.，2013）。

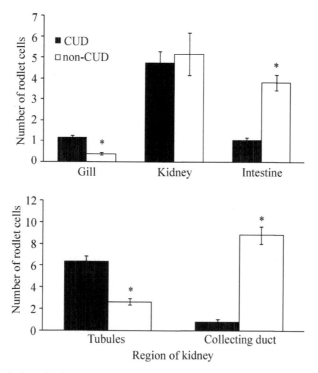

图 3-29 （a）鳃部、肾脏和肠道中杆状细胞的数量（样品面积=66189μm²），和受 CUD 感染和未受 CUD 感染的澳洲龙纹斑的肾脏的（b）第一个近端小管和收集管。数值是以平均±SE（n=5）表示。星号（*）表示任何重大的受 CUD 和未受 CUD 感染之间的变化（方差分析，$P<0.05$）（A G Schultz，2014）。

巨噬细胞病毒引起的高死亡率疾病严重影响水产养殖，在东亚和东南亚最频繁的爆发。作为养殖菜鱼和观赏鱼的鱼苗的国际贸易有助于这些病毒的传播，这些病毒已经蔓延到欧洲和澳洲等地区。巨噬细胞病毒（DGIV）是一种具有传染的脾脏和肾脏坏死性病毒（ISKNV）。鱼被恒温在23±1℃并且通过腹腔内（IP）注射或与含巨噬细胞病毒的澳洲龙纹斑养在一起进行感染。根据 qPCR 测定和组织学观察到的巨噬细胞病毒的包涵体，某一物质被认为对巨噬细胞病毒敏感。在携带病毒的澳洲龙纹斑和黄金鲈鱼，麦克夸里鱼，麦考利鲈鱼，澳洲麦氏鲈鱼以及澳洲龙纹斑之间病毒发生平行传播（图3-30）。这表明，从23℃恒温养殖的感染鱼中脱落的巨噬细胞病毒可以在淡水中存活，随后感染这些幼鱼。此外，巨噬细胞病毒通过腹腔内注射对澳洲龙纹斑高度致病（Rimmer，A. E.，2017）。

在2003年2月，维多利亚水产设施养殖中由一种巨细胞病毒（虹彩病毒科）引起的澳洲龙纹斑苗死亡率为90%。该疾病其与公认的传染性脾肾坏死病毒（ISKNV）和虹

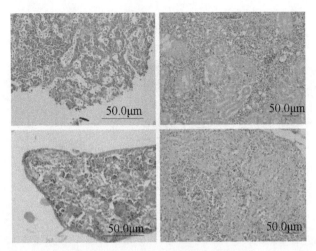

图3-30 来自受DGIV感染鱼中试验的典型的组织病理学损伤（苏木精和伊红染液）范围
(a) 澳洲麦氏鲈的前肾和特征多形性嗜碱性包涵体细胞（箭）不存在普遍的或个别的细胞坏死。
(b) 大范围坏死的澳洲麦氏鲈鱼脾脏与嗜碱性的嗜碱性包涵体细胞（箭）结合在一起。注意有几个大的浅棕色着色黑素噬菌体中心。(c) 澳洲龙纹斑肾大范围坏死的造血组织。红细胞和肾小管未受影响。(d) 澳洲龙纹斑鱼脾具有典型的多形性嗜碱性包涵体细胞（箭）和普遍的白髓坏死，使红细胞稀少（Rimmer, A. E., 2017）。

彩病毒株（DGIV）密切相关，这两个以前都没有在澳洲的养殖或野生鱼类中报道过。对澳洲龙纹斑幼鱼腹腔注射经PCR确定阳性巨细胞病毒DNA的丽丽鱼的组织匀浆过滤液体导致死亡率>90%。澳洲龙纹斑幼鱼和丽丽鱼共栖也会引起死亡；当鱼体被分割时，病毒通过水在两种鱼类之间传播。组织病理学显示病灶与维多利亚爆发的报道相同，在病灶中观测到125~130μnm的二十面体病毒粒子而且大多数暴露的鱼经PCR检测都是阳性的。DNA测序证实在虹彩病毒、传染性脾肾坏死病毒、从丽丽鱼中获得的病毒培养液和出现于实验感染的澳洲龙纹斑中的病毒之间大部分的衣壳蛋白和三磷酸核苷酸序列有99.9到100%的同源性。这些发现证实了澳洲龙纹斑对从东南亚进口的观赏鱼中存在的巨细胞病毒非常敏感（图3-31、图3-32）（Go, J., 2006）。

从2003年因患流行病而死的人工养殖的澳洲龙纹斑中和从亚洲进口的患病丽丽鱼中提取组织通过福尔马林固定、石蜡包埋进行磷酸二氢钙，三磷酸核苷酸和其他病毒基因测序，该组织来自2003年被亚洲进口的患流行病死亡的丽丽鱼。澳洲龙纹斑的虹彩病毒（MCIV）和2004年死在澳洲水族馆里的丽丽鱼的虹彩病毒（DGIV）几乎是完全相同（99.95%），超过4527bp，并且与传染性脾肾坏死病毒（ISKNV）有很高的同源性（99.9%）。这些病毒最有可能是巨细胞病毒属中的单一物种并且可能起源于同一地

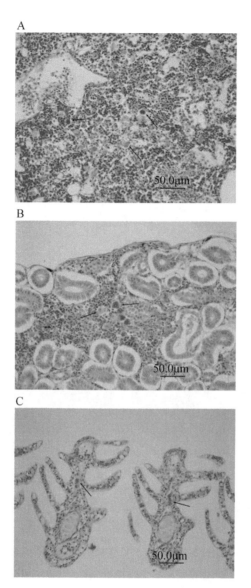

图 3-31 首次受 IP 感染试验的 2 尾澳洲龙纹斑中的嗜碱性的肥大细胞（箭头指示的地方）。苏木精和伊红。A，前肾；B，后肾；C，鳃（Go, J., 2006）。

点。开发了用于特异性 PCR 的引物和用于快速鉴别澳洲龙纹斑的虹彩病毒/丽丽鱼的虹彩病毒/传染性脾肾坏死病毒与红海鲷鱼虹彩病毒（RSIV）（一种应用病原体）的引物（图 3-33）。用这些引物检测出悉尼零售水族馆里丽丽鱼虹彩病毒（DGIV）的患病率为 22%（95%置信区间为 15%~31%）。观赏鱼的全球贸易可能促进巨细胞病毒的传播，并且使遥远生物地理区域的新宿主物种能够出现疾病（Go, J., 2006）。

目前已报道的鱼病毒如造血坏死病毒、虹彩病毒、类似小核糖核酸的病毒和双

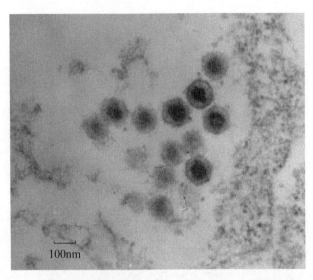

图 3-32 首次受 IP 感染试验的澳洲龙纹斑鱼种的后肾上的二十面体病毒体 from the first IP injection trial（Go, J., 2006）。

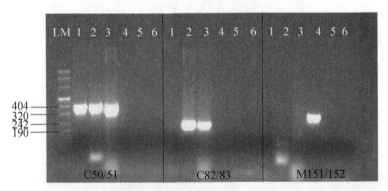

图 3-33 PCR 检测巨噬细胞病毒的检测和对红海鲷病毒的鉴别和动物造血坏死病毒的鉴别。丽丽鱼的虹彩病毒（DGIV-2004），使用病毒基因特异性 MCP 引物 C50/C51 的 PCR 仪测量虹彩病毒（RSIV）和流行性胰坏死病毒（EHNV），MCP 引物适合于虹彩病毒群体内部的病毒 C82/C83 并且 MCP 引物适合于流行性胰坏死病毒内部的病毒 M151/M152. LM，分子大小标准参照物（bp）；在每一板：条带 1，RSIV；条带 2，MCIV；条带 3，DGIV-2004；条带 4，EHNV；条带 5，对照组水样；条带 6，PCR 混合试验组（Go, J., 2006）。

RAN 病毒（birnavirus）等，虹彩病毒（Lancaster M J，2003）和 EHNV 病毒有感染澳洲龙纹斑案例，其中虹彩病毒的感染造成的死亡率很高。病鱼食欲不振，昏睡，感染后 4~7d 死亡。病鱼的脾脏，造血组织，鳃组织和心脏组织内发现嗜碱性溶血颗粒。在澳洲本土养殖过程中，发生了一个严重的案例，该案例中虹彩病毒引起 4~6cm 的鱼苗大量死亡，死亡率在90%以上；温度达到26~27℃时，10~15cm的鱼也受到了感染，死亡率在25%以上；目前没有报告成鱼有临床表现（罗土炎，2015）。

五、有毒有害物质的危害及特征

澳洲龙纹斑是一种标志性的澳洲本土淡水鱼，该物种是澳洲最大河系但也是许多环境污染物最终沉积的墨瑞达令河流域上的本地鱼类。有机三丁基锡（TBT）和二丁基锡（DBT）是常见的淡水和海洋环境污染物，同时众所周知哺乳动物和水生生物对它们具有免疫毒性。在本研究中，三丁基锡和二丁基锡被用作样本抗毒素去评估免疫功能测定的效率（丝裂原刺激的淋巴细胞，在头肾组织的吞噬作用，巨噬细胞吞噬功能和血清溶菌酶活性），并去比较澳洲龙纹斑与其他鱼类的敏感性。在之前报道的虹鳟鱼的亚致死浓度有机锡是澳洲龙纹斑的致死浓度。体内三丁基锡暴露在0.1和0.5mg/kg之间会刺激澳洲龙纹斑的吞噬功能，而三丁基锡在最高浓度2.5mg/kg时减少了淋巴细胞数量并且细胞有丝分裂发生。二丁基锡在澳洲龙纹斑中是更有效的免疫抑制剂，可以造成吞噬活性显著降低和淋巴细胞数显著减少（Harford，A.J.，2010）。

Harford，A.J.（2005）研究了常用农药暴露对澳洲龙纹斑的头肾细胞的吞噬细胞功能和细胞合成的影响。从澳洲龙纹斑中分离出头肾免疫细胞，并且在先前最佳的测量荧光微球的吞噬作用的条件下培养（1×10（6）细胞/ml）。将这些细胞培养物暴露于3种农药中，即有机锡，三丁基锡和二丁基锡，有机氯农药硫丹和有机磷农药毒死蜱。体外有机锡暴露对头肾细胞具有高度地免疫。在最高浓度（即10mg/L）下，硫丹暴露导致澳洲龙纹斑发生吞噬细胞反应的转调。毒死蜱显示出很小的免疫毒性，尽管在澳洲龙纹斑淋巴细胞中呈剂量依赖性降低（Harford，A.J.，2005）。

六、澳洲龙纹斑养殖过程中主要病害的诊断和预防

澳洲龙纹斑养殖过程中主要病害包括水霉病、车轮虫病、小瓜虫病、斜管虫病、盾纤毛虫病、聚缩虫病、细菌性烂鳃病、细菌性烂尾病、病毒性疾病。就健康养殖而言，除了把握科学饲养环节之外，重要的工作就是要严格防控疾病发生与及时的治疗。因此，在澳洲龙纹斑养殖实践中，要注重应用疾病预防措施，注重疾病的辨别与诊断，同时实施有效的防治。在此过程中要因地制宜制定防治方案，既能有效防控，又能防止环境污染。在实践中，总结了澳洲龙纹斑养殖过程中9种常见病诊断方法，预防措施，治疗方案以及要点说明等，列入表3-7之中（罗土炎，2015）。

表3-7 澳洲龙纹斑养殖过程中主要病害的诊断方法和预防措施（罗土炎，2015）

鱼病类型	诊断方法	预防措施	治疗方案	注意事项
水霉病	病鱼发病早期肉眼不易察觉，随着水霉菌在伤口处大量繁殖，外菌丝会形成形似灰白色棉絮。	(1) 清除池底过多淤泥及腐烂物质，使用生石灰彻底清塘，保持水质优良和养殖环境清洁；(2) 在放养、捕捞、运输过程中，仔细操作，避免鱼体受伤；(3) 及时清除死掉的鱼卵及鱼尸体。	(1) 使用食盐和小苏打混合液 400mg/L 水体全池泼洒；新洁尔灭 5mg/L 水体全池泼洒；水霉净（水杨酸）0.3mg/L 水体全池泼洒，每天 1 次，连用 3 次。(2) 在上述药剂处理的同时，在投喂的饲料中添加适量抗菌类药物，连续投喂 5~7d，以防继发细菌感染。水霉病一旦发生，很难治疗，所以要以预防为主，防治结合。	尽量避免鱼体受伤，特别要注意选别、搬动等操作
车轮虫病	目检：鱼成群绕池狂游，呈"跑马"症状；昂头浮游，皮肤出现蓝灰色斑块；鱼体破烂不堪。镜检：镜检确诊病原微生物。	(1) 彻底清塘，杀灭水体中和底泥中的病原；(2) 鱼苗、鱼种下塘前使用 2%~3% 食盐溶液消毒 10~15min。	采用 0.05~0.1mg/kg 的车轮净，药浴 16h，严重时重复药浴 16h，同时添加抗生素，如 0.5~1mg/kg 的 10% 氟苯尼考药物制剂，预防细菌感染并发症发生。	有机质过多时注意该病发生
小瓜虫病	目检：病鱼体表、鳍条、鳃上均发现有 0.5~1mm 的白色小点状突起，个别重度病鱼全身皮肤和鳍条布满白点，全身体表如同覆盖着一层白色薄膜，黏液增多，体色暗淡无光。病鱼消瘦，多漂浮于水面不游动或缓慢游动。病鱼经常呈浮头状漂浮于水面。镜检：用载玻片刮取病鱼尾鳍或体表白点涂片，可发现小瓜虫成虫在活动。	采用生石灰清塘、鱼种消毒、合理放养等措施防止小瓜虫病的发生与传播。	根据小瓜虫的生活史（一半的时间处于营养体状态），通过调节水体温度来调节小瓜虫的生长周期，可进行有效治疗；用 1~3mg/kg 的戊二醛和 1~2mg/kg 的鱼康乐浸泡病鱼 5d 即可杀灭小瓜虫，感染严重时注意添加抗生素，如 0.5~1mg/kg 的 10% 氟苯尼考药物制剂，预防细菌并发症发生。	(1) 多发生 pH 变化高的水质中，例如从高 pH 迅速降到低 pH 等；(2) 该病多发生在下雨天

(续表)

鱼病类型	诊断方法	预防措施	治疗方案	注意事项
斜管虫病	目检：斜管虫寄生在鱼的鳃及皮肤上，大量寄生时可引起黏液增多、呼吸困难、游动缓慢而死。常在鳃上与其他寄生虫病并发。 镜检：必须用显微镜进行检查确诊。	放养前用生石灰彻底清塘。鱼种放养前用3%～4%的食盐溶液浸浴5min。保持水质良好。	使用1～3mg/kg的戊二醛和1～2mg/kg的鱼康乐浸泡2d即可杀灭斜管虫。	多发生在有机质含量高的水质中
盾纤毛虫病	目检：病鱼发病初期出现口唇部、鳃盖、鳍边缘发白的症状，严重者出现体表溃烂和充血，眼睛充血和皮下肌肉组织溃疡。 剖检：发现腹腔积水，肝脏充血，脾脏花状，肾脏暗红，肠中有大量黄色黏液。 镜检：确认盾纤毛虫病。	盾纤毛虫感染目前还没有很好的治疗方法，预防措施为彻底清塘，保持水质良好。放养密度不宜过大，4～6cm的鱼苗，每立方水体放养量不超过200尾。调节水质，定期泼洒益生微生物。	用鱼康乐拌料口服（0.5～1g/kg），连用5d，对盾纤毛虫防治有一定的效果。	多发生在有机质含量高的水质中
聚缩虫病	目检：病鱼食欲减退，甚至拒食，行动迟钝，病鱼常聚集到池壁或固定物摩擦体表，体表黏液增多。发病初期体表，尾鳍或背鳍边缘出现发白的斑点，有絮状物附着，严重者出现背鳍和尾鳍隆起，充血和溃烂。 镜检：确认聚缩虫病。	（1）保持水质良好。定期使用含氯消毒剂进行水体消毒。饵料中定期添加Vc、电解多维和免疫多糖，提高鱼体的免疫力； （2）放养前用生石灰彻底清塘。鱼种放养前用3%～4%的食盐溶液浸浴5min。	使用0.05～0.1mg/kg的水体纤虫净能有效将其杀灭。	多发生在有机质含量高的水质中
细菌性烂鳃病	单独在池边或池角游动，体色加深，胸鳍基部充血；病鱼鳃丝腐烂带有污泥，不易洗去，鳃盖骨的内表皮往往充血，中间部分的表皮常腐蚀成一个圆形不规则的透明小窗（俗称开天窗）。 镜检：病变区域的细胞组织呈现不同程度的腐烂、溃烂和"侵蚀性"出血等症状。	（1）保持池塘养殖水良好状况，合理投料，确保饲料中不含有病原菌；（2）及时排污换水，定期使用底质改良剂、EM等微生态制剂调节池塘水质；（3）在疾病流行季节，可用生石灰彻底清塘消毒，或用15～20mg/L的生石灰全池泼洒进行预防，每隔10d使用1次。	（1）0.5～1.0mg/L苯扎溴铵水溶液全池泼洒，2～3d使用1次，连用2～3次；（2）0.8～1.5mg/L的聚维酮碘水溶液全池泼洒；（3）拌料投喂1～2g/kg的抗菌药物土霉素或2～3g/kg 10%的氟苯尼考，连喂5～7d。	夏季是该病的流行季节

(续表)

鱼病类型	诊断方法	预防措施	治疗方案	注意事项
细菌性烂尾病	烂尾病发病早期鱼尾柄处苍白，黏液流失，尾鳍慢慢被蛀蚀，然后尾柄尾鳍及尾鳍柄处充血，尾鳍鳍条开始溃烂。严重时鱼尾尾鳍大部分或全部断裂，尾柄肌肉出血、溃烂，骨骼外露。	（1）保持池塘良好水质，养殖过程细心操作，尽量避免鱼体受伤，定期水体消毒；（2）鱼苗、鱼种下塘前使用2%～3%的食盐溶液或8～10mg/L高锰酸钾水溶液浸洗10～15min进行鱼体消毒。	用1%高碘酸钠溶液1.5～2.0mg/L水体全池泼洒，2～3d使用1次，连用2次；内服抗菌药物盐酸土霉素1～2g/kg饲料投喂。	养殖过程中尽量避免鱼体损伤
病毒性疾病	目检：食欲不振，昏睡，感染后4~7d死亡。镜检：病鱼的脾脏、造血组织、鳃组织和心脏组织内发现嗜碱性溶血颗粒。	切断虹彩病毒和EHNV病毒来源，新购买的鱼苗先放到暂养池养殖10d左右，确保没有感染后转移到精养池。	一旦发现有鱼苗出现类似的症状，立刻根据鱼苗的大小调节水体温度，尽量避免鱼苗的大量死亡。	避免引入病原

（罗钦、刘洋、饶秋华）

第二节 研究论文摘编

74. Draft Genome Sequence of Streptococcus sp. X13SY08, Isolated from Murray Cod (Maccullochella peelii peelii)[①]

[英文摘要]：Streptococcus sp. X13SY08, isolated from freshwater Murray cod fish, likely presents a novel species of Streptococcus. Here, we present an annotated draft genome sequence of this species, which will improve our understanding of its physiology and pathogenesis.

[译文]：

从澳洲龙纹斑中分离出来的链球菌的基因组序列 sp. X13SY08

从淡水澳洲龙纹斑中分离出来的链球菌 sp. X13SY08 可能呈现出一种新型的链球菌。在这里，我们提出了一个注释该物种的基因组序列草图，这将改善我们对其生理和发病机制的理解。

参考文献：Liu, Y., Zeng, R., Weng, B., *et al.* 2016. Draft genome sequence of streptococcus sp. x13sy08, isolated from murray cod (maccullochella peelii peelii) [J]. Ge-

---

① 研究论文摘编序号续上一章，全书同。

nome Announcements, 4 (1), e01470-15.

(罗钦、李冬梅、黄惠珍　译)

75. Characterisation of 'swollen yolk-sac syndrome' in the Australian freshwater fish Murray cod, Maccullochella peelii peelii, and associated nutritional implications for large scale aquaculture

[英文摘要]: The Murray cod is artificially propagated by harvesting naturally spawned fertilised eggs laid in nest boxes by captive, pond-reared broodfish, in mid to late southern spring and early summer. From 1988-1993, a condition referred to as 'swollen yolk-sac syndrome' (SYSS) accounted for increasingly significant mortalities of Murray cod eggs and newly hatched larvae, often resulting in total mortality of eggs and larvae during incubation and larval rearing stages.

[译文]：

澳洲淡水鱼鳕鱼"卵黄囊肿胀综合症"的特性及其对大规模水产养殖的相关营养影响

在南部春季中后期和夏季初期，进行澳洲龙纹斑人工繁育，通过收集捕获的养在池塘的亲鱼在产卵箱中的受精卵进行自然孵化。从1988—1993年，被称为"卵黄囊肿胀综合症"（SYSS）的疾病导致了澳洲龙纹斑卵和新孵出幼鱼的死亡率越来越高，卵和幼鱼的死亡通常发生在孵化和苗种培育阶段。

参考文献：Gunasekera, R. M., Gooley, G. J., Silva, S. S. D. 1998. Characterisation of 'swollen yolk-sac syndrome' in the australian freshwater fish murray cod, maccullochella peelii peelii, and associated nutritional implications for large scale aquaculture [J]. Aquaculture, 169 (1-2), 69-85.

(罗钦、李冬梅、黄惠珍　译)

76. Emergence of epizootic ulcerative syndrome in native fish of the Murray-Darling River System, Australia: hosts, distribution and possible vectors

[英文摘要]: Epizootic ulcerative syndrome (EUS) is a fish disease of international significance and reportable to the Office International des Epizootics. In June 2010, bony herring Nematalosa erebi, golden perch Macquaria ambigua, Murray cod Maccullochella peelii and spangled perch Leiopotherapon unicolor with severe ulcers were sampled from the Murray-Darling River System (MDRS) between Bourke and Brewarrina, New South Wales Australia. Histopathology and polymerase chain reaction identified the fungus-like oomycete Aphanomyces invadans, the causative agent of EUS. Apart from one previous record in N. erebi, EUS has been recorded in the wild only from coastal drainages in Australia. This study is the first published account of A. invadans in the wild fish populations of the MDRS, and is the first confirmed record of EUS in M. ambigua, M. peelii and L. unicolor. Ulcerated carp Cyprinus carpio collected at the time of the same epizootic were not found to be infected by EUS, supporting previous accounts of resistance against the disease by this species. The lack of previous clinical evidence, the large number of new hosts (n=3), the geographic extent (200 km) of this epizootic, the severity of ulceration and apparent high pathogenicity suggest a relatively recent invasion by A. invadans. The epizootic and associated environmental factors are documented and discussed within the context of possible vectors for its entry into the MDRS and recommendations regarding continued surveillance, research and biosecurity are made.

[译文]:

在澳洲墨瑞-达令河流域本地鱼类中体外寄生溃疡综合症的出现：宿主、分布和可能的载体

体外寄生溃疡综合症（EUS）是一种全球性的鱼类疾病并且可向国际兽疫局报告。在2010年6月，从澳洲新南威尔士州的布鲁克和布雷沃里纳之间的墨瑞-达令河流域（MDRS）中的有严重溃疡的骨鲱鱼，黄金鲈鱼，澳洲龙纹斑和金色鲻鲈鱼中取样。组织病理学和聚合酶链反应识别出了真菌媒介丝囊霉卵菌，即体外寄生溃疡综合症的病原体。除了之前在骨鲱鱼上的一个记录之外，体外寄生溃疡综合症在野外仅在澳洲沿海流域被记录。本研究首次报道了真菌媒介丝囊霉在墨瑞-达令河流域的野生鱼群中，并且首次确认在黄金鲈鱼，澳洲龙纹斑和金色鲻鲈鱼中体外寄生溃疡综合症的记录。在同一流行期间收集的溃疡鲤鱼没有发现感染体外寄生溃疡综合症，支持先前关于该鱼类对这种疾病免疫的说法。由于缺乏之前的临床证据，在200km范围内这种流行病的大量新宿主（n=3）严重的溃疡和明显的高致病性表明相对近期被真菌媒介丝囊霉入侵。本流行病和相关的环境因素在其可能进入墨瑞-达令河流域的环境中进行了记录和讨论，并制定关于持续监测，研究和生物安全的建议。

参考文献：Boys, C. A., Rowland, S. J., Gabor, M., et al. 2012. Emergence of epizootic ulcerative syndrome in native fish of the murray-darling river system, australia: hosts, distribution and possible vectors [J]. Plos One, 7 (4), e35568.

（罗钦、李冬梅、黄惠珍　译）

77. The pathology of chronic erosive dermatopathy in Murray cod, Maccullochella peelii peelii (Mitchell)

[英文摘要]：Chronic erosive dermatopathy (CED) is a disease of intensively farmed Murray cod in Australia that has been reported in association with the use of groundwater (mechanically extracted from shallow boreholes) supplies. CED results in focal ulceration of the skin overlying sensory canals of the head and flanks. Trials were conducted at an affected fish farm to study the development of the condition, both in Murray cod and in goldfish, and also to assess the reported recovery of lesions when affected fish were transferred to river water. Grossly, lesions began after 2~3 weeks with degeneration of tissue at the periphery of pores communicating with the sensory canals. Widening of these pores along the axis of the canals resulted from a loss of tissue covering the canal. Histopathologically, hyperplasia of the canal epithelial lining was seen after 3 weeks in borehole water and subsequent necrosis and sloughing of this tissue resulted in the loss of the canal roof. Canal regeneration occurred when fish were transferred from borehole water into river water. The lack of lesions in other organs and the pattern of lesion development support exposure to waterborne factors as the most likely aetiology.

[译文]：

澳洲龙纹斑中慢性腐皮病的病理学研究

慢性腐皮病（CED）是一种集约化养殖澳洲澳洲龙纹斑的疾病，据报道其与使用地下水（从浅水井机械提取）养殖有关。慢性腐皮病导致鱼头部和两侧的感觉神经上的皮肤局部溃疡。在受感染的鱼场进行了试验，研究澳洲龙纹斑和金鱼病情的发展，并评估了当受感染的鱼类被转为河水养殖时病灶可痊愈的报道。大体上，2~3周后病变出现，伴随着感觉气管毛孔周围的组织退化。这些毛孔随着气管的轴线扩张是由气管上的组织缺失造成。在组织病理学上，井水养殖3周后观察到管道上皮层增生，并且这个组织随后坏死和脱落导致通道顶层损毁。当鱼从井水转到河水中养殖时发生管道再生。其他器官没有病变和病变发展模式支持暴露在水性因素是最有可能的病因。

参考文献：Baily, J. E., Bretherton, M. J., Gavine, F. M., *et al*. 2005. The pathology of chronic erosive dermatopathy in murray cod, maccullochella peelii peelii (mitchell) [J]. Journal of Fish Diseases, 28 (1), 3.

（罗钦、李冬梅、黄惠珍 译）

78. Groundwater pre-treatment prevents the onset of chronic ulcerative dermatopathy in juvenile Murray cod, Maccullochella peelii peelii (Mitchell)

[英文摘要]: Chronic ulcerative dermatopathy (CUD) is a disfiguring condition affecting Murray cod cultured in untreated groundwater. This study sought to further investigate the possible etiology of the syndrome and determine whether groundwater pre-treatment could suppress the development of CUD in juvenile Murray cod. Electrolyte enrichment or pre-treatment with UV irradiation delayed the onset of CUD. In contrast, pre-conditioning of groundwater either in a vegetated earthen pond or in the presence of artificial macrophytes drastically reduced both the incidence and severity of CUD, with more than 90% of fish exhibiting no visual signs. Haematology and blood parameters were assessed for future diagnostic potential, but no changes in blood parameters were observed, even in advanced CUD-affected fish. This paper identified that the treatment of groundwater via an earthen pond and in the presence of an artificial macrophyte are two highly effective methods of preventing CUD arising in juvenile Murray cod.

[译文]:

地下水预处理预防了澳洲龙纹斑幼鱼慢性溃疡性皮肤病的发生

慢性溃疡性皮肤病（CUD）是一种对在未处理的地下水中培养的澳洲龙纹斑造成腐皮影响的病。本研究旨在进一步研究该症状可能的病因并且确定地下水预处理是否能抑制慢性溃疡性皮肤病在墨瑞鳕幼鱼中的发展。电解质富集或紫外线照射预处理可以延迟慢性溃疡性皮肤病的发生。相比之下，经过覆盖着植被的土池或当代大片的人工造林预处理的地下水大大降低了慢性溃疡性皮肤病的发病率和严重程度，超过90%的鱼没有出现看得见的病症。血液学和血液参数被用来进行对未来诊断潜力的评估，但是即使在晚期受慢性溃疡性皮肤病感染的鱼中也没有观察到血液参数变化。本文认为，流经土池和当代大片的人工造林的地下水处理方法是两种预防在墨瑞鳕幼鱼中发生慢性溃疡性皮肤病的高效方法。

参考文献: Aarong, S., Sarahl, S., Paull, J., et al. 2011. Groundwater pre-treatment prevents the onset of chronic ulcerative dermatopathy in juvenile murray cod, maccullochella peelii peelii (mitchell) [J]. Aquaculture, 312 (1), 19-25.

（罗钦、李冬梅、黄惠珍 译）

79. Osmoregulatory balance in Murray cod, Maccullochella peelii peelii (Mitchell), affected with chronic ulcerative dermatopathy

[英文摘要]: This study examined the osmoregulatory capability of Murray cod, Maccullochella peelii peelii, affected by chronic ulcerative dermatopathy (CUD) in intensive aquaculture. This condition appears to arise only in facilities utilizing groundwater, with the causative agent suggested to be a water-borne factor. Healthy Murray cod (~700 g) were transferred to a CUD-affected farm to monitor the progression of the syndrome and began to show signs of CUD after approximately five months. In order to evaluate possible effects of CUD on osmoregulation; plasma electrolyte concentrations, osmolality, and Na^+, K^+-ATPase activities were measured, and gill histology and immunohistochemistry were analyzed. Plasma electrolyte concentrations and osmolality of CUD-affected Murray cod were consistent with reference values determined for non CUD-affected fish. A greater number of gill mucous cells were observed in Murray cod cultured at the CUD-affected farm compared to non CUD-affected fish. We also found an un-identified cell type that was present solely in the gills of CUD-affected Murray cod. Gill Na^+, K^+-ATPase activity was significantly higher in severely CUD-affected Murray cod compared to individuals transferred to the CUD-affected farm. While there appeared to be some minor changes in the gills of CUD-affected fish, this study demonstrated that Murray cod were able to effectively osmoregulate, although, perhaps at an energetic cost.

[译文]：

澳洲龙纹斑中渗透压调节平衡对慢性溃疡性皮肤病的影响

本研究调查了在集约化水产养殖中受慢性溃疡性皮肤病（CUD）影响的澳洲龙纹斑中的渗透压调节能力。这种情况似乎只出现在利用地下水的设施养殖中，病原体是通过水传播的。将健康的澳洲龙纹斑（约700g）转移到受慢性溃疡性皮肤病感染的养殖场中，监测该症状的进展，并且约五个月后开始出现慢性溃疡性皮肤病的症状。为了评估慢性溃疡性皮肤病对渗透调节的可能影响；测量血浆电解质浓度，渗透压，Na^+，K^+-三磷酸核苷酸活性，并且分析鳃部组织学和免疫组织化学。受慢性溃疡性皮肤病感染的澳洲龙纹斑的血浆电解质浓度和渗透压被确定与非受慢性溃疡性皮肤病感染的鱼的参考值一致。与非受慢性溃疡性皮肤病感染的鱼相比，在受慢性溃疡性皮肤病感染的养殖场中培养的澳洲龙纹斑观察到更多数量的鳃黏液细胞。我们还发现一种未确定的细胞类型，仅存在于受慢性溃疡性皮肤病感染的澳洲龙纹斑的鳃中。与转移到受慢性溃疡性皮肤病感染的养殖场的个体相比，受严重的慢性溃疡性皮肤病感染的澳洲龙纹斑的鳃中Na^+，K^+-三磷酸核苷酸活性显着升高。虽然在受慢性溃疡性皮肤病感染的鱼的鳃部显示有一些微小的变化，本研究表明澳洲龙纹斑能够有效地渗透调节，尽管可能是个能量消耗。

参考文献：Schultz, A G, Healy, J M, Jones, P L, et al. 2008. Osmoregulatory balance in Murray cod, Maccullochella peelii peelii (Mitchell), affected with chronic ulcerative dermatopathy [J]. Aquaculture, 280, 45-52.

（罗钦、李冬梅、黄惠珍　译）

80. Rodlet cells in Murray cod, Maccullochella peelii peelii (Mitchell), affected with chronic ulcerative dermatopathy

[英文摘要]：We have previously identified an unknown cell type in the gills of Murray cod affected with chronic ulcerative dermatopathy (CUD), a condition that causes severe erosion of epidermis surrounding cephalic and lateral line sensory canals. The condition arises in aquaculture facilities that utilize groundwater, with the cause of the condition suggested to be an unknown contaminant (s). Light and transmission electron microscopy were used to characterize and quantify the unknown cells in CUD-affected Murray cod. The cells were identified as rodlet cells and were characterized by their oval or round shape, basally located nucleus, thick fibrillar capsule surrounding the cell, and multiple rodlet sacs containing a central electron-dense core within the cell. Rodlet cells were present in the gills, kidney and intestine of non-CUD-affected and CUD-affected Murray cod; however, differences in the numbers were observed between the groups of fish. A significantly greater number of rodlet cells were observed in the gills and collecting ducts of CUD-affected fish. This is the first report of rodlet cells in Murray cod, and we suggest that the increased rodlet cell numbers in CUD-affected Murray cod may be in response to unknown water contaminant (s) present in the groundwater that give rise to CUD.

[译文]：

澳洲龙纹斑中罗氏细胞受慢性溃疡性皮肤病的影响

我们之前在受慢性溃疡性皮肤病（CUD）感染的澳洲龙纹斑鳃中鉴定出一种未知的细胞种类，慢性溃疡性皮肤病是一种导致头部周围的表皮和侧感线严重侵蚀的情况。这种情况出现在利用地下水的水产设施养殖中，原因可能是一种未知污染物。光镜和透射电镜被用于受慢性溃疡性皮肤病感染的澳洲龙纹斑中未知细胞的形态描述和量化。细胞被鉴定为罗氏细胞，其特征是椭圆形或圆形，细胞核位于基部，厚的纤维囊包围着细胞，以及在细胞内含有一个中央电子密集核的多个小棒状囊。在非受慢性溃疡性皮肤病感染和受慢性溃疡性皮肤病感染的澳洲龙纹斑鳃，肾和肠中存在罗氏细胞，然而，在鱼群之间罗氏细胞被观察到的数量存在差异。在鱼鳃和受慢性溃疡性皮肤病感染的鱼导管中明显地观测到更多的细胞数。本文首次报道了在澳洲龙纹斑中的罗氏细胞，并且我们认为在受慢性溃疡性皮肤病感染的澳洲龙纹斑中增加的细菌数量可能对使慢性溃疡性皮肤病增加的存在于地下水中的知污染物有反应。

参考文献：A G Schultz, P L Jones, T Toop. 2014. Rodlet cells in murray cod, maccullochella peelii peelii, (mitchell), affected with chronic ulcerative dermatopathy [J]. Journal of Fish Diseases, 37 (3), 219-228.

（罗钦、李冬梅、黄惠珍　译）

81. Gastric cryptosporidiosis in farmed Australian Murray cod, Maccullochella peelii peelii

[英文摘要]：Protozoa morphologically consistent with Cryptosporidium sp. were histologically-evident in the stomach of three cohorts of Murray cod (Maccullochella peelii peelii) from a single aquaculture facility in Australia. One cohort was asymptomatic. A second cohort had shown low-level mortalities, reduced appetite, and slowed growth, and fluid distension of the stomach and small intestine. Fish of the third cohort, sampled a year following diagnosis of the second cohort, were of stunted size and had displayed abnormal swimming behaviour. Microscopically, infection was associated with mild to moderate gastric mucosal mononuclear leukocyte infiltrate and occasional gastric mucosal epithelial cell necrosis and sloughing. Cryptosporidium molnari was confirmed in one cohort using SSU rDNA phylogenetic reconstruction with all available Cryptosporidium genotypes in fish. This study is the first to report Cryptosporidium sp. infection in Murray cod. Associated pathology and clinical signs highlight the possibility of production losses for the affected industry.

[译文]：

人工养殖澳洲澳洲龙纹斑中的胃隐孢子虫病

来自一个澳洲水产设施养殖场中的3种澳洲龙纹斑胃中的原生动物形态与隐孢子虫形态在组织结构上明显一致。一组无症状。第二组群显示死亡率低，食欲降低，生长慢，胃和小肠肿胀充液。第三组病鱼是在诊断出第二组病鱼一年后取样，其显示鱼体发育不良并且游动异常。在显微镜下，传染病与轻度至中度的胃黏膜单核白细胞渗透及偶尔的胃黏膜上皮细胞坏死和脱落有关。使用分子生物学鉴定的系统发育重建与在鱼中所有可利用的隐孢子虫基因型证实了隐孢子虫属于同一种群。本研究首次报道了隐孢子虫感染澳洲龙纹斑。相关的病理和临床症状突显了对受影响行业生产的损失的可能性。

参考文献：Rona Barugahare, Joy A. Becker, Matt Landos, et al. 2011. Corrigendum to "gastric cryptosporidiosis in farmed australian murray cod, maccullochella peelii peelii" [J].

Aquaculture, 318 (3-4), 483-483.

(罗钦、李冬梅、黄惠珍 译)

82. Evidence of multiple species of Chilodonella (Protozoa, Ciliophora) infecting Australian farmed freshwater fishes

[英文摘要]: Parasitic Chilodonella species, Chilodonella piscicola and Chilodonella hexasticha, cause considerable economic losses globally to freshwater farmed fish production. Some genetic studies of Chilodonella spp. have indicated that many species within the genus may form cryptic species complexes. To understand the diversity of Chilodonella spp. infecting Australian freshwater farmed fish, specimens were isolated from infected barramundi (Lates calcarifer) and Murray cod (Maccullochella peelii) from fish farms in tropical north Queensland (QLD), temperate Victoria (Vic) and New South Wales (NSW) for genetic and morphological analysis. Parasites were stained and measured for morphological description and comparative phylogenetic analyses were performed using the mitochondrial small subunit (mtSSU) rDNA marker. Morphological analyses revealed four distinct morphotypes of Chilodonella infecting farmed barramundi and Murray Cod. Three putative species were isolated from barramundi (Chilodonella hexasticha, C. acuta and C. uncinata) and one from Murray cod (C. piscicola). However, phylogenetic analyses detected only three distinct genotypes, with the putative C. hexasticha and C. piscicola sharing 100% sequence identity. This suggests that Australian isolates of C. hexasticha and C. piscicola could represent the same species and may exhibit phenotypic plasticity. Further molecular analysis, including isolates from the type localities, should be performed to support or refute the synonymy of these species.

[译文]:

多种斜管虫属（原生动物，纤毛虫门）感染澳洲养殖淡水鱼的证据

寄生的斜管虫种，斜管虫属和斜管虫在全球对淡水鱼养殖造成了相当大的经济损失。一些斜管虫种的遗传研究已经表明许多种斜管虫在属内可能形成隐蔽种复合物。为了了解感染在澳洲人工养殖淡水鱼中的斜管虫种的多样性，从热带北昆士兰（QLD），温带维多利亚州（Vic）和新南威尔士州（NSW）鱼场的受感染的的澳洲肺鱼（尖吻鲈）和澳洲龙纹斑中分离出来样品，用于遗传和形态学分析。寄生虫被染色并且对该虫进行形态学描述和发育系统测量，发育系统的比较分析采用线粒体小亚基（mtSSU）rDNA进行标记。形态分析显示有四种不同形态的斜管虫感染了人工养殖的尖吻鲈鱼和澳洲龙纹斑。从尖吻鲈鱼中分离出3种虫（斜管虫，斜管蚧和斜管线虫）和在澳洲龙纹斑中分离出1种（斜管虫鱼蛭）。然而，系统发育分析仅检测到3种不同的基因型，其中斜管虫和斜管虫鱼蛭共享100%序列一致性。这表明在澳洲分离出的斜管虫和斜管虫鱼蛭是同一属并且表型可塑性。将来对包括特定地区分离物的分子分析，主要执行支持或反驳这些种类的同物异名。

参考文献：Gomes, G. B., Miller, T. L., Vaughan, D. B., et al. 2017. Evidence of multiple species of chilodonella, (protozoa, ciliophora) infecting australian farmed freshwater fishes [J]. Veterinary Parasitology, 237, 8-16.

（罗钦、李冬梅、黄惠珍 译）

83. Experimental Infection of Australian Freshwater Fish with Epizootic Haematopoietic Necrosis Virus（EHNV）

[英文摘要]：The ranavirus, epizootic hematopoietic necrosis virus (EHNV), is endemic to southern Australia with natural outbreaks resulting in mass mortality events in wild Redfin Perch Perca fluviatilis (also known as Eurasian Perch) and less severe disease in farmed Rainbow Trout Oncorhynchus mykiss. To further investigate the host range for EHNV, 12 ecologically or economically important freshwater fish species from southeastern Australia were exposed experimentally to the virus. A bath-challenge model at 18±3℃ was employed with limited use of intraperitoneal inoculation to determine if a species was likely to be susceptible to EHNV. Of the species tested, Murray-Darling Rainbowfish Melanotaenia fluviatilis and Dewfish Tandanus tandanus (also known as Freshwater Catfish) were considered to be potentially susceptible species. EHNV was isolated from approximately 7% of surviving Eastern Mosquitofish Gambusia holbrooki, indicating this widespread alien fish species is a potential carrier. The infection of Silver Perch Bidyanus bidyanus and Macquarie Perch Macquaria australasica and the lack of infection in Murray Cod Maccullochella peelii peelii and Golden Perch Macquaria ambigua ambigua after exposure to EHNV via water confirmed earlier data from Langdon (1989). Five other species of native fish were potentially not susceptible to the virus or the fish were able to recover during the standard 35-d postchallenge observation period. Overall, it appeared that EHNV was less virulent in the present experimental model than in previous studies, but the reasons for this were not identified.

[译文]：

流行性造血坏死病毒（EHNV）感染澳洲淡水鱼的实验

蛙科病毒属，流行性造血坏死病毒（EHNV）是澳洲南部特有的疫情，自然暴发导致野生红鳍鲈鱼（也称为欧亚鲈鱼）大规模死亡和人工养殖的彩虹鳟鱼轻微的生病。为了进一步调查 EHNV 的宿主范围，来自澳洲东南部的 12 种具有重要生态或经济效益的淡水鱼在实验中接触病毒。在 18±3℃ 下采用 bath-challenge 模型有限的使用腹膜内接种来确定某种鱼类是否可能对流行性造血坏死病毒敏感。在所测试的鱼类中，墨累-达令彩虹鱼和澳洲鲶鱼（又称淡水鲶鱼）被认为是潜在的易感染鱼类。从约 7% 的幸存的东部食蚊鱼中分离出流行性造血坏死病毒，表明这种分布广的外来鱼类是潜在的携带者。银鲈和澳洲麦氏鲈感染病毒以及澳洲龙纹斑和黄金鲈鱼没有感染病毒，在来自伦敦（1989）的通过水接触流行性造血坏死病毒的早期实验数据后就被证实了。另外五种本地鱼类也可能对病毒不敏感，或者在受感染后正常养殖 35d 期间内能够观察到鱼类康复。总之，在当前实验模型中流行性造血坏死病毒的毒性比早期实验小，但造成这样的原因尚未被确定。

参考文献：Becker, J. A., Tweedie, A., Gilligan, D., et al. 2013. Experimental infection of australian freshwater fish with epizootic haematopoietic necrosis virus (ehnv) [J]. Journal of Aquatic Animal Health, 25 (1), 66-76.

（罗钦、李冬梅、黄惠珍　译）

84. Susceptibility of a number of Australian freshwater fishes to dwarf gourami iridovirus (Infectious spleen and kidney necrosis virus)

[英文摘要]：Megalocytiviruses cause high mortality diseases that have seriously impacted aquaculture, with the most frequent outbreaks occurring in East and South-East Asia. The international trade of juvenile fish for food and ornamental aquaculture has aided the spread of these viruses, which have spread to Europe and Australia and other regions. Australian freshwater fishes were examined for susceptibility to infection with the exotic megalocytivirus, dwarf gourami iridovirus (DGIV), which belongs to a group with the type species, Infectious spleen and kidney necrosis virus (ISKNV). Fish were held at 23±1℃ and challenged by intraperitoneal (IP) injection or by cohabitation with Murray cod, Maccullochella peelii (Mitchell) infected with DGIV. A species was deemed to be susceptible to DGIV based on evidence of viral replication, as determined by qPCR, and megalocytic inclusion bodies observed histologically. Horizontal transmission occurred between infected Murray cod and golden perch, Macquaria ambigua (Richardson), Macquarie perch, Macquaria australasica (Cuvier) and Murray cod. This indicated that DGIV shed from infected fish held at 23℃ can survive in fresh water and subsequently infect these naïve fish. Further, DGIV administered

IP was highly pathogenic to golden perch, Macquarie perch and Murray cod. Compared to these species, the susceptibility of southern pygmy perch, Nannoperca australis (Gunther) was lower. Freshwater catfish (dewfish), Tandanus tandanus (Mitchell), were not susceptible under the experimental conditions based on the absence of clinical disease, mortality and virus replication. This study showed the potential risks associated with naive and DGIV-infected fish sharing a common water source.

[译文]：

几种澳洲淡水鱼对虹彩病毒（传染性脾脏和肾脏坏死病毒）的敏感性

巨噬细胞病毒引起的高死亡率疾病严重影响水产养殖，在东亚和东南亚最频繁的暴发。作为养殖菜鱼和观赏鱼的鱼苗的国际贸易有助于这些病毒的传播，这些病毒已经蔓延到欧洲和澳洲等地区。探讨了澳洲的淡水鱼对外来的巨噬细胞病毒（DGIV）感染的敏感性，该病毒是一种具有传染的脾脏和肾脏坏死性病毒（ISKNV）。鱼被恒温在23±1℃并且通过腹腔内（IP）注射或与含巨噬细胞病毒的澳洲龙纹斑养在一起进行感染。根据qPCR测定和组织学观察到的巨噬细胞病毒的包涵体，某一物质被认为对巨噬细胞病毒敏感。在携带病毒的澳洲龙纹斑和黄金鲈鱼，麦克夸里鱼，麦考利鲈鱼，澳洲麦氏鲈鱼以及澳洲龙纹斑之间病毒发生平行传播。这表明，从23℃恒温养殖的感染鱼中脱落的巨噬细胞病毒可以在淡水中存活，随后感染这些幼鱼。此外，巨噬细胞病毒通过腹腔内注射的对黄金鲈鱼，澳洲麦氏鲈鱼和澳洲龙纹斑高度致病。与这些鱼类相比，巨噬细胞病毒对南部矮鲈鱼澳洲矮鲈（Gunther）的敏感性较低。在基于缺乏临床疾病，死亡率和病毒复制基础的实验条件下，淡水鲶鱼（dewfish），澳洲鳗鲶（Mitchell）不易感染。本研究显示幼小的和携带巨噬细胞病毒的鱼苗共用同一水源存在潜在风险。

参考文献：Rimmer, A. E., Whittington, R. J., Tweedie, A., et al. 2017. Susceptibility of a number of australian freshwater fishes to dwarf gourami iridovirus (infectious spleen and kidney necrosis virus) [J]. Journal of Fish Diseases, 40 (3), 293-310.

（罗钦、李冬梅、黄惠珍　译）

85. Experimental transmission and virulence of a megalocytivirus (Family Iridoviridae) of dwarf gourami (Colisa lalia) from Asia in Murray cod (Maccullochella peelii peelii) in Australia

[英文摘要]：In February 2003 there were 90% losses of Murray cod (Maccullochella peelii peelii) fingerlings in a Victorian aquaculture facility. The disease was caused by a megalocytivirus (Family Iridoviridae) closely related to the recognized species Infectious Spleen and

Kidney Necrosis Virus (ISKNV) and strain dwarf gourami iridovirus (DGIV), neither of which had previously been reported from farmed or wild fish in Australia. Experimental transmission trials were undertaken to test the hypothesis that the outbreak could have arisen through introduction of a virus with ornamental gouramis imported from South East Asia. Intraperitoneal injection of Murray cod fingerlings using filtered tissue homogenates from dwarf gourami (Colisa lalia) positive for megalocytivirus DNA by polymerase chain reaction (PCR) resulted in>90% mortality. Mortality was also induced by cohabitating Murray cod fingerlings and dwarf gourami; as the fish were physically separated, the virus spread between the two species via water. Histopathology revealed lesions identical to those reported in the Victorian outbreak, 125-130nm icosahedral virions were observed in lesions and most exposed fish were PCR positive. DNA sequencing confirmed 99.9 to 100% homology of major capsid protein and ATPase nucleotide sequences between DGIV, ISKNV, the viral inoculum obtained from dwarf gourami and virus present in experimentally infected Murray cod. These findings confirm that Murray cod are highly susceptible to a megalocytivirus present in ornamental fish imported from South East Asia. The implications for aquaculture, conservation of native fish and quarantine policy are discussed.

[译文]：

亚洲丽丽鱼的巨细胞属病毒（虹彩病毒科）在澳洲澳洲龙纹斑中的传播和毒性试验

在2003年2月，维多利亚水产设施养殖中的澳洲龙纹斑苗有90%的损失。该疾病是由一种巨细胞病毒（虹彩病毒科）引起的，其与公认的传染性脾肾坏死病毒（ISKNV）和虹彩病毒株（DGIV）密切相关，这两个以前都没有在澳洲的养殖或野生鱼类中报道过。传播试验的进行检验这个假设，即这次病毒暴发可能是通过从东南亚进口一种携带病毒的观赏鱼引进的。对墨瑞鳕幼鱼腹腔注射经PCR确定阳性巨细胞病毒DNA的丽丽鱼的组织匀浆过滤液体导致死亡率>90%。墨瑞鳕幼鱼和丽丽鱼共栖也会引起死亡；当鱼体被分割时，病毒通过水在两种鱼类之间传播。组织病理学显示病灶与维多利亚暴发的报道相同，在病灶中观测到125~130μnm的二十面体病毒粒子而且大多数暴露的鱼经PCR检测都是阳性的。DNA测序证实在虹彩病毒、传染性脾肾坏死病毒、从丽丽鱼中获得的病毒培养液和出现于实验感染的澳洲龙纹斑中的病毒之间大部分的衣壳蛋白和三磷酸核苷酸序列有99.9%到100%的同源性。这些发现证实了澳洲龙纹斑对从东南亚进口的观赏鱼中存在的巨细胞病毒非常敏感。同时讨论了对水产养殖，保护本地鱼种和检疫政策的影响。

参考文献：Go, J., Whittington, R. 2006. Experimental transmission and virulence of a megalocytivirus (family iridoviridae) of dwarf gourami (colisa lalia) from asia in murray cod (maccullochella peelii peelii) in australia [J]. Aquaculture, 258 (1), 140-149.

（罗钦、李冬梅、黄惠珍　译）

86. The molecular epidemiology of iridovirus in Murray cod (Maccullochella peelii peelii) and dwarf gourami (Colisa lalia) from distant biogeographical regions suggests a link between trade in ornamental fish and emerging iridoviral diseases

[英文摘要]：Iridoviruses have emerged over 20 years to cause epizootics in finfish and amphibians in many countries. They may have originated in tropical Asia and spread through trade in farmed food fish or ornamental fish, but this has been difficult to prove. Consequently, MCP, ATPase and other viral genes were sequenced from archival formalin-fixed, paraffin-embedded tissues from farmed Murray cod (Maccullochella peelii peelii) that died during an epizootic in 2003 and from diseased gouramis that had been imported from Asia. There was almost complete homology (99.95%) over 4527 bp between Murray cod iridovirus (MCIV) and an iridovirus (DGIV) present in dwarf gourami (Colisa lalia) that had died in aquarium shops in Australia in 2004, and very high homology with infectious spleen and kidney necrosis virus (ISKNV) (99.9%). These viruses are most likely to be a single species within the genus Megalocytivirus and probably have a common geographic origin. Primers for genus-specific PCR and for rapid discrimination of MCIV/DGIV/ISKNV and red sea bream iridovirus (RSIV), a notifiable pathogen, were developed. These were used in a survey to determine that the prevalence of DGIV infection in diseased gourami in retail aquarium shops in Sydney was 22% (95% confidence limits 15%~31%). The global trade in ornamental fish may facilitate the spread of Megalocytivirus and enable emergence of disease in new host species in distant biogeographic regions.

[译文]：

对澳洲龙纹斑和遥远生物地理区域的丽丽鱼中虹彩病毒的分子流行病学研究表明观赏鱼的贸易与新出现的虹彩病毒病之间有关联

引起许多国家鲸鱼和两栖动物流行病的虹彩病毒已经出现超过 20 年。它们可能起源于热带亚洲，而且通过养殖的商品鱼或观赏鱼的贸易进行传播，但这一点很难证明。因此，从 2003 年因患流行病而死的人工养殖的澳洲龙纹斑中和从亚洲进口的患病丽丽鱼中提取组织通过福尔马林固定、石蜡包埋进行磷酸二氢钙,三磷酸核苷酸和其他病毒基因测序,该组织来自2003年被亚州进口的患流行病死亡的丽丽鱼。澳洲龙纹斑的虹彩病毒(MCIV)和2004年死在澳洲水族馆里的丽丽鱼的虹彩病毒(DGIV)几乎是完全相同(99.95%),超过4527bp,并且与传染性脾肾坏死病毒(ISKNV)有很高的同源性(99.9%)。这些病毒最有可能是巨细胞病毒属中的单一物种并且可能起源于同一地点。开发了用于特异性PCR的引物和用于快速鉴别澳洲龙纹斑的虹彩病毒/丽丽鱼的虹彩病毒/传染性脾肾坏死病毒与红海鲤鱼虹彩病毒(RSIV)(一种应用病原体)的引物。用这些引物检测出悉尼零售水族馆里丽丽鱼虹彩病毒(DGIV)的患病率为22%(95%置信区间为15%~31%)。观赏鱼的全球贸易可能促进巨细胞病毒的传播,并且使遥远生物地理区域的新宿物种能够出现疾病。

参考文献:Go, J., Lancaster, M., Deece, K., et al. 2006. The molecular epidemiology of iridovirus in murray cod (maccullochella peelii peelii) and dwarf gourami (colisa lalia) from distant biogeographical regions suggests a link between trade in ornamental fish and emerging iridoviral diseases [J]. Molecular & Cellular Probes, 20 (3-4), 212.

(罗钦、李冬梅、黄惠珍 译)

87. Effect of in vitro and in vivo organotin exposures on the immune functions of murray cod (Maccullochella peelii peelii).

[英文摘要]:Murray cod (Maccullochella peelii peelii) is an iconic native Australian freshwater fish and an ideal species for ecotoxicological testing of environmental pollutants. The species is indigenous to the Murray-Darling basin, which is the largest river system in Australia but also the ultimate sink for many environmental pollutants. The organotins tributyltin (TBT) and dibutyltin (DBT) are common pollutants of both freshwater and marine environments and are also known for their immunotoxicity in both mammals and aquatic organisms. In this study, TBT and DBT were used as exemplar immunotoxins to assess the efficiency of immune function assays (i. e., mitogen-stimulated lymphoproliferation, phagocytosis in head kidney tissue, and serum lysozyme activity) and to compare the sensitivity of Murray cod to other fish species. The organotins were lethal to Murray cod at concentrations previously reported as sublethal in rainbow trout (i. e., intraperitoneal [i. p.] lethal dose to 75% of the Murray cod [LD75] = 2.5mg/kg DBT and i. p. lethal dose to 100% of the Murray cod [LD100] = 12.5mg/kg TBT and DBT). In vivo TBT exposure at 0.1 and 0.5mg/kg stimulated the phagocytic function of Murray cod ($F=6.89$, $df=18$, $p=0.004$), while the highest concentration of 2. 5mg/kg TBT decreased lymphocyte numbers (F = 7. 92, df = 18, p = 0. 02) and mitogenesis (F = 3. 66, df = 18, p = 0. 035). Dibutyltin was the more potent immunosuppressant in Murray cod, causing significant reductions in phagocytic activity (F = 5. 34, df = 16, p = 0. 013) and lymphocyte numbers (F = 10. 63, df = 16, p = 0. 001).

[译文]：

体内外的有机锡暴露对澳洲龙纹斑免疫功能的影响

澳洲龙纹斑是一种标志性的澳洲本土淡水鱼，是环境污染物生态毒理学测试的理想物种。该物种是澳洲最大河系但也是许多环境污染物最终沉积的墨瑞达令河流域上的本地鱼类。有机三丁基锡（TBT）和二丁基锡（DBT）是常见的淡水和海洋环境污染物，同时众所周知哺乳动物和水生生物对它们具有免疫毒性。在本研究中，三丁基锡和二丁基锡被用作样本抗毒素去评估免疫功能测定的效率（丝裂原刺激的淋巴细胞，在头肾组织的吞噬作用，巨噬细胞吞噬功能和血清溶菌酶活性），并去比较澳洲龙纹斑与其他鱼类的敏感性。在之前报道的虹鳟鱼的有机锡亚致死浓度是澳洲龙纹斑的致死浓度（例如，对澳洲龙纹斑腹腔注射75%的致死剂量［LD75］＝2.5mg/kg 二丁基锡，对澳洲龙纹斑腹腔内注射100%的致死剂量［LD100］＝12.5mg/kg 三丁基锡和二丁基锡）。体内三丁基锡暴露在0.1和0.5mg/kg之间会刺激澳洲龙纹斑的吞噬功能（F=6.89, df=18, p=0.004），而三丁基锡在最高浓度2.5mg/kg时减少了淋巴细胞数量（F=7.92, df=18, p=0.02）并且细胞有丝分裂发生（F=3.66, df=18, p=0.035）。二丁基锡在澳洲龙纹斑中是更有效的免疫抑制剂，可以造成吞噬活性显著降低（F=5.34, df=16, p=0.013）和淋巴细胞数显著减少（F=10.63, df=16, p=0.001）。

参考文献：Harford, A. J., O'Halloran, K., Wright, P. E. 2010. Effect of in vitro and in vivo organotin exposures on the immune functions of murray cod (maccullochella peelii peelii) [J]. Environmental Toxicology & Chemistry, 26 (8), 1649-1656.

（罗钦、李冬梅、黄惠珍　译）

88. The effects of in vitro pesticide exposures on the phagocytic function of four native Australian freshwater fish

[英文摘要]：There are limited data concerning the lethal and sublethal effects of environmental pollutants on Australian freshwater fish and consequently many of the Australian water quality guidelines are based on data from exotic fish species. This study used a flow cytometric assay to assess the effect of in vitro exposures to commonly used pesticides, on the phagocytic function and cellular composition of head kidney cells from four Australian native fish,

i. e. crimson-spotted rainbowfish (Melanotaenia fluviatilis), silver perch (Bidyanus bidyanus), golden perch (Macquaria ambigua) and Murray cod (Maccullochella peelii). Head kidney immune cells were isolated from the four native fish and incubated (1 x 10 (6) cells/mL) under previously optimised conditions to measure the phagocytosis of fluorescent-latex beads. These cell cultures were exposed to three classes of pesticides, i. e. the organotins, tributyltin and dibutyltin, the organochlorine endosulfan and the organophosphate chlorpyrifos. The in vitro organotin exposures were highly immunotoxic to head kidney cells from the Australian fish, although there were some differences in immunotoxic responses between species. At the highest concentration (i. e. 10 mg/L), endosulfan exposure resulted in the modulation of phagocytic responses in all species except for silver perch. Chlorpyrifos displayed little immunotoxicity, although there was a dose-dependent reduction in Murray cod lymphocytes. These studies describe the first investigation of the phagocytic response of Australian freshwater fish immunocytes in the presence of environmental pollutants, and will help to determine appropriate ecotoxicity testing for Australian freshwater environments.

[译文]：

体外农药暴露对4种澳洲本土淡水鱼的吞噬菌细胞功能的影响

关于环境污染物对澳洲淡水鱼的致死和亚致死效应的数据有限，因此许多澳洲的养殖水质是参考国外鱼类的养殖数据。本研究使用流式细胞术检测来评估在体外常用农药暴露对来自4种澳洲本土鱼类中的头肾细胞的吞噬细胞功能和细胞合成的影响，即绯红斑鱼，银鲈，金鲈鱼和澳洲龙纹斑。从4种本土鱼中分离出头肾免疫细胞，并且在先前最佳的测量荧光微球的吞噬作用的条件下培养（1×10（6）细胞/mL）。将这些细胞培养物暴露于三种农药中，即有机锡，三丁基锡和二丁基锡，有机氯农药硫丹和有机磷农药毒死蜱。体外有机锡暴露对来自澳洲鱼类的头肾细胞具有高度地免疫，尽管鱼类之间的免疫毒性反应有一些差异。在最高浓度（即10mg/L）下，硫丹暴露导致除银鲈鱼以外的其他所有鱼类发生吞噬细胞反应的转调。毒死蜱显示出很小的免疫毒性，尽管在澳洲龙纹斑淋巴细胞中呈剂量依赖性降低。本研究首次调查了存在于环境污染物中的澳洲淡水鱼类免疫细胞的吞噬反应，并将有助于确定适当的对澳洲淡水环境的生态毒性测试。

参考文献：Harford, A. J., O'Halloran, K., Wright, P. F. 2005. The effects of in vitro pesticide exposures on the phagocytic function of four native australian freshwater fish [J]. Aquatic Toxicology, 75 (4), 330-342.

（罗钦、李冬梅、黄惠珍 译）

89. 小瓜虫对澳洲龙纹斑不同生长阶段苗种的感染与致病性研究

摘要：为了解澳洲龙纹斑苗种人工感染小瓜虫的情况，本研究以小瓜虫感染3种生长阶段鱼体（平均全长分别为1.2cm、2.3cm和10.7cm），记录其达到全部致死时间（LT100），分别为10d、13d和13d；半数致死时间（LT50）分别为6.4d、8.8d和6.7d；最高感染强度分别为168个、150个和1450个小瓜虫。分析表明其小瓜虫感染率、感染强度和死亡率与感染天数均呈显著正相关（$p<0.01$）。相同感染天数的澳洲龙纹斑驯食前和驯食期鱼苗小瓜虫感染强度差异不显著，鱼种阶段的小瓜虫感染强度则显著高于鱼苗阶段（$p<0.01$）。澳洲龙纹斑鱼苗阶段感染小瓜虫初期呈随机分布，随后均为聚集分布；鱼种阶段感染小瓜虫均呈聚集分布。Logistic模型可很好拟合澳洲龙纹斑苗种小瓜虫感染强度、感染率和死亡率曲线；感染率拐点出现时间均早于感染强度和死亡率拐点时间。本研究结果为澳洲龙纹斑苗种培育中防控小瓜虫病提供了实验依据。

The experimental infection and pathogenicity of Ichthyophthirius multiifliis in the larvae and juvenile of Maccullochella peelii

Abstract: To identify the pathogenicity of Ichthyophthirius multiifliis to larvae and juvenile of Murray cod (Maccullochella peelii), the Murray codat three different growth phase were artificial infected with I. multiifliis. The results showed that 100% lethal time (LT100) of the larvae (average length of 1.2cm), the larvae (average length of 2.3cm), and the juvenile (average length of 10.7cm) of Murray cod were 9, 13 and 13 days; the half-leathal time (LT50) were 6.4, 8.8 and 6.7 days after infected by I. multiifliis, respectively. The highest intensity of infection by I. multiifliis were 168150 and 1450 in the fry of Murray cod, respectively. The infection rate, infection intensity of I. multiitliis and mortality of the fay were significantly and positively correlated with the infection days ($p<0.01$). With the same infection days, there was no significant difference of infection intensity among the larvae and juvenile of Murray cod, but infection intensity of juvenile stage was significantly higher than that of the larvae stage. Infection intensity of I. multiifliisin the larvae of the fish was random distribution in the early period, then became aggregation distribution quickly; and it showed aggregated distribution in all the juvenile stage of Murray cod. The curves of the infection rate and infection intensity of I. multiifliis, and mortality of the fish were well fitted using Logistic regression model, which showed inflexion point of infection rate appeared earlier than inflexion point of the infection intensity and mortality.

参考文献：罗钦，饶秋华，李巍，等．2017．小瓜虫对澳洲龙纹斑不同生长阶段苗种的感染与致病性研究［J］．中国预防兽医学报，39（5），379-383。

90. 不同药物对澳洲龙纹斑幼鱼小瓜虫病的治疗效果

摘要：小瓜虫病是澳洲龙纹斑鱼人工养殖过程中威胁最大的病害，特别是在幼鱼阶段，致死率高达100%。通过两组药物治疗方案，对比和分析治疗澳洲龙纹斑小瓜虫病的效果。结果表明，试验组药物治疗方案（鱼康乐+氟苯尼考+盐酸土霉素药物制剂+活性腐殖酸钠）防治小瓜虫病鱼的鱼苗成活率高达93.3%，而参照组药物（福尔马林+工业盐）防治小瓜虫病鱼苗存活率仅30.0%。试验药物的优化配方为：鱼康乐 $2g \cdot t^{-1}$，氟苯尼考 $3g \cdot t^{-1}$，盐酸土霉素药物制剂 $3g \cdot t^{-1}$，活性腐殖酸钠 $0.8g \cdot t^{-1}$。

The Treatment Effect Analysis on Different Drugs of Lchthyophthirius Multifiliis in Maccullochella Peelii

Abstract：Ichthyophthirius disease is one of the biggest disease threat in process of artificial breeding of Maccullochella peelii, especially in the larval stage, mortality rate of up to 100%. In this study, the two drug regimens, comparison and analysis of the treatment of Maccullochella peelii Ichthyophthirius disease effect, to determine the best treatment and determine optimum drug concentration. Experimental results show, In using experimental drugs for treatment scheme (Fish Kangle + Florfenicol + Pharmaceutical preparation of oxytetracycline hydrochloride+Sodiume humate) of Ichthyophthirius disease fish cure rate was 93.3%, in the use of the control drug group (Faure Marin + Industrial salt) in the treatment of Ichthyophthirius disease of fish survival rate only 30.0%. At the same time, to determine the optimum drug concentration for $2g \cdot t^{-1}$ of Fish Kangle, $3g \cdot t^{-1}$ of Florfenicol, $3g \cdot t^{-1}$ of Pharmaceutical preparation of oxytetracycline hydrochloride and $0.8g \cdot t^{-1}$ of Sodiume humate.

参考文献：罗土炎，罗钦，饶秋华，等．2016．不同药物对澳洲龙纹斑幼鱼小瓜虫病的治疗效果［J］．福建农业学报，31（1）：102-105。

91. 澳洲龙纹斑养殖过程中主要疾病诊断及其防治

摘要：澳洲龙纹斑是从澳洲引进的一种淡水养殖的鱼类新品种，其肉质鲜嫩，营养丰富，商品经济价值高。但澳洲龙纹斑对养殖条件要求严格，特别是在其养殖过程中发生的各种鱼病要注重防治。澳洲龙纹斑常见病害主要为寄生虫病、细菌性疾病、真菌性疾病和病毒性疾病。通过对澳洲龙纹斑上述常见病害的临床症状、病原及其流行特性进行分析，提出针对性的诊断方法与有效防治技术。

Common diseases for Murray cod (Maccullochella peelii peelii) and their control

Abstract: Murray cod, Maccullochella peelii peelii, is a freshwater fish newly introduced from Australia. Itoffers aconsiderable commercial potential because of its highly desirable nutritional value and palatability. However, Murray cod is vulnerable to a variety diseases. Numerous diseases caused by bacteria, fungi, and parasites are frequently found on the fish. Thus, the breeding and cultivation conditions for the aquaculture must be maintained at anexceedingly hygienic level. Through careful analyses on the clinical symptoms, pathogens and the epidemiological characteristics of the susceptable diseases, applicable control measures are proposed. Hopefully, they would help prevent, reduce and/or alleviate the occurrence of the diseases facilitating the development of Murray cod aquaculture.

参考文献：罗土炎，罗钦，涂杰峰，等. 2015. 澳洲龙纹斑养殖过程中主要疾病诊断及其防治 [J]. 福建农业学报（6）：562-566。

92. 重金属离子与消毒药物对澳洲龙纹斑幼鱼的急性毒性

摘要：为了获知若干重金属离子与消毒药物对澳洲龙纹斑幼鱼的急性毒性浓度，本研究选用硫酸铜、硫酸锌、氯化镉、硝酸铅4种重金属离子与二氧化氯、聚维酮碘、高锰酸钾、甲醛、氯化钠5种消毒药物，通过调配不同浓度，对澳洲龙纹斑幼鱼进行急性毒性实验，采用静水法测定9种消毒药物及重金属离子对澳洲龙纹斑幼鱼每隔24h的LC50（半数致死浓度）、LC100（绝对致死浓度）和SC（安全浓度）。结果表明，硫酸铜、硫酸锌、氯化镉、硝酸铅、二氧化氯、聚维酮碘、高锰酸钾、甲醛、氯化钠九种消毒药物对澳洲龙纹斑幼鱼的安全浓度（SC）分别为 0.0601mg·L^{-1}、0.2106mg·L^{-1}、0.0039mg·L^{-1}、0.2892mg·L^{-1}、1.8184mg·L^{-1}、0.2125mg·L^{-1}、0.2812mg·L^{-1}、2.24448mg·L^{-1}、0.8271mg·L^{-1}。综合分析认为，4种重金属离子与59种消毒药物对澳洲龙纹斑的急性毒性的大小顺序为：氯化镉>硫酸铜>硫酸锌>聚维酮碘>高锰酸钾>硝酸铅>氯化钠>二氧化氯>甲醛。

Acute Toxicities of Disinfectants and Metal Ions on juvenile Murray cod, Maccullochella peelii peelii

Abstract: In order to know the concentration of Acute toxicity of several disinfectants and Metal Ions to the juveniles of Maccullochella peelii peelii. The juveniles of Maccullochella peelii peelii were tested for providing a evidence for the culture and control of diseases of Murray cod. A certain concentration gradient of copper sulfate, zinc sulfate, chloride cadmium, lead nitrate, chlorine dioxide, povidone-iodine, potassium permanganate, formaldehyde, sodium chloride was prepared, the hydrostatic method for the determination of several disinfectants and metal ions of Murray cod of LC50 (the half lethal concentration) LC100 (absolute lethal concentration) and safe concentration (SC) every 24h. The results showed: The safe concentration (SC) of copper sulfate, zinc sulfate, chloride cadmium, lead nitrate, chlorine dioxide, povidone-iodine, potassium permanganate, formaldehyde, sodium chloride was 0.0601mg·L^{-1}、0.2106mg·L^{-1}、0.0039mg·L^{-1}、0.2892mg·L^{-1}、1.8184mg·L^{-1}、0.2125mg·L^{-1}、0.2812mg·L^{-1}、2.24448mg·L^{-1}、0.8271mg·L^{-1}. The results indicated that, the acute toxicity intensity of 4 kinds of metal ions and 5 kinds of disinfectants to the juveniles of Murray cod: chloride cadmium>copper sulfate>zinc sulfate>povidone-iodine>potassium permanganate>chloride cadmium>sodium chloride>chlorine dioxide>formaldehyde.

参考文献：罗土炎，罗钦，饶秋华，等.2017.重金属离子与消毒药物对澳洲龙纹斑幼鱼的急性毒性[J].福建农业学报，32（7）：691-696。

93. 一种生态防治澳洲龙纹斑小瓜虫的技术（专利）

摘要：本发明公开了一种生态防治澳洲龙纹斑小瓜虫的技术，在发现澳洲龙纹斑身上出现小瓜虫时，这时在所述澳洲龙纹斑所在的水体中加入海水，用盐度计及时检测，把盐度控制到8‰~10‰，浸泡48h，从而杀死小瓜虫。本发明所述的一种生态防治澳洲龙纹斑小瓜虫的技术，通过对小瓜虫的超微结构、感染机制、致病机理以及免疫反应等做一些深入地基础研究后，发现，小瓜虫生活史分为滋养体、包囊、掠食阶段，其中掠食体阶段虫体较为脆弱，此时改变环境对虫体具有很好的抑制作用，可以避免除虫药物对鱼类的不良影响，同时可以彻底有效的杀死小瓜虫。

参考文献：蔡星，朱永祥，刘大勇，等.2013.一种生态防治澳洲龙纹斑小瓜虫的技术.CN103404467A。

94. 澳洲龙纹斑鱼防治小瓜虫的方法（专利）

摘要：为了提供一种对澳洲龙纹斑鱼防治小瓜虫效果良好、致畸形率低、鱼体康复快的技术，发明人提出了如下的解决方案：一种澳洲龙纹斑鱼防治小瓜虫的方法，包括如下步骤：在养殖澳洲龙纹斑鱼的水体中施用杀虫剂，所述杀虫剂包括活性硅藻土；在养殖澳洲龙纹斑鱼的水体中施用消炎药，所述消炎药包括氟苯尼考药物制剂和盐酸土霉素药物制剂；在养殖澳洲龙纹斑鱼的水体中施用水质改良剂，所述水质改良剂包括活性腐殖酸钠。上述技术方案在药物健康低毒、总成本低的条件下，实现了有效的对澳洲龙纹斑鱼防治小瓜虫的技术效果，具有极佳的推广价值。

参考文献：罗钦，罗土炎，张志灯，等.2017.澳洲龙纹斑鱼防治小瓜虫的方法．CN201510111037.6。

注：无英文摘要。

95. 一种鱼类盾纤虫感染症的治疗方法（专利）

摘要：一种鱼类盾纤虫感染症的治疗方法，包括如下步骤：（1）去除感染盾纤虫的鱼眼，而后分别用双氧水、红药水擦拭感染盾纤虫的鱼眼部，将鱼放入水体；（2）将杀虫剂鱼康乐投喂给鱼类口服，所述的杀虫剂鱼康乐包括活性硅藻土、沸石粉和碳酸氢钙；（3）在水体中施用抗细菌药，所述的抗菌药包括乙酰甲硅预混剂；（4）在水体中施用抗霉菌药，所述的抗霉菌药包括水杨酸；（5）在水体中施用消毒剂，所述的消毒剂包括聚维酮碘。本发明提供的鱼类盾纤虫感染症的治疗方法，对鱼类没有损伤、也不会破坏环境，可以将严重症状的鱼类治愈成活，治愈率高，极具推广价值。

参考文献：罗钦，罗土炎，陈华，等．2017．一种鱼类盾纤虫感染症的治疗方法．CN201710078170.5。

注：无英文摘要。

第四章 澳洲龙纹斑营养与品质

第一节 专题研究概况

澳洲龙纹斑肉质嫩白,细腻而结实,肌间刺很少,味美肉鲜,风味独特,不饱和脂肪酸含量高,营养丰富,味道与海水鱼类似,没有通常鱼类的腥味而有一种特殊的香味,被称作世界上最鲜美的淡水鱼之一。该鱼为白肉鱼,在澳洲排名四大经济鱼类之首,是世界上最大的淡水鱼之一,与宝石鲈、黄金鲈、银鲈等原产于澳洲的淡水鱼并列为澳洲鱼。由于澳洲龙纹斑有特殊的外形以及鲜美口感,在澳大利亚素有"国宝鱼"美称。据对鲜活鱼的测定,鱼肉中含4种香味氨基酸,并含有一定数量的EPA和DHA。在烹饪过程中皮易剥,刺易去,在澳大利亚及其周边国家和地区常以澳洲龙纹斑烤制佳肴或制成鱼排招待宾客。是澳大利亚南部墨瑞河水域具有很高营养价值和经济价值的游钓、商业性生产和资源增殖的地方名贵品种(曹凯德,2001;韩茂森,2003;宋理平,2013;李娴,2013;翁伯琦,2016)。

一、常规营养成分

常规营养成分包括蛋白质、脂质、碳水化合物(糖类)、维生素和矿物质(无机盐)、水、纤维素等7大类。鲜活澳洲龙纹斑常规营养成分的组成和含量见表4-1。3种生长阶段的澳洲龙纹斑干样的营养成分含量见表4-2(韩茂森,2003;宋理平,2013;曹凯德,2001;宋理平,2013;罗钦,2015)。

澳洲龙纹斑的水分含量高于宝石鲈、鳜、鳙和翘嘴红鲌等其他经济价值较高的7种淡水鱼类水分含量(67.43%~80.30%),低于草鱼(81.59%)和尼罗罗非鱼(80.85%)。肌肉粗蛋白含量高于青鱼(15.94%)、草鱼(15.94%)和尼罗罗非鱼(15.38%),而低于宝石鲈、鳜和翘嘴红鲌等6种鱼类的粗蛋白(16.52%~19.16%)

（表4-3）。粗脂肪含量高于鳜、鲭和草鱼等4种鱼类粗脂肪的含量（0.62%~1.50%），低于宝石鲈、鲤和黄颡鱼等其他5种鱼类粗脂肪的含量（1.75%~12.65）（宋理平，2013）。

表4-1 鲜活澳洲龙纹斑常规营养物质含量

名称	含量	备注
含肉率（%）	52	
蛋白质（%）	15.96~20	
脂含量（%）	1~1.68	
脂肪酸（mg/g）	4.7	
n-3（mg/g）	1.9	（韩茂森，2003）
D-6（mg/g）	0.3	（宋理平，2013）
钾（K）（mg/g）	0.39	
钠（Na）（mg/g）	0.53	
钙（Ca）（mg/g）	0.14	
镁（Mg）（mg/g）	0.29	
纤维（mg/g）	0.39	（曹凯德，2001）
总醇量（mg/g）	1	
水分（%）	80.75	（宋理平，2013）
粗灰分（%）	0.94	

表4-2 3种生长阶段的澳洲龙纹斑干样的部分营养物质含量（罗钦，2015）

项目	澳洲龙纹斑亲鱼	澳洲龙纹斑商品鱼	澳洲龙纹斑幼鱼
水分/%	9.6	8.9	2.8
脂肪/%	19.2	8.4	7.7
蛋白/%	68.87	78.73	81.81

表4-3 澳洲龙纹斑与几种常见经济鱼类肌肉鱼样的常规营养成分的比较（宋理平，2013）

鱼名	水分	粗蛋白	粗脂肪	粗灰分
澳洲龙纹斑	80.75	15.96	1.68	0.94
宝石鲈	67.43	19.16	12.65	1.12

(续表)

鱼名	水分	粗蛋白	粗脂肪	粗灰分
鳜	79.76	17.56	1.50	1.06
鳙	80.18	16.95	0.74	2.08
鲤	79.58	16.52	2.06	1.18
青鱼	79.63	15.94	0.62	1.23
草鱼	81.59	15.94	0.62	1.22
尼罗罗非鱼	80.85	15.38	1.75	1.07
黄颡鱼	77.63	16.56	4.34	1.24
翘嘴红鲌	80.30	16.70	2.48	1.08

常规营养成分评价主要是评价蛋白质、脂肪、灰分及水分的组成及含量，通常认为干物质含量越高，其总养分含量也就越高，而蛋白质和脂肪的组成及含量则是评价食品营养价值的重点。在人类的饮食结构中鱼类是重要的肉食提供者，鱼肉为人类提供了高品质蛋白质和脂肪。由表4-3可知，澳洲龙纹斑肌肉含水量与尼罗罗非鱼相近，低于草鱼；通过与其他经济鱼类蛋白质和脂肪含量的比较，澳洲龙纹斑是一种蛋白质含量丰富而脂肪含量适中的淡水经济鱼类（宋理平，2013）。

二、氨基酸组成

Rasanthi M.（1999）研究了澳洲龙纹斑鱼卵（未受精和/或受精）从卵发育到卵黄囊全部被吸收成幼鱼时总氨基酸（蛋白质结合+游离）和游离氨基酸（FAA）的变化和发育的关系。在整个发育过程中，游离氨基酸占总氨基酸库的比例很小（0.19%，在两种鱼类受精卵中）。在整个发育过程中，氨基酸库中九种必需氨基酸和八种非必需氨基酸被定量分析。总氨基酸含量从新孵出的鱼发育成（在20±1℃）卵黄囊全吸收的幼鱼时是降低的。总体而言，必需氨基酸含量和非必需氨基酸含量的变化随着氨基酸库变化。在澳洲龙纹斑中，除了两个非必需氨基酸（天冬氨酸和甘氨酸）外，氨基酸含量普遍下降。相比之下，游离氨基酸随着发育而增加，在必需氨基酸和非必需氨基酸这两个方面都反映出这一变化。定性地说，澳洲龙纹斑卵中主要的游离氨基酸是丙氨酸，赖氨酸，亮氨酸和丝氨酸。因为卵蛋白和总氨基酸的含量随着发育而下降，所以结论是卵黄蛋白的降解速率高于胚胎发生过程中的合成代谢和分解代谢。数据还表明，在淡水鱼类中游离氨基酸未必是胚胎发生过程中的主要能量来源（Rasanthi M.，1999）。

对3种不同生长阶段的氨基酸组成含量进行比较分析，3种不同生长阶段的澳洲龙纹斑鱼体肌肉均含有17种氨基酸,但在氨基酸含量上有所差异,氨基酸总含量最高为商品鱼70.96%、其次为幼鱼69.24%、亲鱼最低为65.36%。3种不同生长阶段均以谷氨酸的含量最高,其次为天门冬氨酸、赖氨酸、甘氨酸、亮氨酸、精氨酸、丙氨酸等,并且含量均高于4.00%。3种不同生长阶段的必需氨基酸均以赖氨酸含量最高、甲硫氨酸含量最低,赖氨酸是人体必需氨基酸之一,可促进人体发育、增强免疫力;非必需氨基酸均以谷氨酸含量最高、胱氨酸含量最低,谷氨酸在人体内具有促进红细胞生成、改善脑细胞营养及活跃思维等作用。3种不同生长阶段的7种必需氨基酸含量的高低顺序基本一致,唯一差别是商品鱼的苯丙氨酸含量(2.88%)高于异亮氨酸的含量(2.86%),但含量的差值很小,仅为0.02%,由此可知,澳洲龙纹斑在生长过程中,必须氨基酸的组成成分和比例基本不变(罗钦,2015)。

对3种不同生长阶段的氨基酸组成比例进行统计分析,结果显示必需氨基酸占17种氨基酸总量的比例 EAA/TAA 中最高为商品鱼36.56%、其次为亲鱼36.20%、最低为幼鱼38.39%;非必需氨基酸占17种氨基酸总量的比例 EAA/NEAA 中最高为亲鱼63.80%、其次为商品鱼为63.44%、最低为幼鱼61.61%。由此可知,3种不同生长阶段的氨基酸组成比例中亲鱼和商品鱼的组成比例很接近,而幼鱼的氨基酸组成比例则有所变化,在必需氨基酸中,其 EAA/TAA 值则比亲鱼以及商品鱼低2%左右,而在非必需氨基酸中,EAA/NEAA 比值又高于亲鱼与商品鱼2%左右(罗钦,2015)。

表4-4 3种不同生长阶段澳洲龙纹斑肌肉中氨基酸种类及含量(罗钦,2015)

氨基酸种类	澳洲龙纹斑亲鱼		澳洲龙纹斑商品鱼		澳洲龙纹斑幼鱼	
	含量(%)	组成比例(%)	含量(%)	组成比例(%)	含量(%)	组成比例(%)
甲硫(蛋)氨酸	1.73	7.33	1.72	6.63	1.70	6.40
苯丙氨酸	2.57	10.84	2.88	11.09	2.92	10.97
异亮氨酸	2.61	11.03	2.86	11.01	3.03	11.40
缬草氨酸	2.96	12.50	3.24	12.48	3.38	12.72
苏氨酸	3.09	13.07	3.35	12.91	3.41	12.83
亮氨酸	4.91	20.74	5.40	20.81	5.63	21.19
赖氨酸	5.79	24.48	6.51	25.08	6.51	24.51
必需氨基酸总量(EAA)	23.66	100.00	25.94	100.00	26.58	100.00
胱氨酸	0.44	1.05	0.46	1.03	0.53	1.25
组氨酸	1.49	3.58	1.59	3.53	1.48	3.47
酪氨酸	2.05	4.91	2.22	4.94	2.30	5.39
丝氨酸	3.02	7.25	3.26	7.24	3.21	7.52

(续表)

氨基酸种类	澳洲龙纹斑亲鱼 含量(%)	澳洲龙纹斑亲鱼 组成比例(%)	澳洲龙纹斑商品鱼 含量(%)	澳洲龙纹斑商品鱼 组成比例(%)	澳洲龙纹斑幼鱼 含量(%)	澳洲龙纹斑幼鱼 组成比例(%)
脯氨酸	3.32	7.97	3.52	7.82	2.96	6.95
丙氨酸	4.42	10.61	4.80	10.65	4.44	10.42
精氨酸	4.63	11.10	4.99	11.08	4.61	10.80
甘氨酸	5.74	13.76	5.97	13.27	4.51	10.57
天门冬氨酸	6.48	15.54	7.14	15.86	7.32	17.16
谷氨酸	10.10	24.22	11.07	24.59	11.29	26.47
非必需氨基酸总量（NEAA）	41.70	100.00	45.02	100.00	42.66	100.00
17种氨基酸总量（TAA）	65.36	/	70.96	/	69.24	/
EAA/NEAA	36.20	/	36.56	/	38.39	/
EAA/TAA	63.80	/	63.44	/	61.61	/

注：氨基酸含量为扣除了所有水分的含量。

澳洲龙纹斑肌肉中检测出谷氨酸、天门冬氨酸、丙氨酸、精氨酸和甘氨酸5种鲜味氨基酸。鲜味氨基酸的总量决定了鱼肉味道的鲜美程度。由表4-5可知，根据澳洲龙纹斑与几种经济鱼类鲜味氨基酸总量对比分析，发现澳洲龙纹斑肌肉鲜味氨基酸总量（7.34%）高于宝石鲈（6.99%）。尼罗罗非鱼，黄颡鱼，斑点叉尾鲴鱼；稍次于鳜，异育银鲫和泥鳅含量相近（宋理平，2013）。

表4-5 澳洲龙纹斑肌肉中鲜味氨基酸的组成及含量与常见经济鱼类的比较%（宋理平，2013）

项目	谷氨酸	天门冬氨酸	丙氨酸	精氨酸	甘氨酸	纤维氨基酸总量
澳洲龙纹斑	2.73	1.73	0.99	0.76	0.76	7.34
宝石鲈	2.35	1.73	1.06	0.83	0.83	6.99
鳜	2.93	1.85	1.09	0.81	0.81	7.76
尼罗罗非鱼	2.41	1.61	0.94	0.73	0.73	6.73
黄颡鱼	2.62	1.61	0.94	0.70	0.70	6.88
斑点叉尾鲴鱼	2.54	1.62	0.93	0.68	0.68	6.80
异育银鲫	2.92	2.01	1.16	1.03	1.03	8.17
泥鳅	2.74	1.93	1.00	0.85	0.85	7.43

注：氨基酸含量为鲜样的含量。

对3种特种水产商品鱼肌肉中呈味氨基酸的分布进行分析，3种商品鱼均含有较高的作为天然的鲜味最强的两种氨基酸，即谷氨酸和天门冬氨酸，鲜味类氨基酸含量最高的为澳洲龙纹斑18.21%、其次为鲟鱼17.38%、最低的为石斑鱼17.15%（表4-6）。3种商品鱼肌肉中甜味氨基酸和鲜味氨基酸的含量之和均超过50%（澳洲龙纹斑57.17%，石斑鱼57.68%，鲟鱼57.85%），并且含量相近，说明3种商品鱼在肉质味道鲜美上相近。石斑鱼肉质细嫩洁白，类似鸡肉，素有"海鸡肉"之称，被港澳地区推为我国四大名鱼之一，是高档筵席必备之佳肴；而鲟鱼在我国古代早已享誉大江南北，如陆玑《诗经·卫风·硕人》疏："（鳠即鲟）大者千余斤，可蒸为臛，又可作炸，鱼籽可为酱。"，《本草纲目．鳠鱼》云："其脂与肉层层相间、肉色白、脂色黄如蜡。其脊骨及鼻，并鳍与鳃，皆脆软可食。其肚及仔盐藏亦佳。其鳔亦可作胶。其肉骨煮炙及作炸皆美。"根据3种鱼的氨基酸组成，可以推测出，澳洲龙纹斑的食用价值、营养价值和经济价值较高，是一种非常值得开发利用的淡水鱼类资源（罗钦，2015）。

表4-6 3种特种水产商品鱼肌肉中呈味氨基酸的分布（罗钦，2015）

氨基酸种类	澳洲龙纹斑商品鱼 含量（%）	澳洲龙纹斑商品鱼 组成比例（%）	石斑鱼商品鱼 含量（%）	石斑鱼商品鱼 组成比例（%）	鲟鱼商品鱼 含量（%）	鲟鱼商品鱼 组成比例（%）
苏氨酸	3.35	/	3.07	/	3.03	/
丝氨酸	3.26	/	2.82	/	3.40	/
谷氨酸	11.07	/	10.39	/	10.74	/
甘氨酸	5.97	/	6.47	/	7.53	/
丙氨酸	4.80	/	5.06	/	4.93	/
脯氨酸	3.52	/	3.67	/	4.09	/
赖氨酸	6.51	/	5.93	/	5.88	/
甜味类氨基酸含量	38.47	38.80	37.42	39.55	39.61	40.21
苯丙氨酸	2.88	/	2.63	/	3.22	/
精氨酸	4.99	/	4.91	/	5.14	/
异亮氨酸	2.86	/	2.59	/	2.76	/
缬草氨酸	3.24	/	3.07	/	2.91	/
亮氨酸	5.40	/	4.95	/	4.93	/
甲硫（蛋）氨酸	1.72	/	2.03	/	1.84	/
组氨酸	1.59	/	1.36	/	1.68	/
苦味类氨基酸含量	22.66	22.86	21.54	22.76	22.48	22.81
组氨酸	1.59	/	1.36	/	1.68	/
天门冬氨酸	7.14	/	6.77	/	6.64	/

(续表)

氨基酸种类	澳洲龙纹斑商品鱼		石斑鱼商品鱼		鲟鱼商品鱼	
	含量(%)	组成比例(%)	含量(%)	组成比例(%)	含量(%)	组成比例(%)
谷氨酸	11.07	/	10.39	/	10.74	/
酸味类氨基酸含量	19.80	19.97	18.51	19.56	19.06	19.34
天门冬氨酸	7.14	/	6.77	/	6.64	/
谷氨酸	11.07	/	10.39	/	10.74	/
鲜味类氨基酸含量	18.21	18.37	17.15	18.13	17.38	17.64
氨基酸总量	70.96	100.00	68.25	100.00	71.13	100.00

注：氨基酸含量为扣除了所有水分的含量。

三、氨基酸营养品质评价

蛋白质和氨基酸的含量是食品营养价值高低的重要评判指标之一。通常情况下婴儿对必需氨基酸的需要量较其他年龄段的要高，因此，1973年FAO/WHO制定了以婴儿需要量为最低限度的评分标准，而蛋白质评定标准则是参照营养最全面的的鸡蛋蛋白质组成而制定。FAO/WHO标准和鸡蛋蛋白质评定标准被广泛用于食物营养价值的评定。从表4-7可知，澳洲龙纹斑所含必需氨基酸总量为2776mg/个N，占总氨基酸含量44.04%，高于FAO/WHO的35.38%标准，其必需氨基酸的组成符合FAO/WHO的标准。以蛋白质的氨基酸评分为标准，澳洲龙纹斑限制性氨基酸为缬氨酸；以化学评分为标准，第一限制性氨基酸是蛋氨酸和胱氨酸，第二限制性氨基酸是缬氨酸。澳洲龙纹斑肌肉中赖氨酸的含量高于FAO/WHO标准和鸡蛋蛋白质标准，分别为二者的1.83倍和1.41倍。通常情况下，评价食物蛋白质营养价值是以鸡蛋蛋白质必需氨基酸组成为参评标准的必需氨基酸指数为指标的，澳洲龙纹斑的必需氨基酸指数为95.43（宋理平，2013）。

表4-7 澳洲龙纹斑肌肉中必需氨基酸组成的评价（宋理平，2013）

项目	澳洲龙纹斑(mg/gN)	鸡蛋蛋白(mg/gN)	FAO评分模式(mg/gN)	氨基酸评分	化学评分
苏氨酸	309	292	250	1.24	1.06
缬氨酸	282	411	310	0.91	0.69
异亮氨酸	282	331	250	1.13	0.85

（续表）

项目	澳洲龙纹斑（mg/gN）	鸡蛋蛋白（mg/gN）	FAO 评分模式（mg/gN）	氨基酸评分	化学评分
亮氨酸	521	534	440	1.18	0.98
赖氨酸	623	441	340	1.83	1.41
蛋+酪氨酸	251	386	220	1.14	0.65
苯丙+酪氨酸	509	565	380	1.34	0.90
必需氨基酸总量	2776	2960	2190		
必需氨基酸指数	95.43				
必需氨基酸占氨基酸总量（%）	44.04	48.08	35.38		

营养价值较高的食物蛋白质不仅所含的 EAA 种类要齐全，而且 EAA 之间的比例也要适宜，最好能与人体需要相符合，这样 EAA 吸收最完全，营养价值最高。从表 4-8 分析结果表明，3 种商品鱼肌肉中的 E/T、E/N 依次为：澳洲龙纹斑商品鱼（40.34%、67.62%）>石斑鱼商品鱼（39.27%、64.68%）>鲟鱼商品鱼（37.92%、61.08%），高于 FAO/WHO 的值，低于鸡蛋的值，但是比较接近 FAO/WHO。根据 FAO/WHO 模式，澳洲龙纹斑商品鱼超过了 WHO/FAO 提出的 E/T 应为 40%左右和 E/NT 应为 60%以上的参考蛋白质模式标准，说明澳洲龙纹斑商品鱼肌肉中的蛋白质氨基酸不仅种类组成合理，与人体需要相符合（罗钦，2015）。

表 4-8 3 种特种水产商品鱼肌肉中的 EAA 含量与鸡蛋蛋白、FAO/WHO 标准模式的比较（罗钦，2015）

氨基酸	澳洲龙纹斑商品鱼/(mg·g^{-1}N)	石斑鱼商品鱼/(mg·g^{-1}N)	鲟鱼商品鱼/(mg·g^{-1}N)	鸡蛋蛋白/(mg·g^{-1}N)	FAO/WHO
Ile	225	253	238	331	250
Leu	426	483	426	534	440
Lys	514	579	508	441	340
Met +Cys	172	243	194	386	220
Phe+ Tyr	402	457	451	565	380
Thr	264	300	262	292	250
Val	256	300	252	411	310
TEAA	2260	2615	2331	2960	2190
E/T（%）	40.34	39.27	37.92	48.92	35.73
E/N（%）	67.62	64.68	61.08	95.76	55.58

注：TEAA：The total of essential amino acid（Contain Cys and Tyr）；E/T：TEAA/TAA；E/N：TEAA/（TAA-TEAA）。

食物蛋白质是否拥有较高营养价值的评定标准不仅要考虑必需氨基酸种类是否齐全，还要考虑其必须氨基酸比例是否与人体需求相符，以利于人体最完全地吸收食物中的必需氨基酸。各种蛋白质所含氨基酸的种类和含量是存在差异的，有些蛋白质含人体内所需的全部氨基酸且含量充足，因而其营养价值高；有些蛋白质缺乏人体所需某种必需氨基酸或氨基酸含量不足，而降低了其营养价值。蛋白质的营养价值取决于其氨基酸组成和含量。通常情况下，膳食蛋白质中缺乏的一般都是人体必需氨基酸，尤其是赖氨酸和含硫氨基酸最易缺乏。通过对澳洲龙纹斑FAO/WHO标准、氨基酸评分（AAS）、化学评分（CS）和必需氨基酸指数（EAAI）评价的研究发现，澳洲龙纹斑肌肉的氨基酸组成与含量较接近人体必需氨基酸需求（宋理平，2013）。

澳洲龙纹斑、石斑鱼、鲟鱼3种商品鱼肌肉中甜味氨基酸和鲜味氨基酸的含量之和均超过50%（澳洲龙纹斑57.17%，石斑鱼57.68%，鲟鱼57.85%），其反映了3种商品鱼肉质味道鲜美度是处在同一等级上；同时，澳洲龙纹斑商品鱼（E/T比值为40.34、E/NT比值为67.62）符合WHO/FAO提出的E/T应为40%左右和E/NT应为60%以上的参考蛋白质模式标准，说明澳洲龙纹斑营养品质高，蛋白质氨基酸种类组成合理，符合人体需求（罗钦，2015）。

澳洲龙纹斑肌肉中含有丰富的谷氨酸、丙氨酸、精氨酸、甘氨酸和天门冬氨酸5种鲜味氨基酸，较大多数淡水经济鱼类鲜味氨基酸含量高，这是其肉质更为鲜美的原因之一。谷氨酸不仅是提高食品鲜味的鲜味氨基酸，还是参与脑组织生化代谢中多种生理活性物质合成的重要氨基酸。赖氨酸具有促进钙的吸收和钙蓄积的功能，还有增进食欲、促进婴幼儿的生长和发育的作用。精氨酸对成人来说虽然不是必需氨基酸，但在有些情况如机体发育不成熟或在严重应激条件下，如果缺乏精氨酸，机体便不能维持正氮平衡与正常的生理功能，因此被定义为人体条件性必需氨基酸，对机体有很多生化和治疗作用；对于婴幼儿来说精氨酸是生长所必需的氨基酸，精氨酸还具有促进伤口愈合的功能。澳洲龙纹斑肌肉中谷氨酸、赖氨酸、精氨酸含量较高（分别为2.73%、1.59%、1.13%）。因此，澳洲龙纹斑有较高的营养价值，具备一定保健功效（宋理平，2013）。

澳洲龙纹斑所含丰富的赖氨酸可以弥补谷物食品中的赖氨酸含量不足，从而提高人体对蛋白质的利用。以蛋白质的氨基酸评分（AAS）为标准，澳洲龙纹斑的限制性氨基酸为缬氨酸；以化学评分（CS）为标准，蛋氨酸和胱氨酸为澳洲龙纹斑第一限制性氨基酸，缬氨酸为第二限制性氨基酸。根据FAO/WHO标准，质量较好的蛋白质所含氨基酸中必需氨基酸占氨基酸总量的 40%左右。本研究发现，澳洲龙纹斑 WEAA/WTAA 为 39.93%，符合 FAO/WHO 的评价标准。亮氨酸、异亮氨酸和缬氨酸不仅作为支链氨基酸影响蛋白质的合成，还具有增强免疫力、调节激素代谢等功能，其总含量也影响食物的营养价值。澳洲龙纹斑肌肉中亮氨酸、异亮氨酸和缬氨酸含量丰富，使澳洲龙纹斑更具食用价值（宋理平，2013）。

四、脂肪酸组成分析

Gunasekera R. M（1999）研究了澳洲龙纹斑从卵发育到卵黄囊全部被吸收的幼鱼阶段的脂肪酸的变化。澳洲龙纹斑的多不饱和脂肪酸占总脂肪中 19 种定量脂肪酸的 50% 以上。在发育全阶段，澳洲龙纹斑中含量最高的脂肪酸依次为二十二碳六烯酸，花生四烯酸，油酸和棕榈酸，在大多数情况下都超过 100μg/mg 总油脂；在澳洲龙纹斑卵中，二十二碳六烯酸和二十碳五烯酸的比例为 7.3∶1，发育阶段几乎保持不变，并且显著高于一般鱼卵的 2∶1 比例。在澳洲龙纹斑发育过程中，棕榈油酸显著减少而花生四烯酸显著增加。总而言之，澳洲龙纹斑倾向于保存 n-3 和 n-6 系列高度不饱和脂肪酸（HUFA），显示其在首次饲喂中的重要性。这些现象与澳洲龙纹斑的摄食习惯有关，因为它是顶级的淡水肉食性鱼类（Gunasekera R.，M1999）。

Palmeri，G.（2007）研究了人工养殖澳洲龙纹斑不同组织中的脂肪和脂肪酸的分布（图 4-1、图 4-2）。在不同部位的鱼肉中发现脂肪含量的差异，在背/颅部分（P1）中最低，在腹侧/尾部（P8）较高（p<0.05）。后者也测得了最高含量的单不饱和脂肪酸和最低的多不饱和脂肪酸，花生四烯酸，二十二碳六烯酸和 n3/n6 的比例。一般来说，不同部位的鱼肉的脂肪含量与 PUFA 和 MUFA 直接相关。饱和脂肪酸和二十碳五烯酸的含量在不同部位的鱼肉中没有显示出任何明显的趋势，而 DHA 和 ArA 的含量差异显著。这项研究表明，脂肪在澳洲龙纹斑体内的沉积有明显的差异，不同的脂肪酸在不同部位的鱼肉中的沉积也不同（Palmeri，G.，2007）。

在澳洲龙纹斑肌肉中共检测出脂肪酸 19 种，其中饱和脂肪酸 7 种，占总脂肪酸的 24.36%；单不饱和脂肪酸 4 种，占总脂肪酸的 24.36%；多饱和脂肪酸 8 种，占总脂肪酸的 43.44%，不饱和脂肪酸含量（67.80%）远高于饱和脂肪酸含量（31.95%）（表 4-9）。油酸含量最高，为 18.75%，其次为棕榈酸 17.46%、亚油酸 17.28%，最少的是芥酸 0.62%。多不饱和脂肪酸中 EPA 与 DHA 的含量分别为 4.56%和 10.12%（宋理平，2013）。

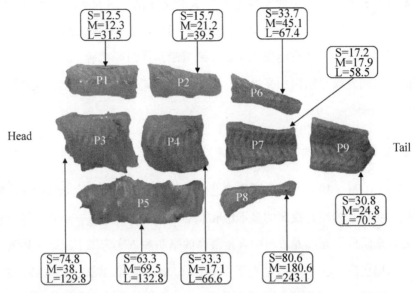

图 4-1 左侧鱼片不同部位的分析。标明了不同规格鱼的每个小的（S）中等的（M）和大的（L）鱼片部位的脂肪含量（mg g^{-1}）（Palmeri，G.，2007）。

表 4-9 澳洲龙纹斑肌肉的脂肪酸组成及含量%（宋理平，2013）

饱和脂肪酸	含量	单不饱和脂肪酸	含量	多不饱和脂肪酸	含量
豆蔻酸（C14：0）	2.08	棕榈油酸（C16：1）	3.92	亚油酸（C18：2）	17.28
十五碳酸（C15：0）	1.26	油酸（C18：1）	18.75	亚麻酸（C18：3）	3.27
棕榈酸（C16：0）	17.46	花生一烯酸（C20：1）	1.07	二十酸二烯酸（C20：2）	1.20
十七碳酸（C17：0）	1.29	芥酸（C22：1）	0.62	二十酸三烯酸（C20：3）	0.73
硬脂酸（C18：0）	6.98	MUFA	24.36	花生四烯酸（C20：4）	2.97
花生酸（C20：0）	1.67			EPA（C20：5）	4.56
山俞酸（C22：0）	1.21			DPA（C22：5）	3.31
SFA	31.95			DHA（C22：6）	10.12
				PUFA	43.44

在澳洲龙纹斑肌肉中共检测出 19 种脂肪酸，其脂肪酸组成总体呈现：多不饱和脂肪＞饱和脂肪酸＞单不饱和脂肪酸。近年来研究发现，多不饱和脂肪酸具有多种生理学功能，它们在降血脂、降血压、抑制血小板凝集、提高生物膜液态性、抗肿瘤和免疫调节等方面发挥着重要作用，摄入多不饱和脂肪酸能显著降低心脑血管疾病的发病率。Mattson（1940）研究报道：单不饱和脂肪酸也具有一定的降血脂作用。EPA 和 DHA 是人体必需的多不饱和脂肪酸，具有许多重要的营养生理学功能。澳洲龙纹斑肌肉中不饱

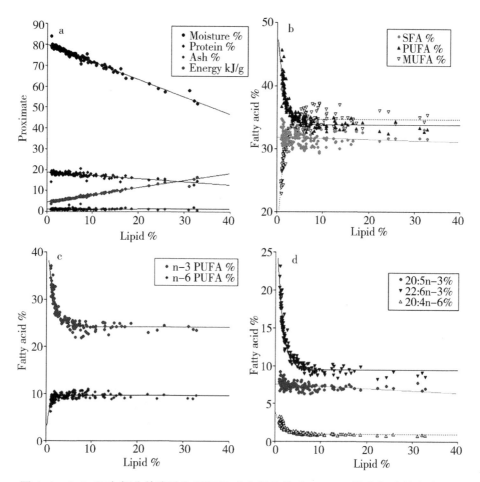

图4-2 （a）肌肉部分的脂肪和近似组成之间的关系，（b）肌肉部分的脂肪和SFA、MUFA和PUFA百分含量之间的关系，（c）肌肉部分的脂肪和n-3和n-6PUFA百分含量之间的关系，和（d）肌肉部分的脂肪和Ara、EPA和DHA百分含量之间的关系。在所有情况下，都给出了最佳匹配的线（Palmeri，G.，2007）。

和脂肪酸含量较高，多不饱和脂肪酸占总脂肪酸含量的43.44%，其中EPA和DHA含量丰富（宋理平，2013）。

通过对澳洲龙纹斑肌肉营养成分分析与品质评价，发现其肌肉中含有丰富、全面的营养物质，蛋白质含量高，氨基酸种类齐全且人体必需氨基酸含量较高，含有大量不饱和脂肪酸，具有较高的营养价值。因此，澳洲龙纹斑具有健康、全面的膳食功能，具有较高的开发价值和广阔的市场前景，作为优质的淡水经济鱼类，值得大力推广养殖（宋理平，2013）。

五、停餐对品质的影响

Palmeri, G.（2008）评估了经过停餐0周，2周和4周后密集养殖的澳洲龙纹斑的生物学，营养和感官变化（图4-3、图4-4）。与未停餐的鱼相比，停餐2周和4周的鱼体重有显著的减轻，分别为4.1%和9.1%，而内脏周围脂肪含量没有改变。与未停餐的鱼相比，停餐时间对饱和脂肪酸，单不饱和脂肪酸和高度不饱和脂肪酸的浓度没有显著的影响，而多不饱和脂肪酸，n-3和n-3高度不饱和脂肪酸显著高于对照组（$p<0.05$）。消费者因偏好停餐鱼的味道而能够发现停餐鱼和未停餐鱼之间的差异（$p<0.05$），但是，消费者无法区分停餐了2周和4周的鱼。这项研究表明，2周的停餐期对于改善养殖的澳洲龙纹斑的最终感官质量是必要的和足够的。因此，通过停餐可改善食用肉的营养品质（Palmeri, G., 2008）。

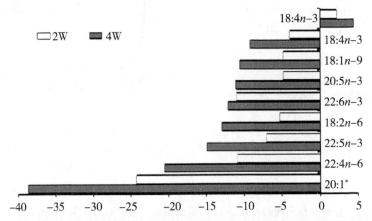

图4-3 评估了停餐2到4周后整个鱼片的总脂肪酸损失（mg）。注意亚麻油酸（18：4n-3）的含量。20：1代表20：1同分异构体的总和（20：1n-9和20：1n-11）（Palmeri, G., 2008）。

将人工养殖的澳洲龙纹斑，置于干净的水系中停餐18d，鱼失去的体重从6%降至14%，体重减轻与停餐/饥饿的天数高度相关。停餐18d后，与对照组相比，条件因子和肝细胞性指数显著下降（$p<0.05$）。鱼肉脂肪含量在试验期间没有变化。在试验过程中，二十碳五烯酸降低，二十二碳五烯酸增加（$p<0.05$），而二十二碳六烯酸没有显著性变化。停餐对挥发性风味化合物组合物的改善有积极的贡献，显著（$p<0.05$）减少总挥发性醛类化合物，并增加总挥发性碳氢化合物。由于在停餐过程的最后阶段样品之间没有发现重大差异，因此，12d是获得改进人工饲养的澳洲龙纹斑的挥发性化合物的最短持续时间，同时体重减轻为最少（图4-5、图4-6）（Palmeri, G., 2009）。

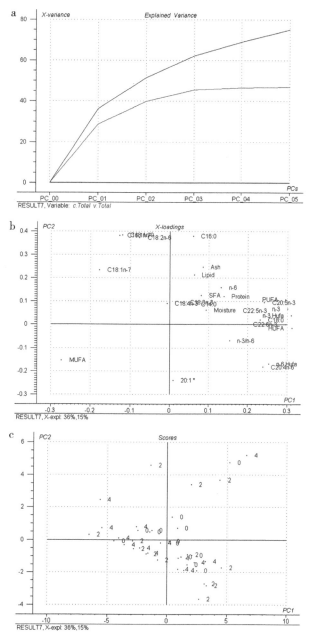

图 4-4 （a）-（c）澳洲龙纹斑净化 0/2 和 4 周后的主成分分析（%）和脂肪酸组成（%）的变化、荷载和记数检查的解释图（Palmeri，G.，2008）。

Palmeri，G.（2010）用菜籽油（CO）或鱼油（FO）饲料调控喂养两组澳洲龙纹斑 90d。用菜籽油饲料喂养的鱼转为 FO 饲料喂养 90d。进行不同时间（0，5，10 和 15d）的停餐，以减少停餐开始时 CO 喂鱼的初始脂肪量，从而加速鱼油类脂肪酸的回

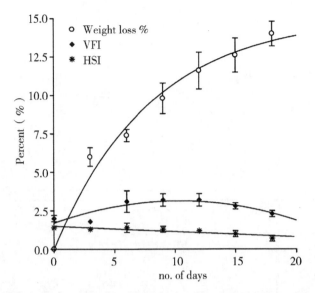

图4-5 标绘指数、线性和二阶多项式与时间（天数）和饥饿的澳洲龙纹斑中失重%、比肝重（HSI）和内脏脂肪指数（VFI）之间的关系图（Palmeri, G., 2009）。

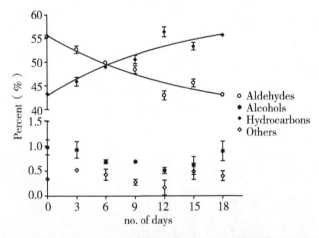

图4-6 净化时间（天数）与澳洲龙纹斑中的主要挥发性风味化合物之间的关系图（Palmeri, G., 2009）。

收率。在停餐过程中，鱼的总脂肪量没有显著减少，无论是鱼肉还是全身。在调理期结束时，饲料中的脂肪酸组成与鱼组织中的相对应。除了n-6多不饱和脂肪酸外，这些差异在重新喂养后趋于稳定。然而，停餐对鱼的最终脂肪酸组成没有影响。因此，澳洲龙纹斑在经受了长达15d的停餐后，脂肪量没有明显减少。(Palmeri, G., 2010)。

（罗钦、潘葳、黄敏敏）

第二节 研究论文摘编

96. Biometric, nutritional and sensory changes in intensively farmed Murray cod (Maccullochella peelii peelii, Mitchell) following different purging times[①].

[英文摘要]: A period of purging before harvesting is common practice in intensive aquaculture to eliminate any possible off flavours from the fish. The present study was conducted to evaluate the biometrical, nutritional and sensory changes in intensively farmed Murray cod (Maccullochella peelii peelii) after 0, 2 and 4 weeks of purging. After the main biometric parameters were recorded, fish were analysed for proximate, fatty acid composition and flavour volatile compounds. A consumer preference test (triangle test) was also conducted to identify sensorial differences that may affect the consumer acceptability of the product. Fish purged for 2 and 4 weeks had a significant weight loss of 4.1% and 9.1%, respectively, compared to unpurged fish, whilst perivisceral fat content did not change. The concentration of saturated (SFA), monounsaturated (MUFA) and highly unsaturated (HUFA) fatty acids were not significantly affected by purging time, while polyunsaturated fatty acids (PUFA), n-3 and n-3 HUFA were significantly higher ($P<0.05$) in purged fish compared to unpurged fish. Consumers were able to detect differences between the purged and unpurged fish ($P<0.05$) preferring the taste of the purged fish. However, consumers were unable to distinguish between fish purged for 2 and 4 weeks. This study showed that a 2 weeks purging period was necessary and sufficient to ameliorate the final organoleptic quality of farmed Murray cod. With such a strategy the nutritional qualities of edible flesh are improved while the unavoidable body weight loss is limited.

[译文]:

在不同的停餐时间之后,集中养殖的澳洲龙纹斑的生物特征,营养和感官变化。

在集约化水产养殖过程中,收获前进行停餐是消除鱼类任何可能异味的常见做法。本研究旨在评估经过停餐0周、2周和4周后密集养殖的澳洲龙纹斑的生物学,营养和感官变化。记录主要生物特征参数后,分析鱼的近似脂肪酸组成和风味挥发性化合物。还进行消费者偏好测试(三角测试)以识别可能影响消费者对产品可接受度的感官差

---

① 研究论文摘编序号续上一章,全书同。

异。与未停餐的鱼相比，停餐2周和4周的鱼体重有显著的减轻，分别为4.1%和9.1%，而内脏周围脂肪含量没有改变。与未停餐的鱼相比，停餐时间对饱和脂肪酸（SFA）、单不饱和脂肪酸（MUFA）和高度不饱和脂肪酸（HUFA）的浓度没有显著的影响，而多不饱和脂肪酸（PUFA），n-3和n-3HUFA显著高于对照组（p<0.05）。消费者因偏好停餐鱼的味道而能够发现停餐鱼和未停餐鱼之间的差异（p<0.05），但是，消费者无法区分停餐了2周和4周的鱼。这项研究表明，2周的停餐期对于改善养殖的澳洲龙纹斑的最终感官质量是必要的和足够的。通过这样的策略，可食用肉的营养品质得到改善，而不可避免的体重损失是有限的。

参考文献：Palmeri, G., Turchini, G. M., Caprino, F., et al. 2008. Biometric, nutritional and sensory changes in intensively farmed murray cod (maccullochella peelii peelii, mitchell) following different purging times [J]. Food Chemistry, 107 (4), 1605-1615.

（罗钦、李冬梅、黄惠珍　译）

97. The amino acid profiles in developing eggs and larvae of the freshwater Percichthyid fishes, trout cod, Maccullochella macquariensis and Murray cod, Maccullochella peelii peelii

[英文摘要]：Results on changes in the total amino acids (protein bound + free) and the free amino acids (FAA) in relation to development, from egg (unfertilised and/or fertilised) to yolk-sac resorbed larva, before first feeding, in two Percichthyid fish, trout cod, Maccullochella macquariensis and Murray cod, Maccullochella peelii peelii, which lay demersal, adhesive eggs, are presented. Throughout development, the FAA accounted for only a small proportion (0.19 % in fertilised eggs of both species) of the total amino acid pool. Nine essential amino acids (EAA) and eight non-essential amino acids (NEAA) were quantified in the amino acid pool at all stages of development. In both species, the total amino acid content decreased during the transformation (at $20\pm1℃$) from newly hatched larva to yolk-sac resorbed larva. Overall, the changes in the TEAA and TNEAA reflected that of the amino acid pool. In trout cod, all but one EAA (lysine) and two NEAA (cysteine and glycine) decreased with ontogeny, from fertilised egg to yolk-sac resorbed larva. In Murray cod, however, the exceptions to the general decline were two NEAA (aspartic acid and glycine). In contrast, the FAA increased with development, the changes being reflected in both FEAA and FNEAA. Qualitatively, the predominant free amino acids in trout cod and Murray cod eggs were alanine, lysine, leucine and serine. Because the egg protein and the total amino acid contents declined with development, it is concluded that the rate of breakdown of yolk protein was higher than the anabolic and catabolic processes during embryogenesis. Data also suggest that in freshwater fish FAA are an unlikely primary energy substrate during embryogenesis.

[译文]：

在真鲈科淡水鳟鳕鱼和澳洲龙纹斑中氨基酸从卵发育到幼鱼的变化

在首次饲喂前，两种真鲈科鳟鳕鱼和澳洲龙纹斑沉性的黏性的鱼卵（未受精和/或受精）从卵发育到卵黄囊全部被吸收成幼鱼时总氨基酸（蛋白质结合+游离）和游离氨基酸（FAA）的变化和发育的关系。在整个发育过程中，游离氨基酸占总氨基酸库的比例很小（0.19%，在两种鱼类受精卵中）。在整个发育过程中，氨基酸库中九种必需氨基酸和八种非必需氨基酸被定量分析。在两种鱼类中，总氨基酸含量从新孵出的鱼发育成（在20±1℃）卵黄囊全吸收的幼鱼时是降低的。总体而言，必需氨基酸含量和非必需氨基酸含量的变化随着氨基酸库变化。在鳟鳕鱼中，除了一个必需氨基酸（赖氨酸）和两个非必需氨基酸（半胱氨酸和甘氨酸）外，总氨基酸含量随着受精卵发育成卵黄囊全吸收的幼鱼而减少。然而，在澳洲龙纹斑中，除了两个非必需氨基酸（天冬氨酸和甘氨酸）外，氨基酸含量普遍下降。相比之下，游离氨基酸随着发育而增加，在必需氨基酸和非必需氨基酸这两个方面都反映出这一变化。定性地说，鳟鳕鱼和澳洲龙纹斑卵中主要的游离氨基酸是丙氨酸，赖氨酸，亮氨酸和丝氨酸。因为卵蛋白和总氨基酸的含量随着发育而下降，所以结论是卵黄蛋白的降解速率高于胚胎发生过程中的合成代谢和分解代谢。数据还表明，在淡水鱼类中游离氨基酸未必是胚胎发生过程中的主要能量来源。

参考文献：Rasanthi M. Gunasekera, Sena S. De Silva, Brett A. Ingram. 1999. The amino acid profiles in developing eggs and larvae of the freshwater percichthyid fishes, trout cod, maccullochella macquariensis, and murray cod, maccullochella peelii peelii [J]. Aquatic Living Resources, 12 (4), 255-261.

（罗钦、李冬梅、黄惠珍　译）

98. Early ontogeny-related changes of the fatty acid composition in the Percichthyid fishes trout cod, Maccullochella macquariensis and Murray cod, M. peelii peelii

[英文摘要]：Changes in the fatty acid profiles of the Percichthyid fish trout cod, Maccullochella macquarlensis (Cuvier), and Murray cod, M. peelii peelii (Mitchell), two Australian native freshwater fish species, were investigated during early development from egg to yolk-sac-resorbed larval stage. In the two Percichthyid fishes polyunsaturated fatty acids (PU-

FA) accounted for more than 50% of the 19 quantified fatty acids in total lipid. The fatty acids that occurred in the highest abundance in both trout cod and Murray cod, in all developmental stages, in order, were docosahexaenoic acid [DHA 22∶6 (n-3)], arachidonic acid [AA 20∶4 (n-6)], oleic acid [18∶1 (n-9)] and palmitic acid (16∶0), all of which exceeded 100 mu g per mg total lipid in most instances: The ratio of 22∶6 (n-3) to eicosapentaenoic acid-20∶5 (n-3) in eggs of trout cod and Murray cod was 5.4∶1 and 7.3∶1, respectively, and remained almost unchanged through development, and was considerably higher than the 2∶1 ratio generally reported for fish eggs. In trout cod, 11 of the 19 fatty acids in total lipid decreased during the transformation from egg to yolk-sac-resorbed larva. In Murray cod, only 16∶1 (n-7) showed a significant decrease whilst 20∶4 (n-6) increased significantly with development. Overall, there was a tendency in both species to conserve n-3 and n-6 series highly unsaturated fatty acids (HUFA), suggesting their essentiality in first feeding larvae. These observations are discussed in relation to the feeding habits of trout cod and Murray cod, which are top order, freshwater carnivores.

[译文]：

在真鲈科鱼鳟鳕鱼和澳洲龙纹斑中早期个体发育的脂肪酸组成的变化

在早期发育过程中，澳洲本土的淡水鳟鳕鱼和澳洲龙纹斑从卵发育到卵黄囊全部被吸收的幼鱼阶段的脂肪酸的变化。在两种真鲈科鱼中多不饱和脂肪酸（PUFA）占总脂肪中19种定量脂肪酸的50%以上。在发育全阶段，鳟鳕鱼和澳洲龙纹斑中含量最高的脂肪酸依次为二十二碳六烯酸［DHA22∶6（n-3）］，花生四烯酸［AA20∶4（n-6）］，油酸［18∶1（n-9）］和棕榈酸（16∶0），在大多数情况下都超过100μg/mg总油脂：在鳟鳕鱼卵和澳洲龙纹斑卵中，22∶6（n-3）和二十碳五烯酸20∶5（n-3）的比例分别为5.4∶1和7.3∶1，发育阶段几乎保持不变，并且显著高于一般鱼卵的2∶1比例。在鳟雪鱼中，从卵发育到卵黄囊全部被吸收的幼鱼阶段总脂肪中19种定量脂肪酸中的11种脂肪酸是减少的。在澳洲龙纹斑发育过程中，只有16∶1（n-7）显著减少而20∶4（n-6）显著增加。总而言之，两种鱼类都倾向于保存n-3和n-6系列高度不饱和脂肪酸（HUFA），显示其在首次饲喂中的重要性。这些现象与鳟鳕鱼和澳洲龙纹斑的摄食习惯有关，它们都是顶级的淡水肉食性鱼类。

参考文献：Gunasekera R. M, De Silva S. S, Ingram B. A. 1999. Early ontogeny-related changes of the fatty acid composition in the percichthyid fishes trout cod, maccullochella macquariensis and murray cod, m. peelii peelii [J]. Aquatic Living Resources, 12 (3),

219-227.

(罗钦、李冬梅、黄惠珍 译)

99. Lipid characterisation and distribution in the fillet of the farmed Australian native fish, Murray cod (Maccullochella peelii peelii)

[英文摘要]: The objective of this study was to determine the distribution pattern of lipids and fatty acids in different tissues of farmed Murray cod (Maccullochella peelii peelii). Differences in lipid content were found amongst different portions of the fillet, being lowest in the dorsal/cranial portion (P1) and highest in the more ventral/caudal portion (P8) (P<0.05). The latter also recorded the highest amount of monounsaturated fatty acid (MUFA) and the lowest in polyunsaturated fatty acids (PUFA), arachidonic acid, 20:4 n﹣6 (ArA), docosahexaenoic acid, 22:6 n﹣3, (DHA) and the n3/n6 ratio. In general, lipid content in the different fillet portions was inversely correlated to PUFA and directly to MUFA. Contents of saturated fatty acids (SFA) and eicosapentaenoic acid, 20:5 n﹣3 (EPA) did not show any discernible trends in the different fillet portions, while significant differences in contents of DHA and ArA were observed. This study shows that lipid deposition in Murray cod varies markedly and that different fatty acids are deposited differently throughout the fillet.

[译文]:

澳洲本土澳洲龙纹斑肉中的脂肪特性及分布

本研究目的是确定人工养殖澳洲龙纹斑（Maccullochella peelii peelii）不同组织中的脂肪和脂肪酸的分布。在不同部位的鱼肉中发现脂肪含量的差异，在背/颅部分（P1）中最低，在腹侧/尾部（P8）较高（p<0.05）。后者也测得了最高含量的单不饱和脂肪酸（MUFA）和最低的多不饱和脂肪酸（PUFA），花生四烯酸［20:4n﹣6（ArA）］，二十二碳六烯酸、二十六碳六烯酸［22:6n﹣3（DHA）］和n3/n6的比例。一般来说，不同部位的鱼肉的脂肪含量与PUFA和MUFA直接相关。饱和脂肪酸（SFA）和二十碳五烯酸［20:5n﹣3（EPA）］的含量在不同部位的鱼肉中没有显示出任何明显的趋势，而DHA和ArA的含量差异显著。这项研究表明，脂肪在澳洲龙纹斑体内的沉积有明显的差异，不同的脂肪酸在不同部位的鱼肉中的沉积也不同。

参考文献：Palmeri, G., Turchini, G. M., De Silva, S. S. 2007. Lipid characterisation and distribution in the fillet of the farmed australian native fish, murray cod (maccullochella peelii peelii) [J]. Food Chemistry, 102 (3), 796-807.

(罗钦、李冬梅、黄惠珍 译)

100. Biometric, nutritional and sensory characteristic modifications in farmed Murray cod (Maccullochella peelii peelii) during the purging process

[英文摘要]: Intensively farmed, market-size Murray cod (~ 600g), were purged (transferred into a clean water system and starved) and sampled at three day intervals for a total of 18 days (D0, D3, D6, D9, D12, D15 and D18). Purged fish lost from 6% (D3) to 14% (D18) body weight, and the weight loss was highly correlated to the number of days of purging/starvation. Condition factor and Hepatosomatic Index decreased significantly ($P < 0.05$) only after 18 days of purging compared to the control (D0). Fillet lipid content (%) did not vary during the trial. Eicosapentaenoic acid (EPA: 20: 5 n-3) decreased and docosapentaenoic acid (DPA: 22: 5 n-3) increased ($P<0.05$) during the trial, while docosahexaenoic acid (DHA: 22: 6 n-3) did not show any significant variation. Purging contributed positively to the improvement of the volatile flavour compound composition, with a significant ($P<0.05$) reduction in total volatile aldehydes and an increase in total volatile hydrocarbons. Since no major differences were found between samples during the last stages of the purging process (D12, D15 and D 18), it is possible to conclude that, under these experimental conditions, 12 days is the minimum duration to obtain an improvement in the volatile compound profile of intensively farmed Murray cod whilst keeping body weight loss to a minimum.

[译文]:
停餐过程对饲养澳洲龙纹斑的生物特征、营养和感官特征的变化

市场规模化密集养殖的澳洲龙纹斑（约600g），置于干净的水系中并且不投喂饲料，以三天的间隔进行停餐和取样，共计18d（D0，D3、D6、D9、D12、D15和D18）。停餐鱼失去的体重从6%（D3）降至14%（D18），体重减轻与停餐/饥饿的天数高度相关。在停餐18d后，与对照组（D0）相比，条件因子和肝细胞性指数显著下降（$p<0.05$）。鱼肉脂肪含量（%）在试验期间没有变化。在试验过程中，二十碳五烯酸（EPA: 20: 5n-3）降低，二十二碳五烯酸（DPA: 22: 5n-3）增加（$p<0.05$），而二十二碳六烯酸（DHA: 22: 6n-3）没有显著性变化。停餐对挥发性风味化合物组合物的改善有积极的贡献，显著（$p<0.05$）减少总挥发性醛类化合物，并增加总挥发性碳氢化合物。由于在停餐过程的最后阶段（D12，D15和D18）样品之间没有发现重大差异，

因此可以得出结论，在这些实验条件下，12d 是获得改进人工饲养的澳洲龙纹斑的挥发性化合物的最短持续时间，同时体重减轻为最少。

参考文献：Palmeri, G., Turchini, G. M., Marriott, P. J., *et al.* 2009. Biometric, nutritional and sensory characteristic modifications in farmed murray cod (maccullochella peelii peelii) during the purging process [J]. Aquaculture, 287 (3-4), 354-360.

（罗钦、李冬梅、黄惠珍　译）

101. Short-term food deprivation does not improve the efficacy of a fish oil finishing strategy in Murray cod

[英文摘要]：Two groups of fish (Maccullochella peelii peelii) were fed for a 90-day conditioning period on a canola oil diet (CO) or a fish oil diet (FO). Canola oil diet fed fish were then shifted to the FO diet for a 90-day finishing period. A variable period of starvation (0, 5, 10 and 15 days) was introduced to reduce the initial lipid level of CO fed fish at the beginning of the finishing period and therefore accelerate the rate of recovery of FO-like fatty acids. During starvation, fish did not show significant reduction in total lipid content, either in the fillet or whole body. At the end of the conditioning period, fatty acid composition of the diet was mirrored in fish tissues. These differences came close to levelling out following re-feeding, with the exception of n-6 polyunsaturated fatty acids (PUFA). However, no effects of the starvation periods on the final fatty acid make-up of fish were recorded. The results of this trial show that Murray cod, when subjected to a starvation period of up to 15 days, does not lose an appreciable quantity of lipid and, therefore, the tested starvation approach to reduce the initial level of lipid has to be considered unsuccessful.

[译文]：

短期停餐并不能提升澳洲龙纹斑鱼油精加工的品质

用菜籽油（CO）或鱼油（FO）饲料调控喂养两组澳洲龙纹斑 90d。用菜籽油饲料喂养的鱼转为 FO 饲料喂养 90d。进行不同时间（0，5，10 和 15d）的停餐，以减少停餐开始时 CO 喂鱼的初始脂肪量，从而加速鱼油类脂肪酸的回收率。在停餐过程中，鱼的总脂肪量没有显著减少，无论是鱼肉还是全身。在调理期结束时，饲料中的脂肪酸组成与鱼组织中的相对应。除了 n-6 多不饱和脂肪酸（PUFA）外，这些差异在重新喂养后趋于稳定。然而，停餐对鱼的最终脂肪酸组成没有影响。这项试验的结果表明，澳洲龙纹斑在经受了长达 15d 的停餐后，脂肪量没有明显减少，因此，经测试采用停餐的方法来降低最初的脂肪是失败的。

参考文献：Palmeri, G., Turchini, G. M., Silva, S. S. D. 2010. Short-term food deprivation does not improve the efficacy of a fish oil finishing strategy in murray cod [J]. Aquaculture Nutrition, 15 (6), 657-666.

（罗钦、李冬梅、黄惠珍　译）

102. Hypoxia, Blackwater and Fish Kills: Experimental Lethal Oxygen Thresholds in Juvenile Predatory Lowland River Fishes

[英文摘要]：Hypoxia represents a growing threat to biodiversity in freshwater ecosystems. Here, aquatic surface respiration (ASR) and oxygen thresholds required for survival in freshwater and simulated blackwater are evaluated for four lowland river fishes native to the Murray-Darling Basin (MDB), Australia. Juvenile stages of predatory species including golden perch Macquaria ambigua, silver perch Bidyanus bidyanus, Murray cod Maccullochella peelii, and eel-tailed catfish Tandanus tandanus were exposed to experimental conditions of nitrogen-induced hypoxia in freshwater and hypoxic blackwater simulations using dried river red gum Eucalyptus camaldulensis leaf litter. Australia's largest freshwater fish, M. peelii, was the most sensitive to hypoxia but given that we evaluated tolerances of juveniles (0.99±0.04g; mean mass±SE), the low tolerance of this species could not be attributed to its large maximum attainable body mass (>100000g). Concentrations of dissolved oxygen causing 50% mortality (LC50) in freshwater ranged from 0.25±0.06mg l^{-1} in T. tandanus to 1.58±0.01mg l^{-1} in M. peelii over 48 h at 25~26℃. Logistic models predicted that first mortalities may start at oxygen concentrations ranging from 2.4mg l^{-1} to 3.1mg l^{-1} in T. tandanus and M. peelii respectively within blackwater simulations. Aquatic surface respiration preceded mortality and this behaviour is documented here for the first time in juveniles of all four species. Despite the natural occurrence of hypoxia and blackwater events in lowland rivers of the MDB, juvenile stages of these large-bodied predators are vulnerable to mortality induced by low oxygen concentration and water chemistry changes associated with the decomposition of organic material. Given the extent of natural flow regime alteration and climate change predictions of rising temperatures and more severe drought and flooding, acute episodes of hypoxia may represent an underappreciated risk to riverine fish communities.

[译文]：

缺氧、黑水和鱼类死亡：幼年捕食性低地河鱼类实验性致死氧阈

　　缺氧对淡水生态系统中的生物多样性构成了越来越大的威胁。在这里，评估澳大利亚墨累-达令盆地（MDB）原产的四种低地河流鱼类在淡水和模拟黑水中生存所需的水生表面呼吸（ASR）和氧气阈值。掠食性物种的幼体阶段（包括黄金鲈鱼，澳洲银鲈，墨瑞鳕鱼和澳洲鳗鲶）暴露于氮诱导缺氧的淡水和使用干河红胶桉树凋落物模拟出缺氧黑水的实验条件。澳大利亚最大的淡水鱼-墨瑞鳕鱼对缺氧最为敏感，但鉴于我们评估了幼体的耐受性（0.99±0.04g；平均质量±标准误差），该物种的低耐受性不能归因于其最大可达量体重（>100000g）。在 25~26℃ 下 48h 内，在淡水中引起50%死亡率（LC50）的溶解氧浓度范围：澳洲鳗鲶是 0.25±0.06mg/L；墨瑞鳕鱼是 1.58±0.01mg/L。逻辑模型预测，澳洲鳗鲶和墨瑞鳕鱼分别在黑水模拟中，第一次死亡可能开始于氧化浓度范围从 2.4mg/L~3.1mg/L。水生表面呼吸先于死亡，这种行为在这里被首次记录在所有四种物种的幼体中。尽管 MDB 低地河流中缺氧和黑水事件自然发生，但这些大型掠食鱼类的幼年阶段易受死亡率（由低氧浓度和有机物分解引起水化学变化而造成死亡）的影响。鉴于自然流态变化、温室效应的气候变化预测、更为严重的干旱和洪水等环境变化程度，缺氧的急性发作可能是一个河流鱼类群落被低估的风险。

参考文献：Small, K., Kopf, R. K., Watts, R. J., et al. 2014. Hypoxia, blackwater and fish kills: experimental lethal oxygen thresholds in juvenile predatory lowland river fishes [J]. Plos One, 9 (4), e94524.

（罗钦、李冬梅、黄惠珍　译）

103. Mortality of larval Murray cod (Maccullochella peelii peelii) and golden perch (Macquaria ambigua) associated with passage through two types of low-head weirs

[英文摘要]：Determining factors responsible for increases in the mortality of freshwater fish larvae are important for the conservation of recruitment processes and for the long-term sustainability of freshwater fish populations. To assess the impact of one such process, Murray cod (Maccullochella peelii peelii Mitchell) and golden perch (Macquaria ambigua Richardson) larvae were arranged into treatment and control groups and passed through different configurations (overshot and undershot) of a low-level weir. Passage through an undershot weir resulted in the death of 95±1% golden perch and 52±13% Murray cod. By comparison, mortality was significantly lower in the overshot treatment and both controls. The relatively large number of undershot weirs within the known distribution of these species could impact upon recruitment over a large scale. It is therefore recommended that water management authorities consider the potential threats of operating undershot gated weirs on the survival of larval fish until further research determines appropriate mitigatory measures for these and other species.

[译文]：

墨瑞鳕幼鱼和黄金鲈幼鱼在通过两种低堰时的死亡率

淡水鱼类幼鱼死亡率增加的决定因素对于淡水鱼类种群的通过补充程序来保护和可持续性是非常重要的。为了评估一个这样的程序的影响，将墨瑞鳕鱼和黄金鲈鱼幼鱼安排在实验组和对照组中，并通过不同结构（上接水的和下接水的）的低堰。通过一个下接水堰时黄金鲈鱼的死亡率为95±1%和墨瑞鳕鱼死亡率为52±13%。相比之下，通过上接水堰的实验组和2个对照组的死亡率明显更低。在这些已知分布的鱼类中，相对众多的下接水堰可能对一个大规模的鱼类补充造成冲击。因此建议水管局考虑下在运营的带闸门的下接水堰对仔鱼存活的潜在威胁，直到将来研究出对这些和其他鱼类的适当的缓解措施。

参考文献：Lee J. Baumgartner, Nathan Reynoldson, Dean M. Gilligan. 2006. Mortality of larval murray cod (maccullochella peelii peelii) and golden perch (macquaria ambigua) associated with passage through two types of low-head weirs [J]. Marine and Freshwater Research, 57 (2), 187-191.

（罗钦、李冬梅、黄惠珍　译）

104. System-Specific Variability in Murray Cod and Golden Perch Maturation and Growth Influences Fisheries Management Options

[英文摘要]：The Murray Cod Maccullochella peelii and Golden Perch Macquaria ambigua are important recreational species in Australia's Murray-Darling Basin (MDB); both species have declined substantially, but recovery is evident in some areas. Minimum length limits (MLLs)—implemented to ensure fish could spawn at least once prior to harvest eligibility—have increased three times in the past decade. We quantified variation in length at 50% maturity (LM50), age at 50% maturity (AM50), and von Bertalanffy growth parameters (k = Brody growth coefficient; L∞ = asymptotic length; t0 = theoretical age at zero length) of these species within two rivers and two reservoirs of the MDB; to investigate whether fish length is a suitable surrogate for AM50 in setting MLLs. Between 2006 and 2013, we collected 1118 Murray Cod and 1742 Golden Perch by electrofishing and gillnetting. Values of k and L∞ were greater for reservoir fish than for riverine fish. For both species, AM50 was generally greater in rivers than in reservoirs; for Murray Cod, LM50 was greater in reservoirs than in rivers. A yield-per-recruit model demonstrated that smaller Murray Cod MLLs would be required for rivers and that an MLL at or below 600 mm (the existing MLL) across all populations could lead to overfishing in some systems. The differences in growth rate and the onset of reproductive maturation between riverine and reservoir populations suggest that system-specific regulations would be more effective at reducing the overfishing risk and meeting fishing quality objectives.

[译文]：

墨瑞鳕鱼和黄金鲈鱼成熟与生长的系统特异性对渔业管理选择的影响

澳洲墨瑞-达令盆地（MDB）中的墨瑞鳕鱼和黄金鲈鱼是重要的休闲渔业；两种鱼类都大幅度下降，但有些地区恢复是明显的。最短的限量（MLL）-确保鱼在适合捕捞前至少产卵一次-在过去十年中增加了三倍。我们量化了50%成熟度（LM50）的长度变化值，50%成熟度的年龄值（AM50），和这些在两个河流和两个墨瑞-达令盆地的水库内的鱼类的贝塔朗菲生长参数（k=布洛迪生长系数；L∞=渐近长度；t0=零长度的理论年龄）；用于调查鱼体长度是否适合用AM50来替代设置MLL。在2006年至2013年期间，我们通过电鱼法和刺网渔法捕获了1118尾墨瑞鳕鱼和1742尾黄金鲈鱼。水库鱼的k值和L∞值比河鱼的大。对于这2种鱼类，河流中的AM50通常大于水库的；对于墨瑞鳕鱼，水库中的LM50比河流的更大。一个每年补充模式证明了河流需要较小的墨瑞鳕鱼MLL，并且在所有种群中，等于MLL或低于600mm（现有MLL）可能导致某些河系过度捕捞。河流与水库之间的生长速度和性成熟时间的差异表明系统特异性的调整对降低过度捕捞风险及达到捕捞质量目标更加有效。

参考文献：Forbes, J. P., Watts, R. J., Robinson, W. A., et al. 2015. System-specific variability in murray cod and golden perch maturation and growth influences fisheries management options [J]. North American Journal of Fisheries Management, 35 (6), 1226-1238.

（罗钦、李冬梅、黄惠珍　译）

105. Cytochemical characterisation of the leucocytes and thrombocytes from Murray cod (Maccullochella peelii peelii, Mitchell)

[英文摘要]：Cytochemistry has proven effective in differentiating specific cell lineages and elucidating their functional properties. This study utilised a range of cytochemical techniques to further investigate the leucocyte populations from Murray cod, an iconic Australian teleost fish species. This analysis provided clear insight into the structure and function of the leucocytes from this fish, which were found to be broadly similar to those of other fish species. However, some important differences were identified in Murray cod, such as the presence of naphthol AS chloroacetate esterase activity in the heterophil population, positive staining for periodic acid Schiff's, alkaline phosphatase and Sudan black B in the lymphocyte population, and a prevalent population of myeloid precursor cells.

[译文]：

来自墨瑞鳕鱼的白细胞和血小板的物理化学表征

细胞化学已被证明能有效区分特异性的细胞谱系并阐明其功能特性。本研究利用一系列细胞化学技术进一步调查墨瑞鳕鱼的白细胞数量，一种标志性的澳洲硬骨鱼类。本研究清楚地了解这种鱼的白细胞的结构和功能，发现它们与其他鱼类大致相似。然而，在墨瑞鳕鱼中鉴定出了一些重大的差异，例如在嗜异细胞种群中出现了萘酚 AS 氯乙酸酯酶活性，过碘酸的正染色，在淋巴细胞种群中出现了碱性磷酸酶和苏丹黑 B，和广泛的骨髓前体细胞。

参考文献：Shigdar, S., Harford, A., Ward, A. C. 2009. Cytochemical characterisation of the leucocytes and thrombocytes from murray cod (maccullochella peelii peelii, mitchell) [J]. Fish Shellfish Immunol, 26 (5), 731–736.

（罗钦、李冬梅、黄惠珍　译）

106. Blood cells of Murray cod Maccullochella peelii peelii (Mitchell)

[英文摘要]：Analysis of the peripheral blood cells of Murray cod Maccullochella peelii peelii identified seven distinct subpopulations including a novel basophilic cell. Haematological reference ranges were established to facilitate future diagnostic blood sampling of this fish.

[译文]：

墨瑞鳕鱼的血细胞研究

对墨瑞鳕鱼外周血细胞的分析确定了包括 1 个新的嗜碱性细胞在内的 7 个不同的亚种。为今后该鱼血样的诊断建立了血液学的参考值范围。

参考文献：Shigdar, S., Cook, D., Jones, P., et al. 2007. Blood cells of murray cod maccullochella peelii peelii (mitchell) [J]. Journal of Fish Biology, 70 (3), 973–980.

（罗钦、李冬梅、黄惠珍　译）

107. 澳洲龙纹斑肌肉氨基酸的分析研究

摘要：为了研究不同生长阶段的澳洲龙纹斑肌肉氨基酸组成的动态变化，同时探讨并比较处于商品鱼阶段的澳洲龙纹斑、石斑鱼、鲟鱼肌肉氨基酸的组成与含量变化，旨在推广澳洲龙纹斑的养殖技术与引导消费。本研究通过氨基酸自动分析仪检测澳洲龙纹斑亲鱼、商品鱼、幼鱼以及石斑鱼商品鱼、鲟鱼商品鱼中肌肉氨基酸含量，系统分析了澳洲龙纹斑不同生长阶段氨基酸组成的变化规律，同时比较了不同种类商品鱼氨基酸组成的含量。结果表明：3 种不同生长阶段的澳洲龙纹斑氨基酸总含量最高为商品鱼 70.96%、其次为幼鱼 69.24%、亲鱼最低为 65.36%（以干基计算），在氨基酸组成中亲鱼和商品鱼的比例较为接近，而幼鱼的氨基酸组成比例则有所变化，在必需氨基酸中，其 EAA/TAA 值则比亲鱼以及商品鱼低 2%左右，而在非必需氨基酸中，EAA/NEAA 比值又高于亲鱼与商品鱼 2%左右；澳洲龙纹斑、石斑鱼、鲟鱼 3 种商品鱼肌肉中甜味氨基酸和鲜味氨基酸的含量之和均超过 50%（澳洲龙纹斑 57.17%，石斑鱼 57.68%，鲟鱼 57.85%），其反映了 3 种商品鱼肉质味道鲜美度是处在同一等级上；同时，澳洲龙纹斑商品鱼（E/T 比值为 40.34、E/NT 比值为 67.62）符合 WHO/FAO 提出的 E/T 应为 40%左右和 E/NT 应为 60%以上的参考蛋白质模式标准。

Amino Acid Analysis of Maccullochella peelii Muscle

Abstract: In order to discuss the change of the amino acids in Maccullochella peelii muscle of different growth stages of composition, and compare the muscle amino acid composition of Maccullochella peelii、Grouper、Sturgeon in marketable fish stage, aims to promote the cultivation and consumption of Maccullochella peelii. In this study, the contents of amino acids in the muscle of Maccullochella peelii parent fish、marketable fish、juvenile fish and Grouper marketable fish、Sturgeon marketable fish were detected by the automatic amino acid analyzer, and analyses the amino acid composition change of Maccullochella peelii at different growth stages, and compare the content of amino acid composition in different kinds of marketable fish. The results show that: the highest total amino acid content in the 3 kinds of different growth stages of Maccullochella peelii was marketable fish 70.96%、followed was juvenile fish 69.24%、the lowest was parent fish 65.36% (dry basis), composition of amino acid in parent fish and marketable fish were very close, and the proportion of juveniles in the essential amino acids of EAA/TAA lower than them about 2% amino acids, but in essential amino acid EAA/NEAA higher than them about 2%; the content of the sweet amino acids and tasty amino acids in 3 kinds marketable fish were more than 50% (Maccullochella peelii 57.17%, Grouper 57.68%, Sturgeon 57.85%), meat taste delicious are about the same. At the same time, the Maccullochella peelii marketable fish (E/T 40.34, E/NT 67.62) accorded with the standard reference model for the protein of WHO/FAO that E/T should be about 40% and

E/NT should be more than 60%.

参考文献：罗钦，罗土炎，颜孙安，等.2015.澳洲龙纹斑肌肉氨基酸的分析研究[J].福建农业学报，30（5）：446-451。

108. 虫纹鳕鲈肌肉营养成分分析与品质评价

摘要：用常规方法测定、分析了澳洲龙纹斑肌肉中营养成分组成与含量。结果显示，澳洲龙纹斑肌肉中水分含量为80.75%，蛋白质为15.96%，粗脂肪为1.68%，灰分为0.94%。肌肉中测出氨基酸17种（未测色氨酸），其中包括人体必需氨基酸7种（占氨基酸总量的39.93%），必需氨基酸组成基本符合FAO/WHO标准。澳洲龙纹斑的主要限制性氨基酸为蛋氨酸、胱氨酸和缬氨酸，其必需氨基酸指数（EAAI）为95.43，鲜味氨基酸占氨基酸总量的45.59%。脂肪酸中多不饱和脂肪酸含量为43.44%，其中EPA与DHA的总含量为14.68%。实验结果表明，澳洲龙纹斑具有较高的食用价值与保健作用。

参考文献：宋理平，冒树泉，胡斌，等.2013.虫纹鳕鲈肌肉营养成分分析与品质评价[J].饲料工业，34（16）：42-45。

注：无英文摘要。

第五章 澳洲龙纹斑饲料与制备

第一节 专题研究概况

澳洲龙纹斑是澳洲暖水性、肉食性、夜行性的淡水养殖鱼类中个体最大的种类，也是深受养殖者和消费者喜爱的白肉鱼，素有澳洲"国宝鱼"之美称。1999年，我国台湾从澳洲引进澳洲龙纹斑鱼苗养殖，大陆自2001年从澳洲引进，经过多年的研究，近几年取得了澳洲龙纹斑规模化繁养殖技术的突破，其养殖得以蓬勃发展。澳洲龙纹斑各生长发育阶段摄食对象有所不同，5~8mm的幼鱼主要摄食浮游动物，15~20mm的鱼苗即可捕食水中昆虫、线虫、红虫、小型甲壳类等，成鱼主要捕食小型鱼类、甲壳类，也捕食蛙类、水禽、乌龟等，在无其他饲料处于饥饿状态时，也会互相残食，养殖条件下，在规格达25~30mm阶段的幼鱼，就可以驯化摄食配合饲料（Gunasekera *et al*, 2000; Rowland, 2004; Ingram, 2009）。

随着澳洲龙纹斑养殖业的发展，迫切需要产业化开发高效、低成本、环境友好型澳洲龙纹斑专用配合饲料，以推进其"标准化、规模化、集约化、产业化"健康养殖的持续发展。配合饲料的质量不仅决定了澳洲龙纹斑生长性能、养殖效益，而且还影响澳洲龙纹斑产品质量与安全。开发的澳洲龙纹斑系列配合饲料，不仅要满足澳洲龙纹斑的营养需求和摄食习性，实现其高效利用，而且更要达到对人类、澳洲龙纹斑和环境的友好，以推进澳洲龙纹斑健康养殖的持续发展。目前，我国对高效环境友好型澳洲龙纹斑系列配合饲料尚在研发之中，建立澳洲龙纹斑配合饲料质量评价体系对指导优质澳洲龙纹斑配合饲料的开发具有重要意义。

一、澳洲龙纹斑的营养需求

澳洲龙纹斑的营养需求是开发其配合饲料的基础。目前，国内外有关澳洲龙纹斑营养需求研究较少,开展了澳洲龙纹斑的营养成分分析、蛋白质营养需求和脂类营养需求等研究工作。鱼体化学组成是在鱼体生长发育过程中,饲料经过消化、吸收并在鱼体内经过代谢、转化沉积的结果,其含量不仅反映了鱼体生长发育和代谢的情况,也反映出了鱼体对于来自于饲料的营养素需要量的差异。鱼类各生长发育阶段的代谢规律(内因)和饲料营养水平、水温、溶解氧、水域环境(外因)共同影响着鱼体自身的物质和能量代谢,进而影响营养素在鱼体中的沉积以及鱼类对饲料中营养素的需要量,因此,在探明某种鱼的鱼体营养成分的种类与含量,不仅可以作为其营养品质评价的基础数据,而且可以为鱼类配合饲料的研发提供理论借鉴和科学依据(Haard,1992; Shearer,1994)。

(一) 蛋白质和氨基酸的营养需求

1. 蛋白质的营养需求

蛋白质是维持澳洲龙纹斑生长、健康、繁殖及其他生命活动所必需的营养素,在澳洲龙纹斑营养上具有非常重要的作用和特殊地位,是其他营养素所无法替代的,必须由饲料供给。作为典型肉食性鱼类的澳洲龙纹斑将摄取的饲料蛋白质在其消化器官内代谢分解成小肽或游离氨基酸,小肽或氨基酸在鱼体内被吸收合成为鱼体蛋白质,用于鱼体生长、修补组织及维持生命。蛋白质是澳洲龙纹斑饲料中所占费用最高的营养素,探明其适宜的营养需求是开发澳洲龙纹斑优质配合饲料需要解决的首要问题。研究表明,澳洲龙纹斑对饲料蛋白质的需求量较高,其适宜需求量主要由蛋白质品质决定,同时也受到澳洲龙纹斑生长阶段、生理状况、养殖密度、养殖模式、环境因子(水温、溶解氧等)、日投喂量、饲料中非蛋白质能量的数量等因素的影响。

迄今为止,有关澳洲龙纹斑蛋白质营养需求研究较少,大多采用测定鱼体蛋白质含量作为其蛋白质营养需求的参考依据之一。饲养实验结果表明,澳洲龙纹斑幼鱼饲料中适宜的蛋白质需求量为50%;在饲料脂肪含量为24%时,澳洲龙纹斑幼鱼饲料中蛋白质含量40%也可以满足其生长发育需要,脂肪含量10%~17%时,蛋白质含量50%的饲料养殖效果更好。这与其他鱼类最佳生长速度的蛋白质需求基本一致(40%~50%),也符合澳洲龙纹斑肉食性鱼类的习性。澳洲龙纹斑饲料中的蛋白质含量维持较为合理的水平才能有效促进鱼体生长,蛋白质含量过低无法满足其快速生长对蛋白质营养的需求,而含量过高则会造成蛋白质吸收效率下降,带来资源浪费和环境污染的问题。综合而言,澳洲龙纹斑幼鱼阶段配合饲料的蛋白质需求量应在45%以上,达到幼鱼快速生长的配合饲料蛋白质含量应在50%左右;养成配合饲料中蛋白质的需求量应在40%以上,达到更佳生长速度的饲料蛋白质水平为45%左右,其中鱼粉对蛋白的贡献率应不低于

50%，且应尽量减少饲料中纤维素的含量，避免食物过快的通过消化道，从而避免营养物质尚未完全吸收就被排出体外。（Gunasekera et al，2000、De Silva et al，2002）

De Silva，S.S.（2002）开展了一个为期56天的澳洲龙纹斑幼鱼的饲养实验，平均重量14.902±020.0402g，饲以五种实验的饲料，一个系列为蛋白质含量40%和脂肪含量为10%，17%和24%（P40L10，P40L17和P40L24），另一个系列为蛋白质含量50%和脂肪含量17%，24%（P50L17和P50L24）。不同饲料幼鱼的特定生长率(%day611)范围从1.18到1.41，除了P40L10外其它饲料在饲养实验之间没有显著性差异。然而，特定生长率在蛋白质水平上随着饲料中油脂含量的增加而增加。与50%蛋白质饲料相比，40%蛋白质饲料的饲料转化率更差，饲料转化率最佳值是1.14，是在喂养P50L17饲料中观测到的。然而，蛋白质效率比在低蛋白饲料饲养的鱼中更好。在不同的饲料实验中，净蛋白利用率并没有显著性差异。如在蛋白质效率比一样的情况下，投喂P40L24饲料的澳洲龙纹斑观测到最高的净蛋白利用率，而投喂P50L24饲料的鱼最低。当两种蛋白质水平上的油脂含量的显著增加时，鱼体油脂含量反映了饲料的油脂含量。然而，随着饲料油脂含量的增加，鱼体脂肪含量并没有增加，并且明显低于全身脂肪含量。在饱和脂肪酸、单一脂肪酸和多不饱和脂肪酸中发现的最高浓度的脂肪酸分别为1602：020、1802：021n-9和2202：026n-3，并且它们中的每一个占比为每个组总数的60%以上。肌肉脂肪酸含量受饲料油脂含量的影响，例如，单脂肪酸的总量（μg02mg61102油脂）范围为7202±025.1（P40L10）至11202±0210（P40L24）和11202±022.8（P50L17）至13202±0211.8（P50L24），n-6系列脂肪酸含量随饲料油脂含量增加而增加，但变化不够明显。最值得注意的是，1802：022n-6随着两种饲料中的油脂水平增加而增加（De Silva，S.S.，2002）。

表5-1 饲养试验的原料与近似组成（De Silva，S.S.，2002）

原料（g/kg）	饲料				
	$P_{40}L_{10}$	$P_{40}L_{17}$	$P_{40}L_{24}$	$P_{50}L_{17}$	$P_{50}L_{24}$
鱼粉	565	580	590	730	745
小麦粉	225	150	80	140	55
α-淀粉	130	120	110	—	—
维生素	30	30	30	30	30
鱼油	25	95	165	75	145
壳聚糖	25	25	25	25	25
近似组成分（as fed basis）					

(续表)

原料（g/kg）	饲料				
	$P_{40}L_{10}$	$P_{40}L_{17}$	$P_{40}L_{24}$	$P_{50}L_{17}$	$P_{50}L_{24}$
%蛋白质	40.9	41.3	40.5	51.1	50.0
%脂肪	9.9	16.8	24.6	16.6	23.8
%灰分	8.4	8.5	8.6	10.1	10.1
%无氮浸出物	27.5	21.2	15.2	22.2	16.1
能量（kJ/g）	16.4	18.2	19.8	20.3	21.8
蛋白能量比（mg·kJ/g）	24.9	22.7	20.4	25.2	22.9

澳洲龙纹斑在测试生长和饲料利用率的56d实验中，其中一式三组澳洲龙纹斑幼鱼饲喂含有40%、45%、50%、55%或60%蛋白质的饲料。最终平均体重，体重增长百分比和生长速率最高的是投喂P50饲料的鱼。饲料转化率和蛋白质效率比最好的也是投喂P50饲料的鱼，但是这些参数与投喂P55和P60饲料的数值差异并不显著。饲料转化率（Y）与饲料中蛋白质含量（X）有关，两者在 $Y=0.004x^2-0.431x+12.305$（$r=0.95$；$P=0.01$）时呈现二阶多项式关系（图5-1）。鱼体水分，蛋白质，脂肪和灰分的比例在用不同饲料养殖中没有差异。蛋白质转化效率（y）与饲料蛋白质（X）含量百分比呈负相关，关系为：$y=62.76-0.62X$（Gunasekera, R. M., 2000）。

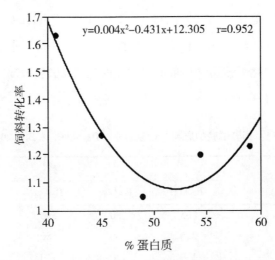

图5-1 饲料转化率（FCR）与等澳洲龙纹斑幼鱼热量的饲料中不同蛋白质百分含量之间的二阶多项式关系。（Gunasekera, R. M., 2000）

Abery, N. W.（2002）评估了作为两种蛋白质来源的血粉和脱脂豆粕替代不同比例的鱼粉对澳洲龙纹斑幼鱼的适用性（图5-2）。在70天养殖实验中，评估9种等氮和等能量饲料对澳洲龙纹斑幼鱼生长性能的影响，以鱼粉为蛋白质的饲料作为对照组，以鱼粉或脱脂豆粕8，16，24和32比例替代鱼粉为蛋白质的饲料作为测试组。使用Cr_2O_3作为标记物测定了饲料干物质的百分比和蛋白质消化率。所有脱脂豆粕处理的饲料的成活率高，但鱼粉理的饲料的鱼的成活率显著低于其他处理的饲料的成活率。投喂脱脂豆粕试验组的生长率（第1天）为1.63±0.06~1.78±0.10，即使在最高比例（32蛋白质来自脱脂豆粕）与对照组（1.65~0.09）的差异也不显著。脱脂豆粕处理的饲料转化率范围为1.36±0.08至1.45±0.07。在大豆粉处理的试验组中，澳洲龙纹斑的蛋白质效率比和蛋白转化效率比较好，大多数脱脂豆粕处理的均优于对照组。在所测试的脱脂豆粕处理的饲料中，饲料干物质百分比和蛋白质消化率的百分比范围分别为70.6±1.46~72.3±1.81和88.6±0.57~90.3±0.17，与对照组比较差异不显著（饲料干物质的百分比为74.3±1.63；蛋白质消化率为91.3±0.55）。讨论了用鱼粉饲养澳洲龙纹斑的成活率和生长速度明显差的原因，以及这些饲料和饲料消化率相对较差的原因。研究表明，澳洲龙纹斑饲料中鱼粉蛋白被替代至32比例不会对鱼体性能和组成造成损害（Abery, N. W., 2002）。

由于不同蛋白源的氨基酸组成和比例存在差异，其对澳洲龙纹斑的营养价值也各异。鱼粉作为一种优质动物蛋白源，在澳洲龙纹斑配合饲料中使用量较大，然而由于鱼粉是一种资源性产品，因其供应量有限而需求量不断增加导致其价格不断攀升，采用其他蛋白源，尤其是植物性蛋白源替代部分鱼粉已成为水产饲料开发研究的热点和前沿。在粗蛋白质为50%，脂肪为14%的基础饲料中，采用脱脂豆粕蛋白和血粉蛋白分别替代鱼粉蛋白8%、16%、24%和32%进行为期70天的饲养实验结果表明，脱脂豆粕蛋白替代32%的鱼粉蛋白对澳洲龙纹斑幼鱼的生长性能、干物质消化率、粗蛋白质的消化率、鱼体健康等与鱼粉组的均差异不显著，但血粉由于适口性等原因，造成澳洲龙纹斑生长较差，这提示在实际生产配方中如选择血粉作为蛋白源应注意多种蛋白源搭配使用，并关注饲料的适口性和氨基酸的平衡问题（Abery et al, 2002）。

测定鱼类对常用饲料原料的消化率是评价其营养成分可利用性的常用手段，也是编制营养全面、成本合理的鱼用饲料配方的必不可少的重要步骤。为了提高筛选澳洲龙纹斑配合饲料蛋白质原料的效率，已有学者开展了各种原料和配合饲料中蛋白质消化率的测定。采用体质量为100.4g的澳洲龙纹斑幼鱼评价其对饲料中鲨鱼粉、豆粕和肉骨粉的表观消化率，结果表明，鲨鱼粉和豆粕的干物质表观消化率分别为73.1±1.58%和70.6±0.82%；粗蛋白的消化率分别为87.5±1.27%和86.5±0.49%；幼鱼对肉骨粉的干

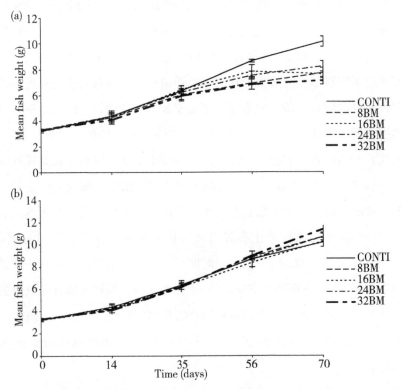

图 5-2 在不同饲料饲养下的澳洲龙纹斑幼鱼的平均体重。(a) 饲喂对照组饲料和豆粕替代饲料的澳洲龙纹斑平均体重的增加；(b) 饲喂对照组饲料和血粉替代饲料的澳洲龙纹斑平均体重的增加 (Abery, N.W., 2002)。

物质表观消化率和粗蛋白消化率较低；澳洲龙纹斑幼鱼对鱼粉的干物质表观消化率为 74.3±1.63%，蛋白质表观消化率达 91.3±0.55%；幼鱼对脱脂豆粕的蛋白质表观消化率为 70.6±1.46%~72.3±1.81%，当蛋白质表观消化率达到 88.6±0.57%~90.3±0.17% 时，均表现出了高的消化率 (Abery et al, 2002)。鱼粉替代技术开发是研发高效、低成本、环境友好型鱼类配合饲料的一个重要方面。植物性蛋白源在不同程度上均存在着氨基酸不平衡、抗营养因子、适口性差及消化率低的特点，采用酶解、发酵、选育、膨化等技术，减少、消除或钝化植物性蛋白源中的抗营养因子，以提高饲料转化率以及干物质和总能的表观消化率 (De Silva et al, 2000)。

2. 氨基酸的营养需求

研究表明，鱼类对蛋白质的营养需求实质是对寡肽和氨基酸的营养需求，特别是对必需氨基酸的营养需求。鱼类的 10 种必需氨基酸是指赖氨酸、色氨酸、蛋氨酸、亮氨酸、组氨酸、异亮氨酸、缬氨酸、苯丙氨酸、精氨酸和苏氨酸等。澳洲龙纹斑配合饲料中必须提供足够、平衡的各种必需氨基酸，以保证其快速、健康生长，并避免必需氨基酸的浪费，以节约饲料成本。尽管目前尚未见有关澳洲龙纹斑的必需氨基酸营养需求研究报道，但有学者分析了不同生长阶段的澳洲龙纹斑肌肉中氨基酸组成及其比例，这可为澳洲龙纹斑配合饲料配方中氨基酸参数的确定提供参考。采用氨基酸自动分析仪检测了澳洲龙纹斑亲鱼、商品鱼、幼鱼中肌肉氨基酸含量。结果表明，3个生长阶段的澳洲龙纹斑肌肉中总氨基酸含量最高为商品鱼（70.96%）、其次为幼鱼（69.24%）、亲鱼最低（65.36%），且亲鱼和商品鱼的氨基酸组成的比例较为接近，幼鱼的氨基酸组成比例则有所变化；必需氨基酸中，其必须氨基酸/总氨基酸值则比亲鱼以及商品鱼低2%左右；非必需氨基酸中，必须氨基酸/非必需氨基酸比值又高于亲鱼与商品鱼2%左右（罗钦等，2015）。采用Agilent1100型液相色谱仪分析了体质量45~50g的澳洲龙纹斑幼鱼肌肉中的氨基酸，检测出氨基酸17种（未测色氨酸），其中包括人体的必需氨基酸7种（占氨基酸总量的39.93%），必须氨基酸组成基本符合FAO/WHO标准，澳洲龙纹斑幼鱼的主要限制性氨基酸为蛋氨酸、胱氨酸和缬氨酸，其必需氨基酸指数为95.43，鲜味氨基酸占氨基酸总量的45.59%。这些必需氨基酸需求量数据的获得，对于澳洲龙纹斑饲料实际生产中配方的氨基酸指标调整提供了重要的参考标准和指导作用（宋理平，2013）。

S S De Silva（2000）评估了富含蛋白质的豆粕，鲨鱼肉骨粉废弃物或肉骨粉饲料投喂澳洲龙纹斑的干物质、蛋白质和能量消化率和氨基酸利用率。饲喂澳洲龙纹斑的对照组饲料的蛋白质含量为50，脂肪含量为10%。试验组饲料由含30%的替代蛋白成分和70%的对照组饲料组成，并且以氧化铬为标记物进行消化率测定。在投喂鲨鱼肉粉（73.16±1.58和87.56±1.27）和豆粕（70.66±0.82和86.56±0.49）这两类饲料，最高的干物质和蛋白质分别被观测到。投喂任一饲料，其能量的差异都不显著。上述结果在干物质和营养成分的物消化率上也都观察到。投喂这两类饲料，其必需氨基酸的差异显著。在短鳍鳗鱼中，投喂肉骨粉饲料（50.56±4.25）的总必需氨基酸含量明显低于其他种饲料。在墨鱼鳕鱼中，投喂肉骨粉饲料的总必需氨基酸含量、总非必需的氨基酸含量和总氨基酸含量（分别为89.7%，82.1%和85.2%）显着高于（总非必需氨基酸除外）其他对照组（分别为84.5%，77.6%和80.6%），肉骨粉饲料的必需氨基酸含量均超过82%。结果显示在不同鱼类中饲料的利用率有差异。蛋白质和必需氨基酸这两个参数之间存在密切相关性，蛋白质提供的氨基酸利用率相当可靠（图5-3）（S S De Silva，2000）。

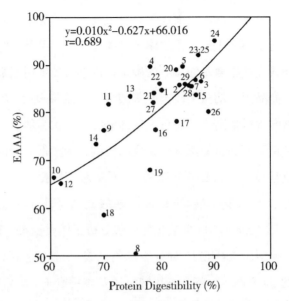

图5-3 蛋白消化率百分比（%）与不同鱼种的饲料和/或原料的必需氨基酸百分比之间的关系，组成了最佳拟合线。数据点显示澳洲龙纹斑（1±4）和短鳍南美鳗鱼（5±8），罗非鱼（9±14），Sadiku & Jauncey（1995）；大西洋鲑鱼（15±19），Anderson etal（1992）；金头海鲷（20±25），Lupatsch etal（1997）；鲤鱼（26±29），Hossain & Jauncey（1989）（Silva, S. S. D., 2000）。

二、脂肪及脂肪酸的营养需求

（一）脂肪的营养需求

脂肪是维持澳洲龙纹斑生长、健康和繁殖等生命活动所必需的营养素，澳洲龙纹斑对脂肪利用率较高，对碳水化合物利用率较低，饲料中添加多不饱和脂肪酸将有助于澳洲龙纹斑获得更好的生长速率和饲料利用效率。因此，脂肪就成为澳洲龙纹斑的重要能量来源之一。饲料中的脂肪含量适宜，澳洲龙纹斑就能充分利用；饲料中脂肪含量不足或缺乏，澳洲龙纹斑摄取的饲料中蛋白质就会有一部分作为能量被消耗掉，饲料蛋白质利用率下降，同时还可发生脂溶性维生素和必需脂肪酸缺乏症，从而影响澳洲龙纹斑生长，造成饲料蛋白质浪费和饵料系数升高，然而饲料中脂肪含量过高时，短时间内可以促进澳洲龙纹斑的生长，降低饲料系数，但长期摄食高脂肪饲料会使澳洲龙纹斑产生代谢系统紊乱，增加体内脂肪含量，导致鱼体脂肪沉积过多，内脏尤其是肝脏脂肪过度聚集，产生脂肪肝，进而影响蛋白质的消化吸收并导致机体抗病力下降。此外，饲料脂肪含量过高也不利于饲料的贮藏和成型加工，因此，只有使用脂肪和蛋白质含量均适宜的饲料才能实现澳洲龙纹斑养殖的最佳效果。研究表明，影响澳洲龙纹斑饲料中脂肪营养需求的主要因素有鱼体大小、鱼的生理状态、脂肪源、饲料组成（特别是蛋白质：脂肪：碳水化合物之比）、水温和水体中饵料生物的种类与含量、摄食时间等。澳洲龙纹斑对脂肪有较高的消化率，尤其是低熔点脂肪，其消化率一般在90%以上。饲料所含的脂肪，不能直接被鱼类吸收，必须经过消化酶分解为甘油和脂肪酸后，才能被吸收，它不仅是鱼类的能源物质，而且作为脂溶性维生素A、D、E、K的载体，促进输送与其吸收。添加适量脂肪可节约蛋白质，增进食欲，提高饲料利用率，在生产上有更重要的意义。有关澳洲龙纹斑饲料中脂肪的营养已经开展了一些探讨。研究表明，澳洲龙纹斑幼鱼蛋白质含量在50%时，获得最大生长速度的脂肪含量为17%，若饲料蛋白质含量为40%时，饲料中脂肪含量为24%可以获得良好的生长效果（De Silva et al，2002）。澳洲龙纹斑养成饲料中蛋白质含量大于45%时，饲料中的脂肪含量以大于15%为宜。综合文献报道，澳洲龙纹斑对饲料中脂肪的含量需求在12%~25%，高含量的脂肪可以提供充分的能量，从而节约蛋白质，促进鱼体快速生长，但是也可能导致脂肪在鱼体肌肉、肝脏和腹腔中的大量累积，对鱼类健康造成一定不利影响。鱼类对饲料脂肪的利用能力，也与脂肪来源及必需脂肪酸的种类有密切关系。鱼类饲料中采用植物油替代鱼油已经是研究热点。采用100%的鳕鱼油、100%的菜籽油、100%的亚麻籽油以及100%的混合油（鳕鱼油：菜籽油为1:1）、100%的混合油（鳕鱼油：亚麻籽油为1:1）为脂肪源，配制成5种等氮等能等脂（脂肪含量17%）的5种饲料在水温22℃条件下饲喂澳洲龙纹斑幼鱼（平均初始体质量为6.45±1.59g）84天，结果表明，鳕鱼油组和100%混合油（鳕鱼油：亚麻籽油为1:1）组的幼鱼最终体质量、特殊增重率和摄食量均显著高于亚麻籽油组，但各组间饲料效率和蛋白质效率差异不显著；且澳洲龙纹斑幼鱼肌肉中的脂肪酸组成与饲料脂肪源的脂肪酸组成较为吻合，饲喂鱼油饲料的澳洲龙纹斑肌肉中脂肪酸中的EPA，ArA和DHA最高，饲喂菜籽油饲料的澳洲龙纹斑肌肉中油酸含量高达每克脂肪中油酸为192.2±10.5mg，饲喂亚麻油饲料的澳洲龙纹斑肌肉中的亚麻酸含量高达每克脂肪中亚麻酸为107.1±6.7mg，本研究表明，澳洲龙纹斑饲料中鱼油可以被100%的菜籽油取代或50%的亚麻籽油取代，对其生长影响不显著（Turchini et al，2006a）。进一步的研究表明，饲喂植物油饲料的澳洲龙纹斑在上市前转喂以鳕鱼油的饲料16w，澳洲龙纹斑肌肉中的脂肪酸组成与含量与一直饲喂鱼油饲料相似（Turchini et al，2006b）。低鱼粉饲料中植物油也是可以部分替代鱼油，只是替代比例不能太大，可替代25%的鱼油（Turchini et al，2007b；Gunasekera et al，2000）。

在23.7±0.802℃下，进行了超过60天的在澳洲龙纹斑幼鱼（25.4±0.8102g）饲料中添加人工养殖鲑鱼内脏中提取的鱼油（TO）的养殖效果实验。使用五种含有同等氮（48%蛋白质），同等脂肪（16%）和同等能量（21.802kJ02g611）的饲料，其中鱼油以25%（0%~100%）的增量替代。在含有50%~75%鲑鱼内脏鱼油的鱼饲料中观察到最好的生长和饲料效率。在每种情况下，通过二阶多项式方程（$p<0.05$）描述了比生长速率（SGR），食物转化率（FCR）和蛋白质效率比（PER）与饲料中鲑鱼内脏鱼油含量的关系：$SGR = -0.44TO^2 + 0.52TO + 1.23$（$r^2 = 0.90$，$p<0.05$）；$FCR = 0.53TO^2 - 0.64TO + 1.21$（$r^2 = 0.95$，$p<0.05$）；和 $PER = -0.73TO^2 + 0.90TO + 1.54$（$r^2 = 0.90$，$p<0.05$）。在不同饲料的饲养中，鱼体和肌肉近似成分的差异显著，在肌肉中发现的油脂比在鱼体中的少。不论饲料如何，在澳洲龙纹斑中发现的最高含量的脂肪酸，均为棕榈酸、油酸、亚油酸和二十碳五烯酸。肌肉的脂肪酸组成反映了饲料中的脂肪酸组成。n-6脂肪酸含量和n-3至n-6比值均显著反映出与生长参数相关，关系如下：SGR和FCR的饮食中n-6的百分比（X）：$SGR = -0.12X^2 + 3.96X - 32.51$（$r^2 = 0.96$）和 $FCR = 0.13X^2 - 4.47X + 39.39$（$r^2 = 0.98$）；和n-3：n-6比值（Z）与SGR，FCR，PER：$SGR = -2.02Z^2 + 5.01Z - 1.74$（$r^2 = 0.88$），$FCR = 2.31Z^2 - 5.70Z + 4.54$（$r^2 = 0.93$）和 $PER = -3.12Z^2 - 7.56Z + 2.80$（$r^2 = 0.88$）。从本研究中可以看出，鲑鱼内脏鱼油可以在澳洲龙纹斑饲料中有效地使用，并且n-3：n-6的比值为1.2时澳洲龙纹斑的生长性能最佳（图5-4至图5-6）（Turchini, G., 2003）。

Turchini GM（2006）研究了投喂含油菜籽或亚麻籽油饲料的澳洲龙纹斑中脂肪酸的代谢。澳洲龙纹斑能够延长和去饱和脂肪酸亚油酸和α-亚麻酸。在投喂油菜籽饲料的鱼中，亚油酸沉积了54.4%，氧化了38.5%，延长和去饱和较高脂肪酸同系物6.4%。在投喂亚麻籽油饲料的鱼中，α-亚麻酸沉积了52.9%，氧化了37%，延长和去饱和了8.6%。与其他淡水鱼相比，n-6脂肪酸对澳洲龙纹斑的作用显得更为重要。同时检测出澳洲龙纹斑利用C18脂肪酸进行能量生产的优先顺序（α-亚麻酸>亚油酸>油酸）。此外，证明在饲料中α-亚麻酸的增加直接造成鱼体去饱和酶活性增加和饱和脂肪酸累积增加。本研究还表明，在尽可能提高鱼类生产长链多元不饱和脂肪酸的能力的情况下，全身脂肪酸平衡法可以被认为是非常适合的和信息丰富的，并且澳洲龙纹斑适合使用食用鱼油替代品（Turchini GM, 2006）。

在投喂25周含有鱼油、卡诺拉油和亚麻籽油的饲料后，转投喂鱼油（完成/洗出）饲料16周，测定澳洲龙纹斑组织中脂肪酸组成的动态变化。在25周的植物油替代饲养之后，在洗出期开始时，澳洲龙纹斑肉中的脂肪酸组成反映了各自的饲养情况。转移到

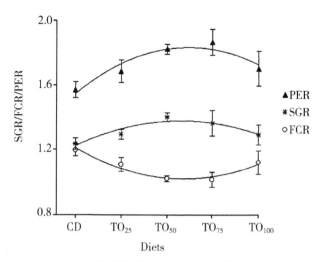

图 5-4 The relationship of t 澳洲龙纹斑幼鱼的平均特定生长率（SGR），饵料转化率（FCR）和蛋白质效率（PER）与不同饲养试验中鲑鱼鱼油（TO）含量的关系。竖条表示均值的标准误差（Turchini, G., 2003）。

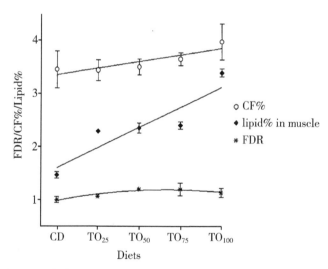

图 5-5 澳洲龙纹斑幼鱼的平均脂肪沉积率（FDR），肥满度系数（CF%）和肌肉的脂肪含量（%）与饲喂鲑鱼鱼油（TO）浓度的关系。竖条表示均值的标准误差（Turchini, G., 2003）。

鱼油饲养后，数量的差异减少，导致鱼油脂肪酸组成的恢复。通过指数方程可以描述脂肪酸的百分比和总积累的变化，并且证明在结束的头几天发生了重大的变化。用稀释模型预测脂肪酸组成。尽管该模型具有一般的可靠性（$Y=0.9234X+0.4260$，$r^2=0.957$，$p<0.001$，其中 X 是脂肪酸的预测百分比；Y 是观察到的脂肪酸百分比），在一些情况下，观测值和预测值的回归比较与权益线明显不同，表明变化率高于预期（即 Y =

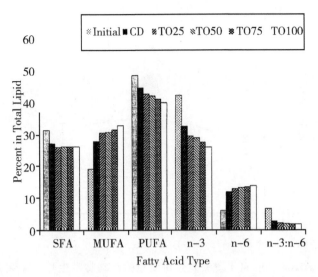

图 5-6 初始鱼样品和试验不同饲料后鱼样品的主要脂肪酸种类（Turchini，G.，2003）。

0.4205X+1.191，$r^2=0.974$，p<0.001，其中 X 为预测百分比的 α-亚麻酸；Y 是观察到的 α-亚麻酸的百分比）。最终，使用距离系数（D），显示先前投喂植物油饲料的鱼的脂肪酸组成，在 70 天或 97 天后，分别恢复到澳洲龙纹斑的鱼肉脂肪酸组成的平均变异性（LO 和 CO 分别）（Turchini，G. M.，2010）。

Turchini，G. M（2010）评估了初始生长和最终阶段澳洲龙纹斑的生长性能。在结束以鱼油为基础的喂养之前，给鱼喂三种中的一种不同脂肪来源的饲料：鱼油、低多不饱和脂肪酸植物油混合物和高不饱和脂肪酸植物油混合物。在生长期结束阶段时，澳洲龙纹斑肌肉的脂肪酸组成反映了各自的饲料；而在结束阶段，差异程度和差异的产生都有所减少。用原来低不饱和脂肪酸植物油喂养的鱼油脂肪酸的恢复更快。在最后冲洗阶段，以投喂植物油为主的饲料的鱼比一直投喂以鱼油为主的饲料的鱼生长更快。这项研究表明，在以植物油为主的养殖鱼类中，高度不饱和脂肪酸可以促进油脂代谢和鱼的新陈代谢，从而促进鱼的生长。这种作用可以称为脂肪补偿生长（Turchini，G. M，2010）。

Francis，D. S.（2006）评估了卡诺拉油和亚麻籽油两种植物油替代鱼油投喂澳洲龙纹斑幼鱼的适用性。目前，澳洲龙纹斑没有专用的商业饲料，而大量使用其他鱼类制定的营养饲料。以五种等氮、等热量、等脂肪的半纯化的实验饲料为研究对象，配方中 17%脂肪来自100%鱼肝油，100%菜籽油，100%亚麻油，1:1的共混物的油菜和鱼肝油和 1:1 的共混物的亚麻籽油和鱼肝油。在 22℃下，将每种饲料每天投喂两次，每组 50

尾澳洲龙纹斑，初始平均重量为6.45±1.59g，持续84天。经鱼肝油和混合油（亚麻籽油：鱼肝油为1:1）处理的最终平均体重、生长速率和日饲料消耗量显著高于亚麻籽油处理的。饲料转化率和蛋白质效率在各处理间差异不显著。试验含有植物油和植物油混合的饲料有较高浓度的n-6脂肪酸，主要是油酸的形成，而在亚麻籽油和混和油（亚麻籽油：鱼肝油为1:1）处理中的n-3脂肪酸的浓度较高。澳洲龙纹斑肉中脂肪酸组成反映了饲料中脂肪的来源。投喂鱼肝油的鱼的鱼肉中EPA、ArA和DHA最高。投喂卡诺拉油饲料的鱼有高浓度的油酸，而投喂亚麻籽油饲料的澳洲龙纹斑肉中有高浓度的α-亚麻酸。目前的研究表明，在澳洲龙纹斑饲养中高达100%的菜籽油和高达50%的亚麻油取代鱼油对生长没有显著的影响（Francis，D. S.，2006）。

众所周知饲料中的脂肪酸可调节鱼类的脂肪酸代谢。然而，鱼类本身把短链脂肪酸转化成有益的长链脂肪酸的能力是不足的，需要减少从饲料中摄入来弥补。研究了饲料中C18多不饱和脂肪酸对脂肪酸代谢的影响。用高含量C18多不饱和脂肪酸的饲料投喂的鱼中的Δ-6去饱和酶活性明显更大。特别是，采用高含量C18多不饱和脂肪酸饲养的鱼，其Δ-6脱氢酶活性对α-亚麻酸比亚油酸起的作用更大。然而，随着饲料中的C18多不饱和脂肪酸的逐渐减少，Δ-6去饱和酶的底物偏好从α-亚麻酸转变为亚油酸。该信息将在养殖澳洲龙纹斑时允许维持n-3长链脂肪酸水平的低鱼油饲料提供宝贵意见（Francis，D. S.，2009）。

在澳洲龙纹斑幼鱼16周的饲养试验中，鱼油和菜籽油每天定时交替使用（amCO：早上菜籽油，下午鱼油，pmCO：早晨鱼油，下午菜籽油）或一系列周期循环使用（2W：菜籽油2周，鱼油2周，4W：菜籽油4周，鱼油4周）。所有试验与最终体重之间没有观察到显著性差异。然而，接受2W处理的鱼比所有其他处理的鱼大。接受2W处理的鱼与对照处理（连续加入50%鱼油和50%菜籽油的鱼）相比，二十碳五烯酸和二十二碳六烯酸（分别为62.1%和24.0%）的总净消失率更低。同样，与CD和pmCO处理相比，amCO处理的二十碳五烯酸总净消失率更低。这些数据表明脂肪酸的利用/保留存在着周期性的机制（图5-7）（Francis，D. S.，2010）。

Varricchio Ettore（2012）研究了在澳洲龙纹斑饲料中添加不同脂肪酸的组合物对饥饿素和瘦素免疫反应细胞在胃肠道和血液中的分布水平。澳洲龙纹斑幼鱼饲喂五种相同能量实验，饲料含有鱼油或以下某一种植物油：橄榄油，葵花籽油，亚麻油，棕榈油，作为饲料添加的脂肪源。胃肠道中胃饥饿素和瘦素细胞的存在和分布都受植物油的影响。胃饥饿素免疫细胞在投喂鱼油的澳洲龙纹斑的胃肠系膜丛中都被发现。瘦素细胞定位于投喂植物油的澳洲龙纹斑肠黏膜。瘦素免疫反应活性在投喂鱼油的澳洲龙纹斑胃肠

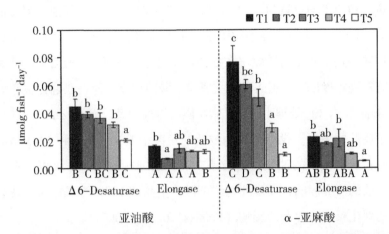

图5-7 体内延长酶和 Δ-6 去饱和酶与亚油酸和 α-亚麻酸（每天鱼 g 的最后产物；平均 ±S. E. M；N=3）的活动图。在柱状图的顶部，小写字母是相同的，表明治疗之间没有显著差异。在柱状图的底部，相同的大写字母表示在试验中没有显著差异（Francis，D. S.，2010）。

系膜丛中检测到。血浆中胃饥饿素和瘦素水平、澳洲龙纹斑的生长速率在不同饲养实验中均无差异（Varricchio Ettore，2012）。

　　澳洲龙纹斑是具有较高养殖潜力的顶级肉食鱼类。目前，没有专门为澳洲龙纹斑制定的商业化的饲料。本研究为澳洲龙纹斑（初始平均体重 80~83.5g）使用两种实验室配制的饲料（DU1 和 DU2；48.9%和 49.1%的蛋白质和 16.9%16.1%的油脂，分别为干物质）和两种原为其他鱼类制定（鲑鱼-CDS 和澳洲肺鱼-CDB）的但现用于喂养澳洲龙纹斑的商业化饲料。两种商业化饲料蛋白质含量较低分别为 46.6%和 44.4%，但是脂肪含量较高（分别为 21.7%和 19.5%）。饲料的能量含量经测定是相似的（约 20~22kj/g）。澳洲龙纹斑的生长性能和饲料利用率在饲养中没有显著差异，但饲喂 DU1 和 DU2 饲料的澳洲龙纹斑的饲料转化率和蛋白质效率比都比较好。鱼体和肌肉中的脂肪沉积显著减少。在用饲料 CDS 喂养的鱼肌肉脂肪酸中，其 n-3、n-6 和多不饱和脂肪酸的含量最低，在用 DU2 和 CDB 喂养的鱼中最高。脂肪酸、棕榈酸和硬脂酸，油酸和棕榈油酸，DHA、EPA、二十二碳五烯酸和亚油酸是饱和脂肪酸，单不饱和脂肪酸，多不饱和脂肪酸的主要脂肪酸，占所有发现的澳洲龙纹斑肌肉中脂肪酸的 80.8%~88.7%。研究表明，澳洲龙纹斑可以饲喂含有 20%大豆粉的饲料（DU2），而不影响生长和鱼体重量。对野生和养殖的澳洲龙纹斑的肌肉组成和脂肪酸组成的差异进行观测，后者差异最明显。野生澳洲龙纹斑相比有较少（p<0.05）饱和脂肪酸，单不饱和脂肪酸，n-3 系列脂肪酸，但肌肉中的 n-6 系列脂肪酸高于养殖鱼。野生鱼也有较低的 n-3/n-6 比例（De

Silva SS，2015）。

（二）必需脂肪酸的营养需求

必需脂肪酸是指机体内不能合成或合成的量不能满足鱼体营养需求但又为鱼体所必需的多不饱和脂肪酸，必需脂肪酸是合成多种生物活性分子的前体，例如前列腺素、白细胞三烯、凝血恶烷类等，必需脂肪酸缺乏，会影响繁殖期间亲鱼的产卵数量和质量以及胚胎的早期发育。必需脂肪酸只有通过摄食才能获得满足机体的营养需求。鱼类的必需脂肪酸通常包括n-3多不饱和脂肪酸中的亚麻酸、二十碳五烯酸、二十二碳六烯酸和n-6多饱和脂肪酸中的亚油酸、花生四烯酸等，对鱼类的生长、免疫和繁殖生理活动具有重要的调控功能。一般认为，饲料中含有1% α-亚麻酸和亚油酸就能满足淡水鱼类必需脂肪酸需要量。研究表明，澳洲龙纹斑对脂肪酸的消化率随碳链延长而降低，随脂肪酸的不饱和度的增加而增强。碳链长度、脂肪酸不饱和度和脂肪酸熔点显著影响脂肪酸的消化率，一般脂肪酸的消化率由大到小依次为多不饱和肪脏酸>单不饱和脂肪酸>饱和脂肪酸和短链>长链脂肪酸（Francis et al.，2007a）。α-亚麻酸和亚油酸总量为51%不变得情况下，α-亚麻酸与亚油酸比值的范围为0.3~2.9内，随着饲料中α-亚麻酸与亚油酸比值的增大，对澳洲龙纹斑幼鱼的生长和肌肉中的脂肪含量影响不显著，但显著影响澳洲龙纹斑幼鱼的肌肉的脂肪酸组成和营养品质，α-亚麻酸与亚油酸比值大（即α-亚麻酸含量高），澳洲龙纹斑幼鱼肌肉中的 EPA 和 DHA 含量高，α-亚麻酸与亚油酸比值小则显著降低肌肉的营养品质。进一步的研究发现，饲料中α-亚麻酸与亚油酸比值会影响澳洲龙纹斑脂肪酸代谢调节，α-亚麻酸更能激化脂肪β-氧化和生物转化，而亚油酸则更能促进脂肪储存，且α-亚麻酸与亚油酸比值高能提高Δ-6去饱和酶的活性，促进肌肉中n-3长链多不饱和脂肪酸的合成（Senadheer et al.，2011）。有关澳洲龙纹斑脂肪酸的营养需求研究较少。目前主要是分析澳洲龙纹斑各组织器官中脂肪酸的种类组成和比例以及部分饲养试验来探讨澳洲龙纹斑的脂肪酸营养需求。已有学者分析了体质量45~50g的澳洲龙纹斑幼鱼肌肉中的脂肪酸组成与含量，共检测出脂肪酸19种，其中饱和脂肪酸7种，占总脂肪酸的31.95%；单不饱和脂肪酸4种，占总脂肪酸的24.36%；多不饱和脂肪酸8种，占43.44%。不饱和脂肪酸含量（67.80%）远高于饱和脂肪酸含量（31.95%）。油酸含量最高，为18.75%，其次为棕榈酸17.46%、亚油酸17.28%，最少的是芥酸0.62%。多不饱和脂肪酸中 EPA 与 DHA 的含量分别为4.56%和10.12%（宋理平等，2013）（Senadheera et al，2010）。

Francis，D. S.（2007）评估了饲养120天菜籽油和亚麻籽油对澳洲龙纹斑幼鱼的影响。在每一个实验中，各投喂五种饲料中的一种，其中鱼油被菜籽油（实验A）或亚麻油（实验 B）按 25%（0~100%）的增量替代。经过不同增量菜籽油和亚麻籽油饲养的幼鱼，其最终重量分别为 112.7g±7.6g~73.8g±9.9g 和 93.9g±3.6g~74.6g±2.2g，鱼的生长随着替代水平的增加呈负增长趋势。鱼肉和肝脏的脂肪酸成分发生了明显的变化，对应饲料中各自的脂肪酸组成。油酸和亚油酸的含量随着油菜籽油含量的增加而增加，与此同时 α-亚麻酸的含量随着亚麻籽油的增加而增加。随着日粮中植物油含量的增加，鱼肉和肝脏中 n-3 高度不饱和脂肪酸的含量降低。结果表明，在低鱼粉日粮中添加植物油替代鱼油对澳洲龙纹斑的影响是有限的（图5-8）（Francis, D.S., 2007）。

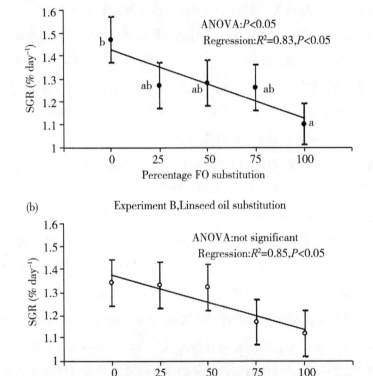

图 5-8 饲喂超过 112d（a）不同等级的菜籽油和（b）亚麻籽油饲料的澳洲龙纹斑幼鱼的特定生长率。方差分析和回归分析作为统计工具对剂量反应试验中的数据进行了比较分析（Francis, D.S., 2007）。

Francis, D.S.（2007）研究了澳洲龙纹斑中 C18 和长链多不饱和脂肪酸的关系。用五分之一的等氮和等能量的半纯化的实验饲养澳洲龙纹斑幼鱼，其喂养的鱼油被替换成按（0, 25, 50, 75, 100%）比例调和的植物油，主要配方与脂肪酸类［饱和脂肪酸、单不饱和脂肪酸，n-3 多不饱和脂肪酸，和 n-6 多不饱和脂肪酸］相匹配。然而，植物油的多不饱和脂肪酸组分以 C18 脂肪酸为主,鱼油的多不饱和脂肪酸组分以 C20 或 C22 脂肪酸为主。一般来说,在五种处理方法的鱼体、鱼肉和肝脏中,都能清楚地反映出饲料脂肪酸组成。脂肪代谢受到饲料脂肪来源改变的影响。随着植物油的替代,C18 多不饱和脂肪酸的饱和度和延伸率增加,并得到更长和更多的不饱和脂肪酸的支持。然而,增加的延长酶和脱氢酶活性不可能弥补因高度不饱和脂肪酸摄入量的下降造成的脂肪酸组成的差距(Francis,D. S.,2007)。

饲喂超过 98 天的不同饲料的澳洲龙纹斑幼鱼的综合评估见图 5-9,相互作用见图 5-10 至图 5-11。

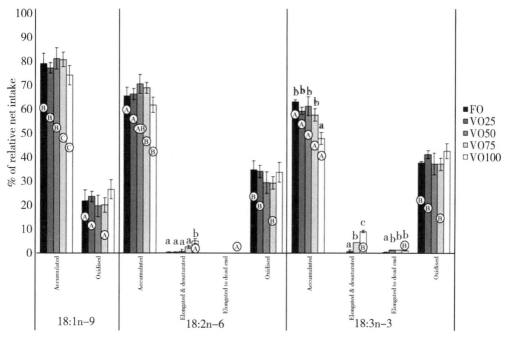

图 5-9 饲喂超过 98d 的不同饲料的澳洲龙纹斑幼鱼的综合评估。数据分别代表进食量 18:1 n-9、18:2 n-6、和 18:3 n-3 的总百分数。小写的上标表示在各饲料组间脂肪酸代谢上具有统计学上的显著差异。在每一组饲养试验中,不同的字母(被圈起)表示在脂肪酸积累、氧化、伸长和饱和度之间的统计差异,以及对代谢最终产物的延伸率(Francis,D. S.,2007)。

鱼油的全球短缺迫使水产养殖饲料企业使用替代油源,其使用对养殖鱼类的最终脂肪酸组成有不利影响。因此,养殖鱼类中脂肪酸代谢的调节是全球研究的核心。本研究旨在评估饲养鱼的各种饲料中 α-亚麻酸/亚油酸的比例。在淡水澳洲龙纹斑上进行了一项饲养试验,投喂以鱼油为对照的饲料或五种不含鱼油而饲料配方含有 α-亚麻酸与亚油酸的比值从 0.3 到 2.9 的饲料,但具有恒定的总 C18 多不饱和脂肪酸(α-亚麻酸+亚

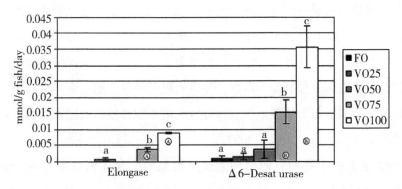

图5-10 饲喂超过98d不同饲料的澳洲龙纹斑幼鱼的延长酶和Δ6-去饱和酶与18：2 n-6的相互作用。在每次饲养试验之间，不同的上标字母（小写）表示统计差异。在每个饲料组中，大写的，圈出的上标的表明酶活性之间有显著的统计学差异（Francis, D. S., 2007）。

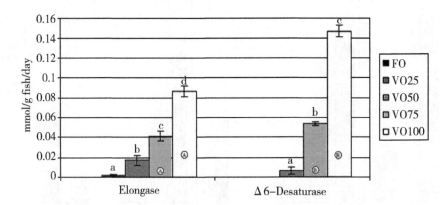

图5-11 饲喂超过98d不同饲料的澳洲龙纹斑幼鱼的延长酶和Δ6-去饱和酶与18：3 n-3的相互作用。在每次饲养试验之间，不同的上标字母（小写）表示统计差异。在每个饲料组中，大写的，圈出的上标的表明酶活性之间有显著的统计学差异（Francis, D. S., 2007）。

油酸）含量。采用全身脂肪酸平衡法对鱼体内脂肪酸代谢进行评价。结果表明，饲料中α-亚麻酸活性更为活跃地被β-氧化和转化，而亚油酸似乎更有效地沉积。亚油酸以恒定水平（约占总摄入量的36%）进行β-氧化成食物供应，而α-亚麻酸则按比例氧化成食物。随着饲料中α-亚麻酸与亚油酸的比例增加，n-3和n-6多不饱和脂肪酸的体内Δ-6去饱和酶活性分别呈上升和下降趋势，清楚表明该酶活性是依赖底物。然而，作用于α-亚麻酸的最大Δ-6去饱和酶活性在3.2186（每尾鱼每天μmolg）的底物水平时达到峰值，这表明额外加入α-亚麻酸不仅浪费，而且在n-3的长链脂肪酸中起到反作用。尽管α-亚麻酸与亚油酸的总供应不变，但两种底物在体内记录的总Δ-6去饱和酶活性与α-亚麻酸与亚油酸的比值同步增加，峰值在1.54，并且一个密切性达3.2倍大的Δ-6去饱和

酶活性对 α-亚麻酸与亚油酸比值被记录（图5-12至图5-15）（Senadheera，S. D.，2011）。

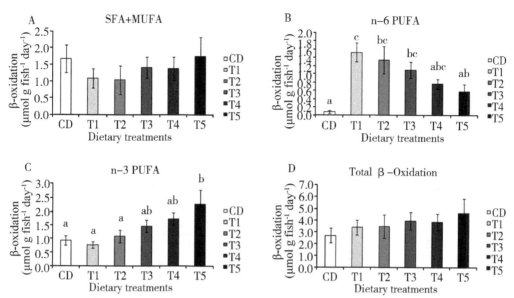

图 5-12　澳洲龙纹斑体内 β-氧化脂肪酸（mol/g·d）：（A）总饱和脂肪酸和单不饱和脂肪酸；（B）n-6 多不饱和脂肪酸；（C）n-3 多不饱和脂肪酸；（D）总脂肪酸。（Senadheera，S. D.，2011）

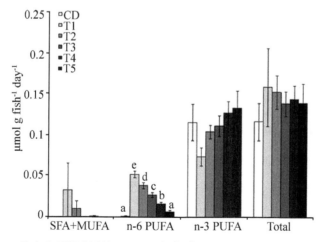

图 5-13　体内中延长酶活性（mol g 鱼$^{-1}$ d^{-1}）（Senadheera，S. D.，2011）。

三、碳水化合物的营养需求

碳水化合物也称糖类是鱼类的脑、鳃组织和红细胞等必需的代谢供能底物之一，与

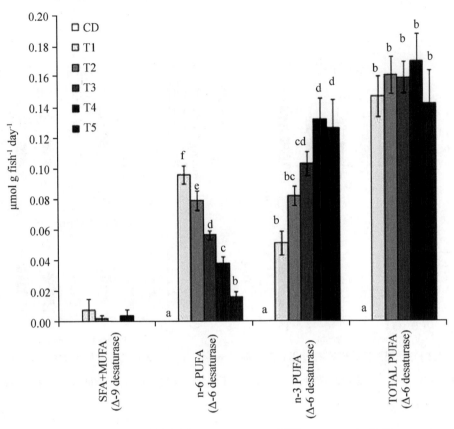

图 5-14 体内总去饱和酶（Δ-9 and Δ-6）活性（μmol g 鱼$^{-1}$d^{-1}）
（Senadheera，S. D.，2011）。

鱼体维持正常的生理功能和存活能力密切相关（Nakano 等，1998）。同时，糖类也是 DNA 和 RNA 的重要组成成分。此外，糖类还是抗体、某些酶和激素的组成成分，参加机体正常的新陈代谢，免疫反应、维持正常的生命活动。研究表明，鱼类主要以蛋白质和脂肪作为能量来源，对糖的利用能力较低，饲料中糖水平超过一定限度会引发鱼类抗病力低、生长缓慢、死亡率高等现象，鱼类被认为具有先天性的"糖尿病体质"（Wilson，1994）。鱼类特别是肉食性鱼类，通常被认为具有葡萄糖不耐受性，当饲料中含有过多的糖类时，它们体内就会长时间维持在高血糖状态，一般认为淡水鱼体内缺乏胰岛素，从而导致其糖代谢能力弱。也有研究表明，鱼体并不缺乏胰岛素，而是缺少相应的胰岛素受体或胰岛素受体功能不全。鱼类营养学研究领域的一个重要课题便是如何提高其对饲料碳水化合物的利用（Kirchner et al，2003）。鱼类配合饲料中使用一定量的碳水化合物，充分发挥其供能和免疫功能，节约蛋白质能耗，增加脂肪的积累，提高机

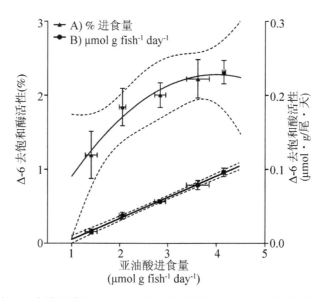

图 5-15 体内 Δ-6 去饱和酶对 18：2n-6 的活性表示为（A）% 进食量和表示为（B）μmol g 鱼$^{-1}$d^{-1}，18：2n-6 和进食量（μmol g 鱼$^{-1}$d^{-1}）的关系。误差线代表 SEM，和 and 虚线代表 95% 回归直线的置信区间。回归方程：（A）Y=−0.1542X^{2}b 1.230X−0.1705；r^2=0.96；检测剩余误差 D'Agostino-Pearson 的当量浓度 P=0.97；（B）Y=0.02851X−0.02327；r^2=0.99；斜度偏差从 0 到 P<0.0001（Senadheera, S. D., 2011）。

体的免疫能力，不但可以缓解目前水产配合饲料行业对鱼粉的过分依赖，减轻氮排泄对养殖水体的污染，而且还可以降低饲料成本，同时有助于配合饲料，特别是膨化配合饲料的制粒，促进鱼类健康养殖业的发展（Polakof et al，2009）。

鱼类对不同来源和种类的碳水化合物利用率各异。鱼类对单、双糖的消化率较高，淀粉次之，纤维素最差，有不少鱼类不能利用纤维素。饲料中碳水化合物含量过高，对鱼类的生长和健康不利。淀粉类既可作为配合饲料的粘结剂，又能在鱼体酶系统的参与下被消化，使其以单糖的形式被吸收供给鱼体生命活动的能量，同时为鱼体新陈代谢形成体脂和合成非必需氨基酸提供原料，节约蛋白质，提高蛋白质的有效利用率。纤维素只作为填充物或载体，起帮助消化和吸收其他营养素的作用，肉食性鱼类消化道中纤维素酶活性极低，不能分解纤维素，纤维素偏高在某种程度上可能反而会起到负面作用。目前，有关澳洲龙纹斑饲料适宜碳水化合物营养需求尚未见报道，但根据鱼类比较营养学的研究结果，肉食性鱼类饲料中的碳水化合物含量不宜超过 20%，为此，建议澳洲龙纹斑适宜的饲料中碳水化合物含量不超过 20%。

四、无机盐的营养需求

无机盐是构成鱼类机体组织的重要成分，有助于维持鱼类的正常生长、健康和繁殖等功能，对机体渗透压、调节机体正常生理代谢、调节酸碱平衡发挥着重要的作用。鱼类除了从饲料中获得矿物元素外，还可以从水环境中吸收矿物质，淡水动物主要通过鳃和体表吸收。镁、钠、钾、铁、锌、铜和硒等矿物元素通常从水中吸收可部分满足鱼类的营养需求，然而磷和硫大部分只能从饲料中吸收补充。矿物质如锌、铁、铜和硒作为金属酶的辅酶，对维持高等脊椎动物免疫系统的细胞功能是至关重要的，但至今未见有关矿物质对澳洲龙纹斑的生长、健康、免疫系统、肌肉品质和繁殖影响的研究报道，也未见澳洲龙纹斑矿物质营养需求的研究报道。澳洲龙纹斑矿物质的营养需求可以参照淡水肉食性鱼类矿物质营养需求，以研发出高效环境友好型澳洲龙纹斑系列配合饲料。

五、维生素的营养需求

维生素是鱼类机体营养素代谢重要的调节和控制因子，是一类含量微小而作用却极大的微量营养素，对维持鱼类的新陈代谢、免疫功能、生长和繁殖是必需的，其主要是通过调节体内物质和能量代谢以及参与氧化还原反应对机体起作用。维生素缺乏时，除导致鱼厌食、新陈代谢受阻、鱼体增重减慢、鱼的抗病力下降外，还会出现一系列缺乏症。然而鱼类体内几乎不能合成任何维生素，都必须从食物中摄取。鱼类需要11种水溶性维生素：硫胺素、核黄素、吡哆醇、泛酸、尼克酸、生物素、叶酸、钴胺素、肌醇、胆碱、抗坏血酸以及4种脂溶性维生素A、维生素D、维生素EE和维生素K。澳洲龙纹斑也同样需要这些维生素。有关澳洲龙纹斑的维生素营养需求尚未见报道，其营养需求可以参照其他肉食性鱼类的维生素营养需求，如大西洋鲑的维生素营养需求。澳洲龙纹斑对维生素需求量受发育阶段、生理状态、饲料组成和品质、饵料生物、养殖模式、环境因素以及营养素间的相互关系等影响，较难准确地测定。

六、澳洲龙纹斑配合饲料研发

澳大利亚集约化养殖澳洲龙纹斑的硬颗粒饲料规格按鱼的大小生产，外观硬颗粒饲料呈咖啡色，光亮，给人的感觉含油很多，嗅之有一股鱼香味，其饲料氨基酸总量接近于43%，饵料配比中要求粗蛋白质不低于50%，并加入添加剂及适量的鱼油。据介绍饲料的配比成分与大西洋鲑鱼相近（韩茂森，2003）。

澳洲龙纹斑21g以后，投喂含蛋白质50%、脂肪10%的饲料比投喂蛋白质60%或55%的饲料生长的更快。但如果饲料中蛋白质低于45%也使澳洲龙纹斑生长缓慢。这表明澳洲龙纹斑对饲料中的蛋白质要求的更高一些而对脂肪的要求相对低一些（曹凯德，2001）。

用水丝蚓、枝角类作养殖效果的对比实验数据记录见表5-2。由表5-2可以看出，前三天投喂水丝蚓的日均增重为0.123g/日。而投喂枝角类的前三天的日均增重为0.140g/日，从3月14日起，投喂水丝蚓组的日均生长增至0.218g/日。但日增重率没有明显的变化；而投喂枝角类的日均增重却减少到0.073g/日，日均增重率由4.72%减少到1.92g/日。由以上数据可知，在培育前期投喂枝角类生长较好，后期投喂水丝蚓组生长较快，因为在澳洲龙纹斑苗种培育的过程中会发生转食现象并且在投喂枝角类三组内发现死亡的鱼亦高于投喂水丝蚓组。到3月22日止，为了保证鱼苗的健康生长，放弃原计划，停止投喂枝角类，全部改投水丝蚓（左瑞华，2001）。

表5-2 不同饵料的养殖效果（g）（左瑞华，2001）

时间	水丝蚓			枝角类		
	体重均值	日均增重	日增重率（%）	体重均值	日均增重	日增重率（%）
10/3	2.48			2.49		
13/3	2.85	0.123	4.62	2.93	0.149	5.41
16/3	3.24	0.163	5.27	3.38	0.150	4.72
19/3	3.80	0.153	4.29	3.70	0.107	3.05
22/3	4.40	0.218	5.28	3.92	0.073	1.98

澳洲龙纹斑在刚开始的3天投喂枝角类的鱼苗生长较快。而后来一段时间投喂水丝蚓组的鱼苗生长较快。这表明澳洲龙纹斑鱼种在各个生长阶段对饵料的营养要求不同，甚至出现食性的转化，如不及时转换饵料，就会错过鱼苗快速生长的高峰期，甚至造成鱼苗的死亡。在培育过程中，如仅使用某一种饵料，就会造成鱼对某种营养成分的缺乏，会抑制其生长。因此在投喂的过程中尽可能多投几种营养物质不同的饵料，以保证营养物质的全面性，促进鱼苗的快速生长（左瑞华，2001）。

鱼苗在培育过程中，摄食率是随着温度的升高而增大，但随着鱼苗生长，摄食率在递渐减小，这是因为鱼苗代谢水平高，生长发育迅速，随着体重的增加就相对减少。由实验结果可以看出，日均增重并不是随着日投饵量升高而呈比例增大，这是因为鱼苗的生长是受多种因素制约的，摄食量增加并不意味饵料的利用率高。相反，适当地降低投饵率，可以提高饵料的转化率。但在饵料不足的情况下，会影响澳洲龙纹斑的迅速生长。就群体而言，会造成澳洲龙纹斑的同步生长受到抑制，从而引起弱肉强食的现象。目前由于对澳洲龙纹斑的许多情况并不是十分了解。因此以充分饵料，换取高生长速度和成活率。以缩短养殖周期，从生产经营策略上来说可能是值得选取的（左瑞华，2001）。

澳洲龙纹斑饵料根据鱼体不同的生长阶段，主要包括以下几种：鲜虾、红虫、冰鲜鱼、颗粒饲料。前期鱼苗在10天之前主要依次投喂红虫、鲜虾和野杂鱼，10天之后转为人工颗粒饲料。颗粒料主要包括鲟鱼颗粒料、大菱鲆颗粒料、海水鱼料。饲料营养蛋白质≥40%、粗脂肪≥8%、粗纤维≤5%、粗灰粉≤18%、食盐≤2%、钙≥2.5%、总磷≥1.2%、赖氨酸≥2.6%。本次试验鱼的生长速度和成活率相对于澳大利亚工厂化养殖模式下偏低，主要是由于对澳洲龙纹斑的认识还有一定的局限性，包括对其前期饵料的优化、后期人工颗粒饲料的选择，以及光线的控制和疾病的预防都有待于进一步的探索（张善发，2009）。

根据澳洲龙纹斑鱼体不同的生长阶段选择饵料，通常前期育苗在10天之前主要依次投喂红虫、鲜虾和野杂鱼，10天之后转为人工颗粒饲料哺，颗粒料主要包括鲟鱼颗粒料、大菱鲆颗粒料、海水鱼料，其中饲料营养蛋白质含量≥40%、粗脂肪含量≥8%、粗纤维含量≤5%、粗灰粉含量≤18%、食盐含量≤2%、钙含量≥2.5%、总磷含量≥1.2%、赖氨酸含量≥2.6%H。每天早晚各投喂2次，投喂量宜为不同发育阶段体重的1.5%~3.5%（郭松，2012）。

澳洲龙纹斑人工配合粉状饲料粗蛋白含量≥45%，脂肪含量≥5.6%，再取花、白鲢新鲜鱼肉，用高锰酸钾或双氧水彻底消毒、冲洗后和粉状饲料专门加工成适口的软颗粒饲料，饲料的大小也随鱼体的大小决定，参考标准如表5-3所示。由于澳洲龙纹斑白天怕光，每天早6：00、晚7：00各投喂一次，观察鱼的摄食情况，适当掌握每天的投饵量，投饵率一般为1%左右，从2011年10月25日放养0.502kg/尾龙纹斑养殖至2012年4月25日，经过6个月的试养，饲料系数为1.75（郭正龙，2012）。

表5-3 澳洲龙纹斑人工配合饲料颗粒大小参考（郭正龙，2012）

鱼体长度（mm）	鱼体重量（g）	饲料直径（mm）
45~75	1.0~5.0	1.5
75~90	5.0~10.0	2.0
90~1105	10.0~15.0	3.0

澳洲龙纹斑在澳大利亚已成功使用配合饲料养殖，在国内尚无成熟的人工饲料配方。Gunasekera 等报道，澳洲龙纹斑配合饲料蛋白质含量为50%时饲料转化率和蛋能比最高。由于澳洲龙纹斑为初次引进养殖。对其饵料要求进行了摸索和尝试。在养殖过程中考虑到澳洲龙纹斑为肉食性鱼类，采用了鲜活饵料和配合饲料相结合的喂养方式。引进鱼苗平均全长4.2cm，平均体重0.86g，开始投喂水丝蚓和绞碎的鲜鱼肉。并逐渐用人工配合开口饵料进行驯化，约1月后喂食饵料时鱼苗集群。此后饵料投喂基本为配合饲料，根据情况经常添加绞碎的鱼肉，春末夏初时在池中投放鲫鱼苗或鲤鱼苗。根据养殖经验和参考文献，加工的配合饲料蛋白质含量为50%，脂肪含量为12%。每天投饵2次，因澳洲龙纹斑不喜强光，投饵时间为清晨和傍晚（李娴，2013）。

澳洲龙纹斑对蛋白的要求较高，鱼苗饲料中的蛋白含量为45%，成鱼的蛋白含量为43%，饲料中的鱼粉需质量好，新鲜度高，目前为定做饲料。饲料中的蛋白含量对鱼的成长影响大，蛋白含量不足会导致鱼群出现抢食打斗现象，同时也导致鱼体抵抗力下降，投饵量可视水质情况适当加减（欧小华，2017）。

（罗钦、林虬、李巍）

第二节　研究论文摘编

109. Performance of juvenile Murray cod, Maccullochella peelii peelii (Mitchell), fed with diets of different protein to energy ratio[①]

[英文摘要]：The results of a 56-day experiment on juvenile Murray cod, Maccullochella peelii peelii, an Australian native fish with a high aquaculture potential, of mean weight 14.9±0.04g, fed with five experimental diets, one a series of 40% protein content and lipid levels of 10, 17 and 24% (P40L10, P40L17 and P40L24), and another of 50% protein and 17 and 24% (P50L17 and P50L24) lipid are presented. The specific growth rate (SGR) (% day^{-1}) of fish maintained on different diets ranged from 1.18 to 1.41, and was not significantly different between dietary treatments, except P40L10 and the rest. However, there was a general tendency for SGR to increase with increasing dietary lipid content at both protein levels. The food conversion ratio (FCR) for the 40% protein series diets were poorer compared with those of the 50% protein diets, and the best FCR of 1.14 was observed with the P50L17

[①] 研究论文摘编序号续上一章，全书同。

diet. The protein efficiency ratio (PER), however, was better in fish reared on low protein diets. The net protein utilization (NPU) also did not differ significantly ($P>0.05$) in relation to dietary treatment. As in the case of PER the highest NPU was observed in Murray cod reared on diet P40L24 and the lowest in fish fed with diet P50L24. The carcass lipid content reflected that of the diets, when significant increases in the lipid content was observed in relation to dietary lipid content at both protein levels. However, body muscle lipid content did not increase with increasing dietary lipid content, and was significantly lower than in the whole body. The fatty acids found in highest concentration amongst the saturates, monoenes and polyunsaturates (PUFAs) were 16∶0, 18∶1n-9 and 22∶6n-3, respectively, and each of these accounted for more than 60% of each of the group's total. The muscle fatty acid content was affected by the dietary lipid content; for example the total amount (in $\mu g\ mg^{-1}$ lipid) of monoenes ranged from 72±5.1 (P40L10) to 112±10 (P40L24) and 112±2.8 (P50L17) to 132±11.8 (P50L24) and the n-6 series fatty acids increased with increasing dietary lipid content, although not always significant. Most notably, 18∶2n-6 increased with the dietary lipid level in both series of diets.

［译文］：

不同蛋白质和能量比的饲料养殖澳洲龙纹斑幼鱼的效果研究

一个为期 56d 的对澳大利亚本土具有较高养殖潜力的澳洲龙纹斑幼鱼的实验，平均重量 14.9±0.04g，饲以五种实验的饲料，一个系列为蛋白质含量 40%和脂肪含量为 10%、17%和 24%（p40L10、p40L17 和 p40L24），另一个系列为蛋白质含量 50%和脂肪含量 17%，24%（p50L17 和 p50L24）。不同饲料幼鱼的特定生长率（SGR）（$\%d^{-1}$）范围从 1.18 到 1.41，除了 p40L10 外其它饲料在饲养实验之间没有显著性差异。然而，特定生长率在蛋白质水平上随着饲料中油脂含量的增加而增加。与 50%蛋白质饲料相比，40%蛋白质饲料的饲料转化率（FCR）更差，饲料转化率最佳值是 1.14，是在喂养 P50L17 饲料中观测到的。然而，蛋白质效率比（PER）在低蛋白饲料饲养的鱼中更好。在不同的饲料实验中，净蛋白利用率（NPU）并没有显著性差异（$p>0.05$）。如在蛋白质效率比 PER 一样的情况下，投喂 P40L24 饲料的澳洲龙纹斑观测到最高的净蛋白利用率 NPU，而投喂 P50L24 饲料的鱼最低。当两种蛋白质水平上的油脂含量的显著增加时，鱼体油脂含量反映了饲料的油脂含量。然而，随着饲料油脂含量的增加，鱼体脂肪含量并没有增加，并且明显低于全身脂肪含量。在饱和脂肪酸、单一脂肪酸和多不饱和脂肪酸（PUFA）中发现的最高浓度的脂肪酸分别为

16∶0、18∶1n-9 和 22∶6n-3，并且它们中的每一个占比为每个组总数的 60% 以上。肌肉脂肪酸含量受饲料油脂含量的影响，例如，单脂肪酸的总量（在油脂中 μg/mg）范围为 72±5.1（P40L10）至 112±10（P40L24）和 112±2.8（P50L17）至 132±11.8（P50L24），n-6 系列脂肪酸含量随饲料油脂含量增加而增加，但变化不够明显。最值得注意的是，18∶2n-6 随着两种饲料中的油脂水平增加而增加。

参考文献：De Silva, S. S., Gunasekera, R. M., Collins, R. A., et al. 2002. Performance of juvenile murray cod, maccullochella peelii peelii (mitchell), fed with diets of different protein to energy ratio [J]. Aquaculture Nutrition, 8 (2), 79-85.

（罗钦、李冬梅、黄惠珍 译）

110. Effect of dietary protein level on growth and food utilization in juvenile Murray cod Maccullochella peelii peelii (Mitchell)

[英文摘要]：Murray cod Maccullochella peelii peelii (Mitchell) is a freshwater Percichthyid fish considered to have high culture potential. Growth and feed utilization were examined in a 56-day experiment, in which triplicate groups of juvenile Murray cod (initial weight 21.5±0.03g) were fed isocalorific diets (gross energy content of about $21kJg^{-1}$) containing 40%, 45%, 50%, 55% or 60% protein (designated P40, P45, etc.). Final mean weight, percentage increase in weight and specific growth rate (SGR; $\%day^{-1}$) were highest in fish fed the P50 diet. Food conversion ratio (FCR; 1.05±0.04) and protein efficiency ratio (PER; 1.98±0.11) were also best in fish on the P50 diet, but the differences in these parameters from the corresponding values on diets P55 and P60 were not always significant. FCR (Y) was related to dietary protein content (X), the relationship being a second-order polynomial, in which $Y=0.004X^2-0.431X+12.305$ (r=0.95; P<0.01). The proportions of carcass moisture, protein, lipid and ash did not differ among the different dietary treatments. The protein conversion efficiency (y) was negatively correlated to percentage dietary protein (X) content, the relationship being：y=62.76-0.62X (r=0.99; P<0.01).

[译文]：

饲料中蛋白质含量对澳洲龙纹斑幼鱼生长和饲料利用的影响

澳洲龙纹斑被认为是具有较高养殖潜力的淡水真鲈科鱼类。在测试生长和饲料利用率的 56d 实验中，其中一式三组澳洲龙纹斑幼鱼（初始重量 21.5±0.03g）饲喂含有 40%、45%、50%、55% 或 60% 蛋白质的饲料（能量约为 21kJ/g）。最终平均体重，体重增长百分比和生长速率（SGR；%/d）最高的是投喂 P50 饲料的鱼。饲料转化率（FCR；1.05±0.04）和蛋白质效率比（PER；1.98±0.11）最好的也是投喂 P50 饲料的鱼，但是这些参数与投喂 P55 和 P60 饲料的数值差异并不显著。FCR（Y）与饲料中蛋白质含量（X）有关，两者在 $Y=0.004X^2-0.431X+12.305$（$r=0.95$；$p<0.01$）时呈现二阶多项式关系。鱼体水分，蛋白质，脂肪和灰分的比例在用不同饲料养殖中没有差异。蛋白质转化效率（y）与饲料蛋白质（X）含量百分比呈负相关，关系为：$y=62.76-0.62X$（$r=0.99$；$p<0.01$）。

参考文献：Gunasekera, R. M., Silva, S. S. D., Collins, R. A., et al. 2000. Effect of dietary protein level on growth and food utilization in juvenile murray cod maccullochella peelii peelii (mitchell) [J]. Aquaculture Research, 31 (2), 181–187.

（罗钦、李冬梅、黄惠珍　译）

111. Growth and nutrient utilization of Murray cod Maccullochella peelii peelii (Mitchell) fingerlings fed diets with varying levels of soybean meal and blood meal

[英文摘要]：The Australian native freshwater fish Murray cod, Maccullochella peelii pellii (Mitchell), currently supports a fledgling inland aquaculture industry, which is thought to have considerable growth potential. The aim of this study was to evaluate the suitability of two alternate protein sources [blood meal (BM)] and defatted soybean meal (SBM) as substitutes for fish meal at various levels of inclusion in diets for juvenile Murray cod. The growth performance of juvenile Murray cod in response to nine isonitrogenous and isocalorific diets (50% protein, 14% lipid, 20.2kJ g^{-1}) consisting of a control diet in which protein was supplied from fish meal, and test diets in which the fish meal protein was substituted at levels of 8%, 16%, 24%, and 32% with BM or SBM was evaluated from a 70-day growth experiment. The per cent apparent dry matter (% ADCdm) and percentage protein digestibility (% ADCp) of the test diets were also determined using Cr_2O_3 as a marker. Survival in all the SBM dietary treatments was high but that of fish on the BM dietary treatments was significantly ($p<0.05$) lower than in all the other dietary treatments. Specific growth rate (% day^{-1}) of Murray cod fed SBM incorporated diets ranged from 1.63P±0.06 to 1.78±0.10 and even at the highest level tested (32% of the dietary protein from SBM) was not significantly different ($p>0.05$) from the fish fed the control diet (1.65±0.09). Feed conversion ratios of the SBM dietary treatments ranged from 1.36±0.08 to 1.45±0.07. The protein efficiency ratios and protein conversion efficiencies of Murray cod in the soybean meal treatments were also good and for a majority of the SBM diets were better than those for the control diet. Per cent ADCdm and ADCp of the SBM diets tested ranged from 70.6±1.46 to 72.3±1.81% and 88.6±0.57 to 90.3±0.17%, respectively, and was not significantly different (p>0.05) from the control diet (% ADCdm 74.3±1.63;% ADCp 91.3±0.55). The reasons for significantly poor survival and growth of Murray cod reared on BM incorporated diets, and relatively poor digestibility of these diets are discussed. The study shows that for Murray cod diets in which fish meal protein is substituted up to 32% performance or carcass composition is not compromised.

[译文]：

在饲料中添加不同比例的豆粕和血粉对澳洲龙纹斑幼鱼的生长和养分利用的研究

澳大利亚本土的淡水澳洲龙纹斑是目前受支持的刚刚起步的内陆水产养殖业，被认为具有相当大的养殖潜力。本研究目的是评估作为两种蛋白质来源［血粉（BM）和脱脂豆粕（SBM）］替代不同比例的鱼粉对澳洲龙纹斑幼鱼的适用性。在70d养殖实验中，评估9种等氮和等能量饲料（50%蛋白，14%脂肪，20.2kJ/g）对澳洲龙纹斑幼鱼生长性能的影响，以鱼粉为蛋白质的饲料作为对照组，以BM或SBM8%，16%，24%和32%比例替代鱼粉为蛋白质的饲料作为测试组。使用Cr_2O_3作为标记物测定了饲料干物质的百分比（% ADCdm）和蛋白质消化率（%ADCp）。所有SBM处理的饲料的成活率高，但BM处理的饲料的鱼的成活率显著（p<0.05）低于其他处理的饲料的成活率。投喂SBM试验组的生长率（%/d）为1.63±0.06～1.78±0.10，即使在最高比例（32%蛋白质来自SBM）与对照组（1.65±0.09）的差异也不显著（p>0.05）。SBM处理的饲料转化率范围为1.36±0.08至1.45±0.07。在大豆粉处理的试验组中，澳洲龙纹斑的蛋白质效率比和蛋白转化效率比较好，大多数SBM处理的均优于对照组。在所测试的SBM处理的饲料中，ADCdm和ADCp的百分比范围分别为70.6±1.46～72.3±1.81%和88.6±0.57～90.3±0.17%，与对照组比较差异不显著（p>0.05）（% ADCdm为74.3±1.63;% ADCp为91.3±0.55）。讨论了用BM饲养澳洲龙纹斑的成活率和生长速度明显差的原因，以及这些饲料和饲料消化率相对较差的原因。研究表明，澳洲龙纹斑饲料中鱼粉蛋白被替代至32%比例不会对鱼体性能和组成造成损害。

参考文献：Abery, N. W., Gunasekera, R. M., De Silva, S. S. 2002. Growth and nutrient utilization of murray cod maccullochella peelii peelii (mitchell) fingerlings fed diets with varying levels of soybean meal and blood meal [J]. Aquaculture Research, 33 (4), 279-289.

（罗钦、李冬梅、黄惠珍　译）

112. Digestibility and amino acid availability of three protein-rich ingredient-incorporated diets by Murray cod Maccullochella peelii peelii (Mitchell) and the Australian shortfin eel Anguilla australis Richardson

[英文摘要]: In this study, the apparent dry matter (ADM), protein (PD) and energy (ED) digestibility, and the amino acid availability (essential, EAAA; non-essential, NEAAA; total, TAAA) of diets incorporated with one of three protein-rich ingredients (soybean meal, shark meat meal waste and meat meal) were evaluated for Murray cod Maccullochella peelii peelii (Mitchell) and the Australian shortfin eel Anguilla australis Richardson. The reference diets (RDs) used for Murray cod and shortfin eel had 50% and 45% protein, and 10% and 15% lipid respectively. The test diets consisted of 30% ingredient and 70% RD, and digestibility estimations were made using chromic oxide as a marker.

In both species, the highest ADM and PD of the test diets was observed for shark meat meal (73.1±1.58 and 87.5±1.27) and soybean meal (70.6±0.82 and 86.5±0.49) diets respectively. The PD of meat meal-incorporated diets was the lowest and, in shortfin eel, significantly so compared with all the experimental diets. For any one species, the ED of the diets did not differ significantly. The above observations were also reflected in dry matter and nutrient digestibility of ingredients. Significant differences ($P < 0.05$) in EAAA among diets were evident in both species. In shortfin eel, TEAAA for the meat meal-incorporated diet (50.5±4.25) was significantly lower than for all the other diets. In Murray cod, TEAAA, TNEAAA and TAAA (89.7%, 82.1% and 85.2% respectively) were significantly higher (except for TNEAAA) for the meat meal-incorporated diet than for the reference diet (84.5%, 77.6% and 80.6% respectively), and all essential amino acids of this diet were available in excess of 82%. The results indicate species differences in the utilization of ingredients. The present data on PD and EAAA, combined with previously published data, indicate a close correlation between these two parameters, suggesting that PD may provide a fairly reliable indication of the amino acid availability.

[译文]:

三种富含蛋白质组分的饲料养殖澳洲龙纹斑和澳洲短鳍南美鳗鱼的消化率和氨基酸利用率

在本研究中,对富含蛋白质的豆粕,鲨鱼肉骨粉废弃物或肉骨粉饲料投喂澳洲龙纹斑和澳洲短鳍南美鳗鱼的干物质（ADM）蛋白质（PD）和能量（ED）消化率和氨基酸利用率（必需的，EAAA；非必需的，NEAAA；总氨基酸，TAAA）进行评估。饲喂澳洲龙纹斑和澳洲短鳍南美鳗鱼的对照组饲料（RD）的蛋白质含量分别为50%和45%，脂肪含量分别为10%和15%。试验组饲料由30%成分和70%RD组成，并且以氧化铬为标记物进行消化率测定。

在投喂鲨鱼肉粉（73.1±1.58 和 87.5±1.27）和豆粕（70.6±0.82 和 86.5±0.49）这两类饲料，最高的干物质和蛋白质分别被观测到。与所有实验饲料相比，投喂含肉骨粉饲料的短鳍鳗鱼的蛋白质是最低的。投喂任一饲料，其能量的差异都不显著。上述结果在干物质和营养成分的物消化率上也都观察到。投喂这两类饲料，其必须氨基酸的差异显著（$p<0.05$）。在短鳍鳗鱼中，投喂肉骨粉饲料（50.5±4.25）的总必需氨基酸含量明显低于其他种饲料。在墨鱼鳕鱼中，投喂肉骨粉饲料的总必需氨基酸含量、总非必的氨基酸含量和总氨基酸含量（分别为89.7%、82.1%和85.2%）显著高于（总非必需氨基酸除外）其他对照组（分别为84.5%、77.6%和80.6%），肉骨粉饲料的必需氨基酸含量均超过82%。结果显示在不同鱼类中饲料的利用率有差异。结合前人研究的数据，本文显示的数据表明蛋白质和必需氨基酸这两个参数之间存在密切相关性，表明蛋白质提供的氨基酸利用率相当可靠。

参 考 文 献：Silva, S. S. D., Gunasekera, R. M., Gooley, G. 2000. Digestibility and amino acid availability of three protein-rich ingredient-incorporated diets by murray cod maccullochella peelii peelii, (mitchell) and the australian shortfin eel anguilla australis, richardson [J]. Aquaculture Research, 31 (2), 195-205.

（罗钦、李冬梅、黄惠珍 译）

113. Performance of intensively farmed Murray cod Maccullochella peelii peelii (Mitchell) fed newly formulated vs. currently used commercial diets, and a comparison of fillet composition of farmed and wild fish

［英文摘要］：Murray cod is a top-order carnivore with high culture potential. Currently, there are no commercial diets formulated specifically for Murray cod. In this study, results of two growth trials on Murray cod （80~83.5-g mean initial weight）, conducted in commercial settings, using two laboratory-formulated diets （DU1 and DU2；48.9% and 49.1% protein, and 16.9% and 16.1% lipid, respectively, on a dry matter basis）, and two commercial diets, formulated for other species （salmon-CD/S and barramundi-CD/B） but used in Murray cod farming are presented. The two commercial diets had less protein （46.6% and

44.4%) but higher lipid (21.7% and 19.5%). The energy content of the feeds tested was similar (about 20~22kJ g^{-1}). The growth performance and feed utilization of Murray cod did not differ significantly amongst the diets, but the food conversion ratio and protein efficiency ratio in fish fed the DU1 and DU2 diets were consistently better. There was significantly less carcass and muscle lipid deposition in fish fed with the latter diets. Of the fatty acids in muscle, the lowest amounts (in μg mg lipid^{-1}) of n-3 (262.5±2.9), n-6 (39.8±0.9) and polyunsaturated fatty acid (PUFA) (302.3±3.8) were observed in fish fed CD/S, and the highest in fish fed DU2 and CD/B. Fatty acids 16∶0 and 18∶0, 18∶1n-9 and 16∶1n-7, and 22∶6n-3, 20∶5n-3, 22∶5n-3 and 18∶2n-6 were the dominant fatty acids amongst the saturates, monoenes and PUFA, respectively, and accounted for 80.8%~88.7% of all identified fatty acids (23) in muscle of Murray cod. The study showed that Murray cod could be cultured successfully on a diet (DU2) containing 20% soybean meal without compromising growth andor carcass quality. Differences in the proximate composition and fatty acid composition of muscle of wild and farmed Murray cod were observed, the most obvious being in the latter. Wild Murray cod had significantly less ($p < 0.05$) saturates (192.6±1.84 vs. 266.3±3.51), monoenes (156.5±8.7 vs. 207.6±6.19), n-3 (145.2±5.24 vs. 261.8±3.2) but higher n-6 (144.3±2.73 vs. 48.3±1.38) in muscle (all values are in μg mg lipid^{-1}) than in farmed fish. Wild fish also had a much lower n-3 to n-6 ratio (1.0±0.03 vs. 5.4±0.09).

[译文]：

新配制饲料与市场通用饲料饲养澳洲龙纹斑的效果比较，并且对养殖和野生的澳洲龙纹斑的肉质成分进行比较

澳洲龙纹斑是具有较高养殖潜力的顶级肉食鱼类。目前，没有专门为澳洲龙纹斑制定的商业化的饲料。本研究为澳洲龙纹斑（初始平均体重80~83.5g）使用两种实验室配制的饲料（DU1和DU2；48.9%和49.1%的蛋白质和16.9%16.1%的油脂，分别为干物质）和两种原为其他鱼类制定（鲑鱼-CD/S和澳洲肺鱼-CD/B）的但现用于喂养澳洲龙纹斑的商业化饲料。两种商业化饲料蛋白质含量较低分别为46.6%和44.4%，但是脂肪含量较高（分别为21.7%和19.5%）。饲料的能量含量经测定是相似的（约20~22kJ/g）。澳洲龙纹斑的生长性能和饲料利用率在饲养中没有显著差异，但饲喂DU1和DU2饲料的澳洲龙纹斑的饲料转化率和蛋白质效率比都比较好。鱼体和肌肉中的脂肪沉积显著减少。在用饲料CD/S喂养的鱼（302.3±3.8）肌肉脂肪酸中，其n-3（262.5±2.9）， n-6（39.8±9）和多不饱和脂肪酸（PUFA）的含量最低（在油脂中 μg/mg），在用 DU2 和 CDB 喂养的鱼中最高。脂肪酸 16：0 和 18：0，18：1n-9 和 16：1n-7，和 22：6n-3，20：5n-3，22：5n-3 和 18：2n-6 是饱和脂肪酸，单不饱和脂肪酸，多不饱和脂肪酸的主要脂肪酸，占所有发现的澳洲龙纹斑肌肉中脂肪酸（23 种）的 80.8%~88.7%。研究表明，澳洲龙纹斑可以饲喂含有 20% 大豆粉的饲料（DU2），而不影响生长和鱼体重量。对野生和养殖的澳洲龙纹斑的肌肉组成和脂肪酸组成的差异进行观测，后者差异最明显。野生澳洲龙纹斑相比有较少（$p<0.05$）饱和脂肪酸（192.6±1.84 对比 266.3±3.51），单不饱和脂肪酸（156.5±8.7 对比 207.6±6.19），n-3（145.2±5.24 对比 261.8±3.2），但肌肉中的 n-6（144.3±2.73 对比 48.3±1.38）（所有值均在油脂中 μg/mg）高于养殖鱼。野生鱼也有较低的 n-3/n-6 比例（1.0±0.03 对比 5.4±0.09）。

参考文献：De Silva S S, Gunasekera R M, Ingram B A. 2015. Performance of intensively farmed murray cod maccullochella peelii peelii (mitchell) fed newly formulated vs. currently used commercial diets, and a comparison of fillet composition of farmed and wild fish [J]. Aquaculture Research, 35 (11), 1039-1052.

（罗钦、李冬梅、黄惠珍　译）

114. Immunohistochemical and immunological detection of ghrelin and leptin in rainbow trout Oncorhynchus mykiss and murray cod Maccullochella peelii peelii as affected by different dietary fatty acids

[英文摘要]：In this study, we report ghrelin and leptin immunoreactive (ir) cells distribution in the gastrointestinal tract and blood ghrelin and leptin levels in rainbow trout (Oncorhynchus mykiss) and Murray cod (Maccullochella peelii peelii) fed diets with different fatty acid compositions. Juvenile rainbow trout and Murray cod were fed five iso-energetic experimental diets containing fish oil (FO) or one of the following vegetable oils (VO): olive oil (OO), sunflower oil (SO), linseed oil (LO), and palm oil (PO); as the added dietary lipid source. The presence and distribution of both ghrelin and leptin ir cells in the gastrointestinal tract were affected by the inclusion of VO. Ghrelin ir cells were found in the gastric glands of rainbow trout and in the mioenteric plexuses of the stomach of Murray cod fed FO. Ghrelin ir cells were localized in the mucosa of the intestine of rainbow trout and Murray cod fed VO. Leptin ir cells were more abundant in the epithelial lining of the mucosa folds and in the glands of the stomach of rainbow trout fed VO. Leptin immunoreactivity was detected in the gastric mioenteric plexus of Murray cod fed FO. No differences were found both in ghrelin and leptin levels in blood plasma or in the growth rates of rainbow trout and Murray cod fed the different experimental diets. These observations suggest that dietary fatty acids play a role in the peripheral feeding regulation.

［译文］：

饲料中不同脂肪酸对彩虹鳟鱼和澳洲龙纹斑的胃饥饿素和瘦素的免疫组化和免疫检测的影响

在本研究中，我们报告彩虹鳟鱼和澳洲龙纹斑饲料中添加不同脂肪酸的组合物对饥饿素和瘦素免疫反应（IR）细胞在胃肠道和血液中的分布水平。幼年彩虹鳟鱼和澳洲龙纹斑饲喂五种相同能量实验，饲料含有鱼油（FO）或以下第一种植物油（VO）：橄榄油（OO），葵花籽油（SO），亚麻油（LO），棕榈油（PO），作为饲料添加的脂肪源。胃肠道中胃饥饿素和瘦素细胞的存在和分布都受VO的影响。胃饥饿素免疫细胞在彩虹鳟鱼的胃腺体中和在投喂FO的澳洲龙纹斑的胃肠系膜丛中都被发现。瘦素细胞定位于投喂VO的彩虹鳟鱼肠黏膜和墨累河鳕鱼肠黏膜。虹鳟黏膜胃黏膜上皮内层和胃腺中瘦素免疫细胞较多。瘦素免疫反应活性在投喂FO的澳洲龙纹斑胃肠系膜丛中检测到。在血浆中胃饥饿素和瘦素水平、虹鳟鱼和澳洲龙纹斑的生长速率的不同饲养实验中均无差异。这些结果表明，饲料脂肪酸在外围饲养调节中起作用。

参考文献：Varricchio Ettore, Russo Finizia, Coccia Elena, et al. 2012. Immunohistochemical and immunological detection of ghrelin and leptin in rainbow trout oncorhynchus mykiss, and murray cod maccullochella peelii peelii, as affected by different dietary fatty acids ［J］. Microscopy Research and Technique, 75（6），771-780.

（罗钦、李冬梅、黄惠珍　译）

115. Effects of Dietary α-Linolenic Acid（18：3n-3）/Linoleic Acid（18：2n-6）Ratio on Fatty Acid Metabolism in Murray Cod（Maccullochella peelii peelii）

［英文摘要］：Global shortages in fish oil are forcing the aquaculture feed industry to use alternative oil sources, the use of which negatively affects the final fatty acid makeup of cultured fish. Thus, the modulation of fatty acid metabolism in cultured fish is the core of an intensive global research effort. The present study aimed to evaluate the effects of various dietary α-linolenic acid（ALA, 18：3n-3）/linoleic acid（LA, 18：2n-6）ratios in cultured fish. A feeding trial was implemented on the freshwater finfish Murray cod, in which fish were fed either a fish oil-based control diet or one of five fish oil-deprived experimental diets formulated to contain an ALA/LA ratio ranging from 0.3 to 2.9, but with a constant total C_{18} PUFA

(ALA+LA) content. The whole-body fatty acid balance method was used to evaluate fish in vivo fatty acid metabolism. The results indicate that dietary ALA was more actively β-oxidized and bioconverted, whereas LA appears to be more efficiently deposited. LA was β-oxidized at a constant level (~36% of net intake) independent of dietary availability, whereas ALA was oxidized proportionally to dietary supply. The in vivo apparent Δ-6 desaturase activity on n-3 and n-6 PUFA exhibited an increasing and decreasing trend, respectively, in conjunction with the increasing dietary ALA/LA ratio, clearly indicating that this enzymatic activity is substrate dependent. However, the maximum Δ-6 desaturase activity acting on ALA peaked at the substrate level of 3.2186 ($\mu mol\ g\ fish^{-1}\ day^{-1}$), suggesting that additional inclusion of ALA is not only wasteful but counterproductive in terms of n-3 LC-PUFA production. Despite a constant total supply of ALA+LA, the recorded total in vivo apparent Δ-6 desaturase activity on both substrates (ALA and LA) increased in synchrony with the ALA/LA ratio, peaking at 1.54, and a 3.2-fold greater Δ-6 desaturase affinity toward ALA over LA was recorded.

[译文]：

饲料中 α-亚麻酸（18：3n-3）/亚油酸（18：2n-6）比例对澳洲龙纹斑脂肪酸代谢的影响

鱼油的全球短缺迫使水产养殖饲料企业使用替代油源，其使用对养殖鱼类的最终脂肪酸组成有不利影响。因此，养殖鱼类中脂肪酸代谢的调节是全球研究的核心。本研究旨在评估饲养鱼的各种饲料中 α-亚麻酸（ALA，18：3n-3）/亚油酸（LA，18：2n-6）的比例。在淡水澳洲龙纹斑上进行了一项饲养试验，投喂以鱼油为对照的饲料或五种不含鱼油而饲料配方含有 ALA/LA 比从 0.3 到 2.9 的饲料，但具有恒定的总 C18 PUFA（ALA+LA）含量。采用全身脂肪酸平衡法对鱼体内脂肪酸代谢进行评价。结果表明，饲料中 ALA 活性更为活跃地被 β-氧化和转化，而 LA 似乎更有效地沉积。LA 以恒定水平（约占总摄入量的 36%）进行 β-氧化成食物供应，而 ALA 则按比例氧化成食物。随着饲料中 ALA/LA 比例增加，n-3 和 n-6PUFA 的体内 Δ-6 去饱和酶活性分别呈上升和下降趋势，清楚表明该酶活性是依赖底物。然而，作用于 ALA 的最大 Δ-6 去饱和酶活性在 3.2186（每尾鱼每日 μmolg）的底物水平时达到峰值，这表明额外加入 ALA 不仅浪费，而且在 n-3 的 LC-PUFA 中起到反作用。尽管 ALA+LA 的总供应不变，但两种底物（ALA 和 LA）在体内记录的总 Δ-6 去饱和酶活性与 ALA/LA 比同步增加，峰值在 1.54，并且一个密切性达 3.2 倍大的 Δ-6 去饱和酶活性对 ALA/LA 被记录。

参考文献：Senadheera, S. D., Turchini, G. M., Thanuthong, T., *et al*. 2011. Effects of dietary α-linolenic acid (18: 3n-3) /linoleic acid (18: 2n-6) ratio on fatty acid metabolism in murray cod (maccullochella peelii peelii) [J]. J Agric Food Chem, 59 (3), 1020-30.

(罗钦、李冬梅、黄惠珍 译)

116. Growth performance, feed efficiency and fatty acid composition of juvenile Murray cod, Maccullochella peelii peelii, fed graded levels of canola and linseed oil

[英文摘要]：In two independent experiments, the effects of dietary inclusion of canola and linseed oil were evaluated in juvenile Murray cod (Maccullochella peelii peelii, Mitchell) over a 112-day period. In each experiment, fish received one of five semi-purified diets in which the dietary fish oil was replaced with canola oil (Experiment A) or linseed oil (Experiment B) in graded increments of 25% (0-100%). Murray cod receiving the graded canola and linseed oil diets ranged in final weight from 112.7±7.6 to 73.8±9.9 g and 93.9±3.6 to 74.6±2.2 g, respectively, and exhibited a negative trend in growth as the inclusion level increased. The fatty acid composition of the fillet and liver were modified extensively to reflect the fatty acid composition of the respective diets. Levels of oleic acid (18：1 n-9) and linoleic acid (18：2 n-6) increased with each level of canola oil inclusion while levels of α-linolenic acid (18：3 n-3) increased with each level of linseed oil inclusion. The concentration of n-3 highly unsaturated fatty acids in the fillet and liver decreased as the amount of vegetable oil in the diets increased. It is shown that the replacement of fish oil with vegetable oils in low fish meal diets for Murray cod is possible to a limited extent. Moreover, this study reaffirms the suggestion for the need to conduct ingredient substitution studies for longer periods and where possible to base the conclusions on regression analysis in addition to anova.

[译文]：

饲养不同梯度的菜籽油和亚麻籽油对澳洲龙纹斑幼鱼生长性能、饵料系数和脂肪酸组成的影响

在为期120d的两个独立的实验中，评估了饲养菜籽油和亚麻籽油对澳洲龙纹斑幼鱼的影响。在每一个实验中，各投投喂五种饲料中的一种，其中鱼油被菜籽油(实验A)或亚麻油（实验B）按25%（0~100%）的增量替代。经过不同增量菜籽油和亚麻籽油饲养的鳕鱼，其最终重量分别为（112.7g±7.6g）~（73.8g±9.9g）和（93.9g±3.6g）~（74.6g±2.2g），鱼的生长随着替代水平的增加呈负增长趋势。鱼肉和肝脏的脂肪酸成分发生了明显的变化，对应饲料中各自的脂肪酸组成。油酸（18：1n-9）和亚油酸（18：2n-6）的含量随着油菜籽油含量的增加而增加，与此同时α-亚麻酸（18：

3n-3) 的含量随着亚麻籽油的增加而增加。随着日粮中植物油含量的增加，鱼肉和肝脏中 n-3 高度不饱和脂肪酸的含量降低。结果表明，在低鱼粉日粮中添加植物油替代鱼油对澳洲龙纹斑的影响是有限的。此外，本研究还重申有必要进行较长时间的成分替代研究，并在可能的基础上进行回归分析和方差分析。

参考文献：Francis, D. S., Turchini, G. M., Jones, P. L., *et al.* 2007. Growth performance, feed efficiency and fatty acid composition of juvenile murray cod, maccullochella peelii peelii, fed graded levels of canola and linseed oil [J]. Aquaculture Nutrition, 13 (5), 335-350.

（罗钦、李冬梅、黄惠珍　译）

117. Dietary lipid source modulates in vivo fatty acid metabolism in the freshwater fish, Murray cod (Maccullochella peelii peelii)

[英文摘要]：The aim of the present investigation was to quantify the fate of C18 and long chain polyunsaturated dietary fatty acids in the freshwater fish, Murray cod, using the in vivo, whole-body fatty acid balance method. Juvenile Murray cod were fed one of five iso-nitrogenous, iso-energetic, semipurified experimental diets in which the dietary fish oil (FO) was replaced (0, 25, 50, 75, and 100%) with a blended vegetable oil (VO), specifically formulated to match the major fatty acid classes [saturated fatty acids, monounsaturated fatty acids, n-3 polyunsaturated fatty acids (PUFA), and n-6 PUFA] of cod liver oil (FO). However, the PUFA fraction of the VO was dominated by C18 fatty acids, while C20/22 fatty acids were prevalent in the FO PUFA fraction. Generally, there was a clear reflection of the dietary fatty acid composition across each of the five treatments in the carcass, fillet, and liver. Lipid metabolism was affected by the modification of the dietary lipid source. The desaturation and elongation of C18 PUFAs increased with vegetable oil substitution, supported by the occurrence of longer and higher desaturated homologous fatty acids. However, increased elongase and desaturase activity is unlikely to fulfill the gap observed in fatty acid composition resulting from decreased highly unsaturated fatty acids intake.

[译文]：

脂肪源在淡水澳洲龙纹斑体内脂肪酸代谢的调节

本研究的目的是利用体内脂肪酸平衡法测定淡水澳洲龙纹斑中 C18 和长链多不饱和脂肪酸的关系。用五分之一的等氮和等能量的半纯化的实验饲养澳洲龙纹斑幼鱼，其喂养的鱼油（FO）被替换成按（0、25、50、75、100%）比例调和的植物油（VO），主要配方与脂肪酸类［饱和脂肪酸、单不饱和脂肪酸，n-3 多不饱和脂肪酸（PUFA），和 n-6 多不饱和脂肪酸］相匹配。

然而，VO 的多不饱和脂肪酸组分以 C18 脂肪酸为主，FO 的多不饱和脂肪酸组分以 C20/22 脂肪酸为主。一般来说，在五种处理方法中鱼体、鱼肉和肝脏中，都能清楚地反映出饲料脂肪酸组成。脂肪代谢受到饲料脂肪来源改变的影响。随着植物油的替代，C18 多不饱和脂肪酸的饱和度和延伸率增加，并得到更长和更多的不饱和脂肪酸的支持。然而，增加的延长酶和脱氢酶活性不可能弥补因高度不饱和脂肪酸摄入量的下降造成的脂肪酸组成的差距。

参考文献：Francis, D. S., Turchini, G. M., And, P. L. J., *et al.* 2007. Dietary lipid source modulates in vivo fatty acid metabolism in the freshwater fish, murray cod (maccullochella peelii peelii) [J]. J Agric Food Chem, 55 (4), 1582-91.

（罗钦、李冬梅、黄惠珍　译）

118. Effects of dietary oil source on growth and fillet fatty acid composition of Murray cod, Maccullochella peelii peelii

［英文摘要］：The Murray cod, an Australian native freshwater fish, supports a relatively small but increasing aquaculture industry in Australia. Presently, there are no dedicated commercial diets available for Murray cod; instead, nutritionally sub-standard feeds formulated for other species are commonly used. The aim of the present investigation was to assess the suitability of two plant based lipid sources, canola oil (CO) and linseed oil (LO), as alternatives to fish oil for juvenile Murray cod. Five iso-nitrogenous, iso-calorific, iso-lipidic semi-purified experimental diets were formulated with 17% lipid originating from 100% cod liver oil (FO), 100% canola oil, 100% linseed oil and 1∶1 blends of canola and cod liver oil (CFO) and 1∶1 blends of linseed and cod liver oil (LFO). Each of the diets was fed to apparent satiation twice daily to triplicate groups of 50 Murray cod with initial mean weights of 6.45±1.59 g for 84 days at 22℃. Final mean weight, specific growth rate and daily feed consumption were significantly higher for the FO and LFO treatments compared to the LO treatment. Feed conversion and protein efficiency ratios were not significantly different amongst treatments. Experimental diets containing vegetable oil and vegetable oil blend (s) had significantly higher concentrations of n-6 fatty acids, predominantly in the form of linoleic acid (LA), while n-3 fatty acids were present in significantly higher concentrations in LO and LFO treatments. The fatty acid composition of Murray cod fillet was reflective of the dietary lipid source. Fillet of fish fed the FO was highest in EPA (20:5n-3), ArA (20:4n-6) and DHA (22:6n-3). Fish fed the CO diet had high concentrations of oleic acid (OlA) (192.2±10.5 mg g lipid^{-1}), while the fillet of Murray cod fed the LO diet was high in α-linolenic acid (LnA) (107.1±6.7 mg g lipid^{-1}). The present study suggests that fish oil can be replaced by up to 100% with canola oil and by up to 50% with linseed oil in Murray cod diets with no significant effect on growth.

[译文]：

饲料中油脂来源对澳洲龙纹斑的生长和鱼肉中脂肪酸组成的影响

澳洲龙纹斑是澳洲本土的淡水鱼，它支持澳洲相对较小但增长的水产养殖业。目前，澳洲龙纹斑没有专用的商业饲料，而广泛使用其他鱼类制定的营养饲料。本研究目的是评估卡诺拉油（CO）和亚麻籽油（LO）两种植物油替代鱼油投喂澳洲龙纹斑幼鱼的适用性。以五种等氮、等热量、等脂肪的半纯化的实验饲料为研究对象，配方中17%脂肪来自100%鱼肝油（FO），100%菜籽油，100%亚麻油，1∶1的共混物的油菜和鱼肝油（CFO）和1∶1的共混物的亚麻籽油和鱼肝油（LFO）。在22℃下，将每种饲料每天投喂两次，每组50尾澳洲龙纹斑，初始平均重量为6.45±1.59g，持续84d。经FO和LFO处理的最终平均体重、生长速率和日饲料消耗量显著高于LO处理的。饲料转化率和蛋白质效率在各处理间差异不显著。试验含有植物油和植物油混合（S）的饲料有较高浓度的n-6脂肪酸，主要是油酸的形成（LA），而在LO和LFO处理中的n-3脂肪酸的浓度较高。澳洲龙纹斑肉中脂肪酸组成反映了饲料中脂肪的来源。投喂FO的鱼的鱼肉中EPA（20∶5n-3）、ArA（6）和DHA（22∶6n-3）最高。投喂CO饲料的鱼有高浓度的油酸（OlA）（每克脂肪中192.2±10.5mg），而投喂LO饲料的澳洲龙纹斑肉中有高浓度的α-亚麻酸（LNA）（每克脂肪中107.1±6.7mg）。目前的研究表明，在澳洲龙纹斑饲养中高达100%的菜籽油和高达50%的亚麻油取代鱼油对生长没有显著的影响。

参考文献：Francis, D. S., Turchini, G. M., Jones, P. L., *et al.* 2006. Effects of dietary oil source on growth and fillet fatty acid composition of murray cod, maccullochella peelii [J]. Aquaculture, 253 (1), 547-556.

（罗钦、李冬梅、黄惠珍　译）

119. Modification of tissue fatty acid composition in Murray cod（Maccullochella peelii peelii, Mitchell）resulting from a shift from vegetable oil diets to a fish oil diet

[英文摘要]：The dynamics of fatty acid composition modifications were examined in tissues of Murray cod fed diets containing fish oil (FO), canola oil (CO) and linseed oil

(LO) for a 25-week period and subsequently transferred to a FO (finishing/wash-out) diet for a further 16 weeks. At the commencement of the wash-out period, following 25 weeks of vegetable oil substitution diets, the fatty acid compositions of Murray cod fillets were reflective of the respective diets. After transfer to the FO diet, differences decreased in quantity and in numerousness, resulting in a revert to the FO fatty acid composition. Changes in percentages of the fatty acids and total accumulation in the fillet could be described by exponential equations and demonstrated that major modifications occurred in the first days of the finishing period. A dilution model was tested to predict fatty acid composition. In spite of a general reliability of the model ($Y = 0.9234X [+] 0.4260$, $R^2 = 0.957$, $P < 0.001$, where X is the predicted percentage of fatty acid; Y the observed percentage of fatty acid), in some instances the regression comparing observed and predicted values was markedly different from the line of equity, indicating that the rate of change was higher than predicted (i.e. $Y = 0.4205X [+] 1.191$, $R^2 = 0.974$, $P < 0.001$, where X is the predicted percentage of [alpha]-linolenic acid; Y the observed percentage of [alpha]-linolenic acid). Ultimately, using the coefficient of distance (D), it was shown that the fatty acid composition of fish previously fed the vegetable oil diets returned to the average variability of the fillet fatty acid composition of Murray cod after 70 or 97 days (LO and CO respectively).

[译文]：

从植物油到鱼油饲养的转变对澳洲龙纹斑组织中脂肪酸组成的影响

在投喂25周含有鱼油（FO）、卡诺拉油（CO）和亚麻籽油（LO）的饲料后，转投喂鱼油FO（完成/洗出）饲料16周，测定澳洲龙纹斑组织中脂肪酸组成的动态变化。在25周的植物油替代饲养之后，在洗出期开始时，澳洲龙纹斑肉中的脂肪酸组成反映了各自的饲养情况。转移到FO饮食后，数量和数量的差异减少，导致FO脂肪酸组成的恢复。通过指数方程可以描述脂肪酸的百分比和总积累的变化，并且证明在结束的头几天发生了重大的变化。用稀释模型预测脂肪酸组成。尽管该模型具有一般的可靠性（$Y = 0.9234X [+] 0.4260$，$r^2 = 0.957$，$p < 0.001$，其中X是脂肪酸的预测百分比；Y是观察到的脂肪酸百分比），在一些情况下，观测值和预测值的回归比较与权益线明显不同，表明变化率高于预期（即$Y = 0.4205X [+] 1.191$，$r^2 = 0.974$，$p < 0.001$，其中X为预测百分比的α-亚麻酸；Y是观察到的α-亚麻酸的百分比）。最终，使用距离系数（D），显示先前投喂植物油饲料的鱼的脂肪酸组成，在70或97d后，分别恢复到澳洲龙纹斑的鱼肉脂肪酸组成的平均变异性（LO和CO分别）。

参考文献：Turchini, G. M., Francis, D. S., De Silva, S. S. 2010. Modification of tissue fatty acid composition in murray cod (maccullochella peelii peelii, mitchell) resulting from a shift from vegetable oil diets to a fish oil diet [J]. Aquaculture Research, 37 (6), 570-585.

(罗钦、李冬梅、黄惠珍 译)

120. Effects of alternate phases of fish oil and vegetable oil-based diets in Murray cod

[英文摘要]：Fish oil (FO) -and canola oil (CO) -based diets were regularly alternated in a daily cycle (amCO：alternation of CO in the morning and FO in the afternoon, and pmCO：alternation of FO in the morning and CO in the afternoon) or in a series of weekly cycles (2W：alternation of 2 weeks on CO and 2 weeks on FO, 4W：alternation of 4 weeks on CO and 4 weeks on FO), over a 16-week period in juvenile Murray cod (Maccullochella peelii peelii). No significant differences were observed between any of the treatments in relation to the final weight. However, fish subjected to the 2W schedule were larger ($P>0.05$) than all other treatments (37.2±0.30 vs. 34.3±0.58 in the control treatment). Fish receiving the 2W treatment had a significantly lower total net disappearance of eicosapentaenoic acid 20：5n-3 (EPA) and docosahexaenoic acid 22：6n-3 (62.1% and 24.0% respectively) compared with the control treatment (fish continuously fed a blend of 50% FO and 50% CO). Likewise, Murray cod receiving the amCO daily schedule had a significantly lower total net disappearance of EPA in comparison with the CD and pmCO treatments. These data point towards the existence of cyclical mechanisms relative to fatty acid utilization/retention.

[译文]：

饲料中鱼油和植物油交替使用阶段对澳洲龙纹斑的影响

在澳洲龙纹斑幼鱼16周的饲养试验中，鱼油（FO）和菜籽油（CO）每天定时交替使用（amCO：早上菜籽油CO，下午鱼油FO，pmCO：早晨鱼油FO，下午菜籽油CO）或一系列周期循环使用（2W：菜籽油CO2周，鱼油FO2周，4W：菜籽油CO4周，鱼油FO4周）。所有试验与最终体重之间没有观察到显着性差异。然而，接受2W处理的鱼比所有其他处理的鱼大（$p>0.05$）（对照处理为37.2±0.30对34.3±0.58）。接受2W处理的鱼与对照处理（连续加入50%FO和50%CO的鱼）相比，二十碳五烯酸20：5n-3（EPA）和二十二碳六烯酸22：6n-3（分别为62.1%和24.0%）的总净消失率更低。同样，与CD和pmCO处理相比，amCO处理的EPA总净消失率更低。这些数据表明脂肪酸的利用/保留存在着周期性的机制。

参考文献：Francis, D. S., Turchini, G. M., Smith, B. K., et al. 2010. Effects of alternate phases of fish oil and vegetable oil-based diets in murray cod [J]. Aquaculture Research, 40 (10), 1123-1134.

（罗钦、李冬梅、黄惠珍　译）

121. Effect of crude oil extracts from trout offal as a replacement for fish oil in the diets of the Australian native fish Murray cod Maccullochella peelii peelii

[英文摘要]：The efficacy of trout oil (TO), extracted from trout offal from the aquaculture industry, was evaluated in juvenile Murray cod Maccullochella peelii peelii (25.4±0.81g) diets in an experiment conducted over 60 days at 23.7±0.8℃. Five isonitrogenous (48% protein), isolipidic (16%) and isoenergetic (21.8kJ g^{-1}) diets, in which the fish oil fraction was replaced in increments of 25% (0-100%), were used. The best growth and feed efficiency was observed in fish fed diets containing 50%~75% TO. The relationship of specific growth rate (SGR), food conversion ratio (FCR) and protein efficiency ratio (PER) to the amount of TO in the diets was described in each case by second-order polynomial equations ($P<0.05$), which were：SGR=-0.44TO^2+0.52TO+1.23 (r^2=0.90, $P<0.05$); FCR=0.53TO^2-0.64TO+1.21 (r^2=0.95, $P<0.05$); and PER=-0.73TO^2+0.90TO+1.54 (r^2=0.90, $P<0.05$). Significant differences in carcass and muscle proximate compositions were noted among the different dietary treatments. Less lipid was found in muscle than in carcass. The fatty acids found in highest amounts in Murray cod, irrespective of the dietary treatment, were palmitic acid (16∶0), oleic acid (18∶1n-9), linoleic acid (18∶2n-6) and eicosapentaenoic acid (20∶5n-3). The fatty acid composition of the muscle reflected that of the diets. Both the n-6 fatty acid content and the n-3 to n-6 ratio were significantly ($P<0.05$) related to growth parameters, the relationships being as follows. Percentage of n-6 in diet (X) to SGR and FCR：SGR=-0.12X^2+3.96X-32.51 (r^2=0.96) and FCR=0.13X^2-4.47X+39.39 (r^2=0.98); and n-3∶n-6 ratio (Z) to SGR, FCR, PER：SGR=-2.02Z^2+5.01Z-1.74 (r^2=0.88), FCR=2.31Z^2-5.70Z+4.54 (r^2=0.93) and PER=-3.12Z^2-7.56Z+2.80 (r^2=0.88) respectively. It is evident from this study that TO could be used effectively in Murray cod diets, and that an n-3∶n-6 ratio of 1.2 results in the best growth performance in Murray cod.

[译文]：

鲑鱼内脏粗提物代替鱼油在澳大利亚本土澳洲龙纹斑饲料饲料中的应用

在23.7±0.8℃下，进行了超过60d的在澳洲龙纹斑幼鱼（25.4±0.81g）饲料中添加人工养殖鲑鱼内脏中提取的鱼油（TO）的养殖效果实验。使用五种含有同等氮（48%蛋白质），同等脂肪（16%）和同等能量（21.8kJ/g）的饲料，其中鱼油以25%（0-100%）的增量替代。在含有50%～75%鱼油 TO 的鱼饲料中观察到最好的生长和饲料效率。在每种情况下，通过二阶多项式方程（p<0.05）描述了比生长速率（SGR），食物转化率（FCR）和蛋白质效率比（PER）与饲料中鱼油 TO 含量的关系：$SGR = -0.44TO^2 + 0.52TO + 1.23$（$r^2 = 0.90$，$p < 0.05$）；$FCR = 0.53TO^2 - 0.64TO + 1.21$（$r^2 = 0.95$，$p < 0.05$）；和 $PER = -0.73TO^2 + 0.90TO + 1.54$（$r^2 = 0.90$，$p < 0.05$）。在不同饲料的饲养中，鱼体和肌肉近似成分的差异显著，在肌肉中发现的油脂比在鱼体中少。不论饲料如何，在澳洲龙纹斑中发现的最高含量的脂肪酸，均为棕榈酸（16:0）、油酸（18:1-9)、亚油酸（18:2-6）和二十碳五烯酸（20:5n-3）。肌肉的脂肪酸组成反映了饲料中的脂肪酸组成。n-6 脂肪酸含量和 n-3 至 n-6 比值均显著反映出（p<0.05）与生长参数相关，关系如下：SGR 和 FCR 的饮食中 n-6 的百分比（X）：$SGR = -0.12X^2 + 3.96X - 32.51$（$r^2 = 0.96$）和 $FCR = 0.13X^2 - 4.47X + 39.39$（$r^2 = 0.98$）；和 N-3：n-6 比值（Z）与 SGR，FCR，PER：$SGR = -2.02Z^2 + 5.01Z - 1.74$（$r^2 = 0.88$），$FCR = 2.31Z^2 - 5.70Z + 4.54$（$r^2 = 0.93$）和 $PER = -3.12Z^2 - 7.56Z + 2.80$（$r^2 = 0.88$）。从本研究中可以看出，鲑鱼内脏鱼油可以在澳洲龙纹斑饲料中有效地使用，并且 n-3:n-6 的比值为 1.2 时澳洲龙纹斑的生长性能最佳。

参考文献：Turchini, G., Gunasekera, R. M., De Silva, S. 2003. Effect of crude oil extracts from trout offal as a replacement for fish oil in the diets of the Australian native fish Murray cod Maccullochella peelii peelii [J]. Aquaculture Research, 34, 697-708.

（罗钦、李冬梅、黄惠珍 译）

122. Fatty acid metabolism in the freshwater fish Murray cod（Maccullochella peelii peelii）deduced by the whole-body fatty acid balance method

［英文摘要］：The whole-body fatty acid balance method was used to investigate the fatty acid metabolism in Murray cod (Maccullochella peelii peelii) fed diets containing canola (CO) or linseed oil (LO). Murray cod were able to elongate and desaturate both 18:2n-6 and 18:3n-3. In fish fed the CO diet, 54.4% of the 18:2n-6 consumed was accumulated, 38.5% oxidized and 6.4% elongated and desaturated to higher homologs. Fish fed the LO diet accumulated 52.9%, oxidized 37% and elongated and desaturated 8.6% of the consumed 18:3n-3. The overall roles of n-6 fatty acids appeared more important in Murray cod compared to other freshwater species. Murray cod also showed a preferential order of utilization of C18 fatty acid for energy production (18∶3n-3>18∶2n-6>18∶1n-9). Moreover, it is demonstrated that an increase in dietary 18∶3n-3 is directly responsible of increased desaturase activity and augmented saturated fatty acid accumulation in the fish body. The present study also suggests that, in the context of the possible maximization of the natural ability of fish to produce long chain polyunsaturated fatty acids, the whole-body approach can be considered well suited and informative and Murray cod is a suited candidate to fish oil replacement for its diets.

[译文]：

采用全身脂肪酸平衡法推导淡水澳洲龙纹斑中脂肪酸的代谢

采用全身脂肪酸平衡法研究投喂含油菜籽（CO）或亚麻籽油（LO）饲料的澳洲龙纹斑（Maccullochella peelii peelii）中脂肪酸的代谢。澳洲龙纹斑能够延长和去饱和脂肪酸18∶2n-6和18∶3n-3。在投喂CO饲料的鱼中，18∶2n-6沉积了54.4%，氧化了38.5%，延长和去饱和较高脂肪酸同系物6.4%。在投喂LO饲料的鱼中，18∶3n-3沉积了52.9%，氧化了37%，延长和去饱和了8.6%。与其他淡水鱼相比，n-6脂肪酸对澳洲龙纹斑的作用显得更为重要。同时检测出澳洲龙纹斑利用C18脂肪酸进行能量生产的优先顺序（18∶3n-3>18∶2n-6>18∶1n-9）。此外，证明在饲料中18∶3n-3的增加直接造成鱼体去饱和酶活性增加和饱和脂肪酸累积增加。本研究还表明，在尽可能提高鱼类生产长链多元不饱和脂肪酸的能力的情况下，全身脂肪酸平衡法可以被认为是非常适合的和信息丰富的，并且澳洲龙纹斑适合使用食用鱼油替代品。

参考文献：Turchini G M, Francis D S, De Silva S S. 2006. Fatty acid metabolism in the freshwater fish murray cod (maccullochella peelii peelii) deduced by the whole-body fatty acid balance method [J]. Comp Biochem Physiol B Biochem Mol Biol., 144 (1), 110-118.

（罗钦、李冬梅、黄惠珍　译）

123. Apparent in Vivo Δ-6 Desaturase Activity, Efficiency, and Affinity Are Affected by Total Dietary C_{18} PUFA in the Freshwater Fish Murray Cod

[英文摘要]：Dietary fatty acids are known to modulate fatty acid metabolism in fish. However, the innate capability of fish to bioconvert short chain fatty acids to health promoting long chain fatty acids (LCPUFA) is insufficient to compensate for a reduced dietary intake. While many studies have focused on the dietary regulation of the fatty acid bioconversion pathways, there is little known regarding the effects of the dietary levels of C18 polyunsaturated fatty acids (PUFA) on fatty acid metabolism. Here, we show a greater degree of apparent enzyme activity (Δ-6 desaturase) in fish fed a diet with higher amounts of dietary C18 PUFA. In particular, fish receiving high amounts of dietary C18 PUFA had a greater amount of Δ-6 desaturase activity acting on 18∶3n-3 than 18∶2n-6. However, with the gradual reduction of dietary C18 PUFA there was a shift in substrate preference of Δ-6 desaturase from 18∶3n-3 to 18∶2n-6. This information will provide valuable insight for the implementation of low fish oil diets, which permit the maintenance of n-3 LCPUFA levels in farmed Murray cod.

[译文]：

淡水鱼澳洲龙纹斑饲料中总 C18 多不饱和脂肪酸对 Δ-6 去饱和酶活性有明显的功效和亲和力

众所周知饲料中的脂肪酸可调节鱼类的脂肪酸代谢。然而，鱼类本身把短链脂肪酸转化成有益的长链脂肪酸（LCPUFA）的能力是不足的，需要减少从饲料中摄入来弥补。然而许多研究集中在脂肪酸生物转化途径的饲养管理方面，人们在饲料中 C18 多不饱和脂肪酸（PUFA）对脂肪酸代谢的影响这方面知之甚少。本文中，我们介绍了用高含量 C18 多不饱和脂肪酸的饲料投喂的鱼中的 Δ-6 去饱和酶活性明显更大。特别是，采用高含量 C18 多不饱和脂肪酸饲养的鱼，其 Δ-6 脱氢酶活性对 18∶3n-3 比 18∶2-6 起的作用更大。然而，随着饲料中的 C18 多不饱和脂肪酸的逐渐减少，Δ-6 去饱和酶的底物偏好从 18∶3n-3 转变为 18∶2n-6。该信息将在养殖澳洲龙纹斑时允许维持 n-3 长链脂肪酸水平的低鱼油饲料提供宝贵意见。

参考文献：Francis, D. S., Peters, D. J., Turchini, G. M. 2009. Apparent in vivo Δ-6 desaturase activity, efficiency, and affinity are affected by total dietary c18 pufa in the freshwater fish murray cod [J]. Journal of Agricultural & Food Chemistry, 57 (10), 4381-90.

（罗钦、李冬梅、黄惠珍 译）

124. Finishing diets stimulate compensatory growth: results of a study on Murray cod, Maccullochella peelii peelii

[英文摘要]：The effective implementation of a finishing strategy (washout) following a grow-out phase on a vegetable oil-based diet requires a period of several weeks. However, fish performance during this final stage has received little attention. As such, in the present study the growth performance during both, the initial grow-out and the final wash-out phases, were evaluated in Murray cod (Maccullochella peelii peelii). Prior to finishing on a fish oil-based diet, fish were fed one of three diets that differed in the lipid source: fish oil, a low polyunsaturated fatty acid (PUFA) vegetable oil mix, and a high PUFA vegetable oil mix. At the end of the grow-out period the fatty acid composition of Murray cod fillets were reflective of the respective diets; whilst, during the finishing period, those differences decreased in degree and occurrence. The restoration of original fatty acid make up was more rapid in fish previously fed with the low PUFA vegetable oil diet. During the final wash-out period, fish previously fed the vegetable oil-based diets grew significantly (P<0.05) faster (1.45±0.03 and 1.43± 0.05, specific growth rate,% day^{-1}) than fish continuously fed with the fish oil-based diet (1.24±0.04). This study suggests that the depauperated levels of highly unsaturated fatty acids in fish previously fed vegetable oil-based diets can positively stimulate lipid metabolism and general fish metabolism, consequently promoting a growth enhancement in fish when reverted to a fish oilbased diet. This effect could be termed lipo-compensatory growth.

[译文]：

结束饲养促进补偿性生长：一项研究澳洲龙纹斑的结果

在以植物油为基础的养殖上的发展阶段之后，有效地实施一种结束策略需要几个星期的时间。在这个最后阶段，鱼的表现几乎没有受到重视。因此，本研究评估初始生长和最终阶段澳洲龙纹斑（Maccullochella peelii peelii）的生长性能。在结束以鱼油为基础的喂养之前，给鱼喂三种中的一种不同脂肪来源的饲料：鱼油、低多不饱和脂肪酸（PUFA）植物油混合物和高不饱和脂肪酸植物油混合物。在生长期结束阶段时，澳洲龙纹斑肉的脂肪酸组成反映了各自的饲料；而在结束阶段，差异程度和差异的产生都有所减少。用原来低PUFA植物油喂养的鱼原脂肪酸的恢复更快。在最后冲洗阶段，以投喂植物油为主的饲料的鱼（1.45±0.03和1.43±0.05，生长速率,%/d）比一直投喂以鱼油为主的饲料的鱼（1.24±0.04）生长更快。这项研究表明，在以植物油为主的养殖鱼类中，高度不饱和脂肪酸可以促进油脂代谢和鱼的新陈代谢，从而促进鱼的生长。这种作用可以称为脂肪补偿生长。

参考文献：TURCHINI, G. M, FRANCIS, D. S, DE SILVA, S. S. 2010. Finishing diets stimulate compensatory growth: results of a study on murray cod, maccullochella peelii peelii [J]. Aquaculture Nutrition, 13 (5), 351-360.

（罗钦、李冬梅、黄惠珍　译）

125. A comparison of reciprocating flow versus constant flow in an integrated, gravel bed, aquaponic test system

[英文摘要]：Murray cod, Maccullochella peelii peelii, and Green oak lettuce, Lactuca sativa, were used to test for differences between two aquaponic flood regimes; reciprocal flow (hydroponic bed was periodically flooded) and constant flow (hydroponic bed was constantly flooded), in a freshwater aquaponic test system, where plant nutrients were supplied from fish wastes, while plants stripped nutrients from the wastewater before it was returned to the fish. The Murray cod had FCRs and biomass gains that were statistically identical in both systems. Lettuce yields were good and a significantly greater amount of both biomass and yield occurred in the constant flow treatment. Constant flow treatments exhibited greater pH buffering capacity, required fewer bicarbonate (buffer) additions to control pH and maintained lower conductivity levels than reciprocal flow controls. Water consumption in the two systems was statistically identical. Overall, results suggest that a constant flow flooding regime is as good as, or better than, a reciprocating flooding regime in the aquaponic test system used.

[译文]：

在综合水体测试系统的砾石床中往复流动与恒定流动的比较

墨瑞鳕鱼和绿栎莴苣被用来测试两种生态防洪系统之间的差异；在淡水鱼菜共生测试系统中，进行相互流动（水培床周期性地淹水）和恒定流动（水培床被不断地淹没），其中的植物营养素是从鱼类废弃物中提供的，与此同时，植物在废水返回到鱼之前从中提取营养物。在两个系统中，墨瑞鳕鱼具有统计学上相同的FCRs（饲料转换率）和生物质吸收。在恒流处理中，莴苣产量良好且生物量和产量均显著增加。恒定流动处理表现出更大的pH缓冲能力，需要更少的碳酸氢盐（缓冲剂）添加物来控制pH并保持比往复流动控制更低的电导率水平。两个系统的耗水量在统计学上是相同的。总体而言，结果表明恒定流动体系与鱼菜共生实验系统中使用的往复流动体系一样好或更好。

参考文献：Lennard, W. A., Leonard, B. V. 2005. A comparison of reciprocating flow versus constant flow in an integrated, gravel bed, aquaponic test system [J]. Aquaculture International, 12 (6), 539-553.

（罗钦、李冬梅、黄惠珍 译）

126. Biomanipulation: a review of biological control measures in eutrophic waters and the potential for Murray cod Maccullochella peelii peelii to promote water quality in temperate Australia

[英文摘要]：Biomanipulation is a method of controlling algal blooms in eutrophic freshwater ecosystems. The most common approach has been to enhance herbivores through a reduction of planktivorous fish and introduction of piscivorous fish. The method was originally intended to reduce grazing pressure on zooplankton, thereby increasing grazing pressure on phytoplankton to increase water clarity and promote the growth of aquatic macrophytes. Biomanipulation has received considerable attention since it was proposed in 1975 where innovative approaches and explanations of the processes have been developed. Although many successful biomanipulation exercises have been conducted internationally, it has received comparatively little attention in the Southern Hemisphere and has not been trialled in the southern temperate climate of South Australia. This is a review to speculate upon the criteria for and against the application of biomanipulation in southern temperate Australia using the native species Murray cod (Maccullochella peelii peelii) and to suggest future research.

［译文］：

生物调控：在富营养化水体中和在有潜力提升墨瑞鳕鱼水质的温带澳洲中的生物调控措施的评论

生物调控是一种控制富营养淡水生态系统中藻类泛滥的方法。最常见的做法是通过减少食用浮游生物的鱼类并引入肉食性鱼类来增加食用浮游植物的鱼类。该方法最初是希望减少在浮游动物上的数量，从而增加浮游植物的数量，使得水分透明度增加和水生植物的生长速度加快。自从1975年提出以来后，生物调控受到了相当大的关注，研发了许多新方法和过程说明。虽然国际上进行了许多成功的生物调控的操作，但在南半球其只受到一点点的关注并且在澳洲南方温带气候地区中没有被实践过。这是一个评论，推测了支持和反对使用本地墨瑞鳕鱼在澳洲南部温带地区进行生物调控的标准，并提出未来的研究的方式。

参考文献：Sierp, M. T., Qin, J. G., Recknagel, F. 2009. Biomanipulation: a review of biological control measures in eutrophic waters and the potential for murray cod maccullochella peelii peelii, to promote water quality in temperate australia [J]. Reviews in Fish Biology & Fisheries, 19 (2), 143-165.

（罗钦、李冬梅、黄惠珍　译）

127. Diet of Murray cod (Maccullochella peelii peelii) (Mitchell) larvae in an Australian lowland river in low flow and high flow years

［英文摘要］：Researchers have hypothesised that influxes of pelagic zooplankton to river channels after floods and high flows are necessary for strong recruitment of some native fish species, including Murray cod (Maccullochella peelii peelii) (Mitchell), in the Murray arling river system, Australia. This study investigated the composition of the diet and gut fullness of drifting Murray cod larvae weekly during two spawning seasons with contrasting flows, to determine if pelagic zooplankton comprised a greater proportion of the gut contents and guts were fuller in a high flow (2000) than in a low flow (2001) year. Gut fullness and yolk levels of 267 larvae were ranked, and prey identified to family level. Approximately 40 and 70% of individuals had been feeding in 2000 and 2001, respectively. Gut fullness increased with declining yolk reserves. Larvae in both the years had an almost exclusively benthic diet, irrespective of the flow conditions at the time. Substantial inundation of dry ground in 2000, albeit restricted to in-channel benches, anastomosing channels and oxbow lakes, did not lead to an influx of pelagic, floodplain-derived zooplankton subsequently exploited by Murray cod larvae. These results have the implications for the management of regulated temperate lowland rivers: high flows cannot automatically be assumed to be beneficial for the fish larvae of all species and their food resources, and caution should be exercised with the timing of flow releases.

[译文]：

澳洲平原河流低流量和高流量年度中的墨瑞鳕幼鱼饵料

研究者们假设，洪水和高流量后河道涌入大量浮游动物，引进一些原生种鱼类是必要的，包括澳大利亚默里河系统的墨瑞鳕鱼（Maccullochella peelii peelii）（Mitchell）。本研究调查了两个不同流量的产卵季节中的墨瑞鳕幼鱼每周的饵料组成和肠道饱满度，以确定浮游动物在肠内容物中的比例是否为高流量年（2000）比低流量年（2001）多。对267个幼鱼的肠道饱满度和卵黄水平含量进行了分级，并对其进行鱼种鉴定。在2000年和2001年，大约分别有40%和70%的鱼在进食。随着卵黄含量的减少，肠道的饱满度增加。无论当时的流动状况如何，两年内的幼鱼全都底栖觅食。2000年大量旱地被洪水淹没，虽然只有长廊、渠道和弓形湖没有大量浮游生物的涌入，但是被洪水淹没带来的的浮游动物被墨瑞鳕幼鱼食用。这些结果对规范温带平原河流的管理有影响：高流量不能想当然地被认为对所有鱼类的幼鱼和它们的饵料来源都是有益的，并且应该谨慎处理流量释放的时间。

参考文献：Kaminskas, S., Humphries, P. 2009. Diet of murray cod (maccullochella peelii peelii) (mitchell) larvae in an australian lowland river in low flow and high flow years [J]. Hydrobiologia, 636 (1), 449-461.

（罗钦、李冬梅、黄惠珍 译）

128. 龙纹斑仔稚鱼配合饲料及其制备方法（专利）

摘要：本发明公开了一种龙纹斑仔稚鱼配合饲料及其制备方法，意在提供一种可满足其生长发育所需，营养均衡、绿色高效的配合饲料；该饲料由进口鱼粉，豆粕，玉米蛋白粉，高筋面粉，乌贼肝脏粉，大豆卵磷脂，复合多维，复合多矿，氯化胆碱，食盐，柠檬酸，乳酸菌，蛋氨酸，维生素C磷酸酯，防霉剂，抗氧化剂组成。本发明提供的龙纹斑仔稚鱼配合饲料，结合澳洲龙纹斑仔稚鱼生长所需，采用"理想氨基酸模式"原理设计开发的营养均衡、环保高效的澳洲龙纹斑仔稚鱼配合饲料。

参考文献：钱晓明，叶建华，董志航，等．2015．龙纹斑仔稚鱼配合饲料及其制备方法．CN104642817A。

注：无英文摘要。

129. 一种虫纹鳕鲈配合饲料及其制备方法（专利）

摘要：本发明涉及一种鱼饲料，特别涉及一种虫纹鳕鲈配合饲料；本发明还涉及一种上述虫纹鳕鲈配合饲料的制备方法。本发明的虫纹鳕鲈配合饲料，包括下列质量份数的各组分：鱼粉21~32份、小杂鱼7~15份，猪肉粉5~14份，豆粕6~16份，四季青6~14份，松花粉3~9份，苍术4~10份，胡桃仁油3~8份，丁香3~7份，槐花2~6份，紫草2~8份，茜草3~9份，三七2~8份，鱼油2~6份，橄榄油1~4份，复合多维1~4份，喂食本发明的虫纹鳕鲈配合饲料后虫纹鳕鲈饲料系数低，虫纹鳕鲈肌肉中粗蛋白含量高、粗脂肪含量低、鲜味氨基酸含量高。

参考文献：王芬，张伟，唐佳，等．2015．一种虫纹鳕鲈配合饲料及其制备方法．CN105211500A。

注：无英文摘要。

130. 一种预防虫纹鳕鲈体表溃疡的饲料及其制备方法（专利）

摘要：本发明涉及一种鱼饲料，特别涉及一种预防虫纹鳕鲈体表溃疡的饲料；本发明还涉及预防虫纹鳕鲈体表溃疡的饲料的制备方法。本发明的预防虫纹鳕鲈体表溃疡的饲料，包括下列质量份数的各组分：鱼粉20~30份、蚯蚓粉8~15份，鱿鱼粉4~10份，米糠6~12份，三白草3~7份，西瓜霜3~8份，蛇蜕2~6份，金沸草3~7份，乌贼骨2~5份，松花粉2~6份，海蛤壳1~5份，炉甘石1~4份，石榴皮2~6份，黄精1~4份，橄榄油1~4份，复合多维1~4份，喂食虫纹鳕鲈后饲料系数低、特定生长率高，抗表面溃疡能力强。

参考文献：王芬，张伟，唐佳，等．2015．一种预防虫纹鳕鲈体表溃疡的饲料及其制备方法．CN105211565A。

注：无英文摘要。

131. 澳洲龙纹斑幼鱼膨化颗粒配合饲料（专利）

摘要：本发明澳洲龙纹斑幼鱼膨化颗粒配合饲料是由各原料组分按膨化颗粒饲料的加工方法而制成的一种混合物，适合于喂养体重在50g到500g幼鱼阶段的澳洲龙纹斑，其各原料组分及其重量百分含量包括：蒸汽烘干鱼粉50%~60%；木薯粉10%~15%；大豆浓缩蛋白6%~10%；豆粕5%~8%；南极磷虾粉2%~5%；冻化鱼油5%~10%；啤酒酵母3~5%；矿物质1.0%~1.4%；磷酸二氢钙0.8%~1.2%；氯化胆碱0.2%~0.4%；复合维生素0.1%~0.3%；螺旋藻0.8%~1.0%；包膜丁酸钠0.1%~0.3%。本发明配合饲料能够提高澳洲龙纹斑的养殖品质，营养全面，适合于幼鱼阶段的澳洲龙纹斑，成本低，诱食性好，水中稳定性好，幼鱼生长快，经济效益好，养殖效果佳。

参考文献：张蕉南，胡兵，李惠，等.2014.澳洲龙纹斑幼鱼膨化颗粒配合饲料.CN104304800A。

注：无英文摘要。

132. 澳洲龙纹斑中成鱼膨化颗粒配合饲料（专利）

摘要：本发明澳洲龙纹斑中成鱼膨化颗粒配合饲料是由各原料组分按膨化颗粒饲料的加工方法而制成的一种混合物，适合于喂养体重大于500g中成鱼阶段的澳洲龙纹斑，其各原料组分及其重量百分含量包括：蒸汽烘干鱼粉45%~55%；木薯粉11%~14%；大豆浓缩蛋白6%~15%；豆粕5%~15%；南极磷虾粉1%~5%；冻化鱼油6%~10%；啤酒酵母4%~6%；矿物质1.5%~2.5%；磷酸二氢钙0.5%~1.5%；氯化胆碱0.2%~0.4%；复合维生素0.1%~0.3%；螺旋藻0.3%~0.5%；包膜丁酸钠0.1%~0.3%。本发明配合饲料能够提高澳洲龙纹斑的养殖品质，营养全面，适合于中成鱼阶段的澳洲龙纹斑，成本低，诱食性好，水中稳定性好，中成鱼生长快，经济效益好，养殖效果佳。

参考文献：胡兵，张蕉南，李惠，等.2014.澳洲龙纹斑中成鱼膨化颗粒配合饲料.CN104286581A。

注：无英文摘要。

133. 澳洲龙纹斑稚鱼膨化颗粒配合饲料（专利）

摘要：本发明澳洲龙纹斑稚鱼膨化颗粒配合饲料是由各原料组分按膨化颗粒饲料的加工方法而制成的一种混合物，适合于喂养体重小于50g稚鱼阶段的澳洲龙纹斑，其各原料组分及其重量百分含量包括：蒸汽烘干鱼粉53%~65%；木薯粉10%~15%；大豆浓缩蛋白5%~8%；豆粕2%~9%；南极磷虾粉2%~5%；冻化鱼油3%~8%；啤酒酵母3%~5%；矿物质0.8%~1.2%；磷酸二氢钙0.8%~1.2%；氯化胆碱0.1%~0.3%；复合维生素0.1%~0.3%；螺旋藻0.8%~1%；包膜丁酸钠0.1%~0.3%。本发明配合饲料能够提高澳洲龙纹斑的养殖品质，营养全面，适合于稚鱼阶段的澳洲龙纹斑，成本低，诱食性好，水中稳定性好，稚鱼生长快，经济效益好，养殖效果佳。

参考文献：林虬，张蕉南，胡兵，等．2014．澳洲龙纹斑稚鱼膨化颗粒配合饲料．CN104286574A。

注：无英文摘要。

134. 澳洲虫纹鳕鲈育成期配合饲料及其制备方法

摘要：本发明公开了一种澳洲虫纹鳕鲈育成期配合饲料及其制备方法，旨在提供一种专门用于澳洲虫纹鳕鲈育成期养殖，并且可显著改善虫纹鳕鲈生长速度，降低环境污染的饲料；该饲料由进口鱼粉20%~40%、虾肉粉5%~10%、鸡肉粉5%~10%、豆粕5%~20%、高筋面粉18%~30%、啤酒酵母2%~8%、乌贼膏2%~8%、鱼油2%~5%、大豆磷脂油2%~5%、维生素预混料0.1%~1.5%、微量元素预混料1%~5%、磷酸二氢钙1%~3%、氯化胆碱0.1%~0.5%、维生素C磷酸酯0.1%~0.5%、特免皇0.1%~0.3%、维生素E 0.03%~0.1%、防霉剂0.01%~0.05%、细麸皮0.5%~1.5%构成；属于鱼饲料技术领域。

参考文献：文远红，张璐，米海峰，等．2014．澳洲虫纹鳕鲈育成期配合饲料及其制备方法．CN103783268A。

注：无英文摘要。

135. 一种澳洲虫纹鳕鲈鱼苗开口配合饲料及其制备方法

摘要：本发明公开了一种澳洲虫纹鳕鲈鱼苗开口配合饲料及其制备方法，旨在提供一种可满足其营养需求、促进其生长发育、增强其非特异免疫力、提高其存活率的绿色健康开口料饲料；该饲料由进口白鱼粉25%~50%、卤虫粉2%~10%、南极磷虾粉3%~10%、乌贼肝脏粉1%~5%、海藻粉5%~15%、啤酒酵母2~5%、α-淀粉3%~10%、高筋面粉10%~20%、大豆卵磷脂1%~5%、鱼油2%~8%、维生素预混料0.2%~1.0%、微量元素预混料1%~5%、磷酸二氢钙1%~5%、枯草芽孢杆菌0.03%~0.1%、维生素C醋酸酯0.1%~0.6%、抗氧化剂0.01%~0.1%、氯化胆碱0.05%~0.5%；属于鱼饲料技术领域。

参考文献：文远红，米海峰，张璐，等．2014．一种澳洲虫纹鳕鲈鱼苗开口配合饲料及其制备方法．CN103798549A。

注：无英文摘要。

136. 澳洲龙纹斑成鱼膨化配合饲料及其制备方法（专利）

摘要：一种澳洲龙纹斑成鱼膨化配合饲料及其制备方法，由下列组份按照重量份数组成：进口鱼粉 60~70 份，特一级面粉 15~18 份，豆粕 1~3 份，啤酒酵母 1~4 份，乌贼肝脏粉 2~4 份，蛋氨酸 0.05~0.2 份，赖氨酸 0.1~0.3 份，甜菜碱 0.2~0.4 份，复合多维 1~3 份，复合多矿 1~3 份，L-抗坏血酸-2-磷酸酯 0.2~0.4 份，磷酸二氢钙 1~3 份，硅型氯化胆碱 0.2~1 份，大豆卵磷脂 2~6 份，进口鱼油 3~7 份，酵母细胞壁 0.1~0.5 份，左旋肉碱 0.1~0.4 份。本发明经过高温膨化后消化率大幅度提高，高温灭杀了大量细菌，避免了疾病的传播，膨化后的饲料紧密，散失率低，避免了饲料散失对水体的污染，且营养成分组成成分丰富。

参考文献：刘大勇，蔡星，秦桂祥，等 . 2013. 澳洲龙纹斑成鱼膨化配合饲料及其制备方法 . CN103250912A。

注：无英文摘要。

137. 一种改善龙纹斑成鱼风味品质的配合饲料及其制备方法（专利）

摘要：本发明公开了一种改善龙纹斑成鱼风味品质的配合饲料及其制备方法，意在提供一种可满足其生长发育所需，风味优良、绿色高效的配合饲料；进口鱼粉 30~50 份，豆粕 20~40 份，玉米蛋白粉 5~10 份，高筋面粉 10~20 份，乌贼肝脏粉 1~3 份，豆油 1~3 份，复合多维 0.2~0.5 份，复合多矿 0.5~1 份，氯化胆碱 0.1~0.3 份，螺旋藻粉 1~3 份，肉碱 0.1~0.4 份，亚油酸 1~3 份，维生素 C 磷酸酯 0.1~0.6 份，防霉剂 0.01~0.1 份，抗氧化剂 0.01~0.1 份；属于鱼饲料技术领域。

参考文献：钱晓明，叶建华，董志航，等 . 2015. 一种改善龙纹斑成鱼风味品质的配合饲料及其制备方法 . CN201510125509.3。

注：无英文摘要。

138. 一种海藻寡糖鱼饲料添加剂（专利）

摘要：一种海藻寡糖鱼饲料添加剂，涉及一种海藻寡糖。针对现有技术的不足，提供能够提高饲料的利用率，促进生长，提高抗病能力，降低成本，同时能够提高鱼质量的一种海藻寡糖鱼饲料添加剂。所述海藻寡糖鱼饲料添加剂的海藻寡糖为龙须菜琼胶寡糖，是利用一株太平洋火色杆菌（Flammeovirga？Pacifica）H2 与龙须菜共培养，通过微生物所产的酶系直接降解龙须菜制备而得的 4~8 糖的混合物。所述太平洋火色杆菌（Flammeovirga？Pacifica）H2 保藏编号为 CCTCC？NO：M2012229。

参考文献：刘洋，曾润颖，罗土炎，等 . 2016. 一种海藻寡糖鱼饲料添加剂 . CN201510894666.0。

注：无英文摘要。

139. 用于生产高质量蛋白浓缩物和脂类的固态发酵系统和方法（专利）

摘要：本发明描述了通过经由固态发酵（SSF）和混合-SSF将植物来源的材料转化为生物可利用的蛋白和脂类，产生高质量蛋白浓缩物（HQPC）和脂类的基于生物的方法，包括因此产生的此类HQPC和脂类作为营养物的用途，包括用作水产养殖食物中的鱼粉代替物的用途。还公开了SSF反应器以及使用所述反应器的方法。

参考文献：詹森·A·布茨马；威廉姆·R·吉本斯；迈克尔·L·布朗. 2016. 用于生产高质量蛋白浓缩物和脂类的固态发酵系统和方法. CN201480055185.4。

注：无英文摘要。

140. 一种生态防治澳洲龙纹斑水霉病的中草药制备方法及应用（专利）

摘要：本发明公开了一种生态防治澳洲龙纹斑水霉病的中草药制备方法及其应用，取大黄干品400g/m³水体，五倍子干品200g/m³水体，加入淡水中，淡水至少没过干品总量的三分之二，煮沸15min后保温15min，然后用纱布滤去药渣，取过滤的药液。本发明通过中草药加盐度调节，可以很好地抑制霉菌的生长。

参考文献：朱永祥，刘大勇，蔡星，等. 2013. 一种生态防治澳洲龙纹斑水霉病的中草药制备方法及应用. CN103272005A。

注：无英文摘要。

第六章　澳洲龙纹斑引种与开发

第一节　专题研究概况

人类的食物几乎完全取自生物资源。全世界大约有 500 万~5000 万个物种，有科学记载的仅 140 万种。随着人口的剧增，人类向自然界索取生物资源的规模越来越大，强度越来越高。然而，无论何国，再大的也不可能拥有所有物种；再小的，也可能有其特有物种。为了全人类的生存和各国的发展，必须共享全球的资源，引种便是捷径之一。拿粮食作物来说，世界主要粮食作物大约 15 种，其中有 7 种作物的产量超过 1000 万吨，即小麦、水稻、玉米、马铃薯、大麦、甘薯及木薯。各国之间对这些物种的良种的交流与引进十分频繁。相互包容和引种是发展农业生产，包括水产生产的必要手段之一（李思发，2005）。

"引种"为褒义词，是人们根据需要而实施的一种有目的、有计划的行动，所期望的是正面效应，即产生经济效益与社会效益。我国是水产生物引种最多的国家之一，引种的途径有作为食用养殖对象引进、作为观赏对象引进、作为生物防控手段引进、作为休闲垂钓和饵料鱼引进、试验目的引进、随压舱水带入、因水利工程的修建而带入和随人工放养放流活动带入。我国虽是世界上水产生物引种后起国家，却已成为引种最多的国家之一。有记载的人为有意引进始自 20 世纪 50 年代后期；70 年代末至 80 年代，我国引进规模逐渐加大，90 年代更趋活跃。据不完全统计，到目前为止，我国从国外引进的水生动物已达 100 种以上。依引进种（含品种）的大类分，我国引进种数的多少依次为：鲈形目鱼类 25 种（其中鲈类 16 种，如大口黑鲈；罗非鱼类 9 种，如尼罗罗非鱼）；鲑形目鱼类 14 种，如虹鳟；鲤形目鱼类 13 种，如露斯坦野鲮，大口牛胭脂鱼；鲇形目鱼类 8 种，如斑点叉尾；鲟形目鱼类 7 种，如匙吻鲟；鲽形目鱼类 4 种，如大菱鲆；脂鲤目鱼类 3 种，如淡水白鲳；贝类 14 种，如海湾扇贝；虾类 6 种，如梵纳对虾；

龟鳖类9种，如鳄龟。

2011年国家出台了《关于加快推进现代化农作物种业发展的意见》（国发〔2011〕8号）文件后，水产新品种引发极大的关注度。水产新品种由农业部公告，具有品种登记号，是科技创新成果和振兴民族水产种业的核心竞争力，是未来保障水产养殖提质增效的关键，根据育种技术手段和来源不同分为选育种、杂交种、引进种和其他类品种。现阶段我国的水产新品种研发虽然具备一定的政策支持、技术保障、社会需求等利好因素刺激，但是依然面临种质资源衰退、重要种源依赖进口、研发体系尚不成熟、推广能力欠缺等因素限制，未来发展道路任重道远（张振东，2015）。

一、我国水产新品种研发基本情况

（一）发展历程

虽然我国水产养殖业发展历史悠久，可以追溯到商周时期，距今约有3000多年，但是品种研发工作从20世纪60年代才开始，起初只是对部分养殖品种进行有意识的人工改良。发展历程大致分为三个阶段：第一阶段，1957年以前，粗放生产苗种阶段。苗种供应主要依靠天然水域采捕培育，1957年育苗采捕量达到234亿尾。第二阶段，突破人工繁殖技术阶段。1958年我国突破"四大家鱼"人工繁殖后，大批野生种和驯养种的人工育苗技术被突破。与此同时，从国外进行引种试养也形成了一定的产业规模。第三阶段，培育研发起步阶段。随着水产养殖业的迅猛发展，20世纪80年代末90年代初由于养殖品种退化而引起的养殖病害频发等问题凸显，迫切需求研发水产新品种。1991国家成立了全国水产原种和良种审定委员会，对水产新品种的审定也正式提上日程。20世纪90年代，共育成兴国红鲤、荷包红鲤、彭泽鲫等多个水产新品种。从此，我国水产新品种研发工作拉开帷幕。

（二）育种研发技术

目前我国水产育种研发技术主要包括选择育种技术、杂交育种技术、细胞工程育种技术和分子育种技术。研究最多、使用最广泛的是群体选择技术。在2000年以后，我国引进并建立了多性状复合技术，迅速在多个养殖品种推广应用。该技术大规模建立家系，对家系和个体标记识别，利用个体本身、同胞、祖先和后代等系谱和测定信息，通过约束极大似然法和最佳线性无偏预测法进行遗传评定，依据综合选择指数选择留种亲本，严格地在家系和个体水平上选种和配种，解决近亲交配及由此导致的种质退化问题。

杂交育种是充分利用种群间的互补效应，尤其是杂交优势。所谓杂交优势，是指不同种群杂交所产生的杂种，其生活力、生长势和生产性能在一定程度上优于两个亲本种群平均值。在水产养殖中，杂交育种有两种方式：种内杂交和种间杂交。种内杂交研究的主要目标为生长速度和抗逆性，目标为培育出高产且抗逆性强的新品种。种间杂交可以用来提高生长速度、调节性别比例、培育不育群体、改善肉质、增强抗病力、增强对极端环境的适应能力、改善重要经济性状等。

细胞工程育种技术和分子育种技术是相对新颖的技术。细胞工程育种技术是在细胞和染色体水平上进行遗传操作改良品种的育种技术，是近年来的研究热点之一。水产生物的细胞工程育种技术研究，目前主要有多倍体育种、雌核发育、单性苗种培育、细胞融合及核质杂交（核移植）等。在分子生物学技术方面，已经对分子辅助育种技术、全基因组育种技术等进行探索和研究，初步建立了相关的技术体系，目前水产养殖生物分子标记的研究多是通过筛选和目标性状相关联的各种分子标记，从而达到间接辅助选育研究的目的。

（三）研发成果

据统计，目前我国已经报道的水产生物有几千种，其中可以养殖的水产种类有170多种，主要养殖经济品种40多种，主要包括"四大家鱼"、鲤、鲫、鲂、鳜、罗非鱼、鳗鲡、鲷、乌鳢、鲈、鲶、黄颡鱼、牙鲆、石斑鱼、鮰、大黄鱼、鲑鳟鱼、大菱鲆、鲟鱼、鲍、对虾、小龙虾（克氏原螯虾）、扇贝、牡蛎、蛤、珍珠贝、鲍鱼、海带、紫菜、裙带菜、海参、海胆、中华鳖等。这些都是目前我国水产新品种研发的主要对象。

截至2015年，国家农业部公告的水产新品种有156个，除了30个引进种外，自主培育的水产新品种有126个，其中选育种76个，占49%；杂交种45个，占29%；其他类5个占3%。除了草鱼意外，重要的养殖种类基本实现新品种的突破。鱼类87个，占56%；虾类14个，占9%；蟹类5个，占3%；贝类21个，占13.5%；藻类21个，占13.5%；其他种类8个，占5%。主要表现出以下特点：（1）是年均育成数量逐渐增多。2000年以前累计完成40个，2001年至2005年完成21个，2006年至2010年完成39个，2011年至2014年完成56个。近年来贝类与藻类品种数量增加明显。42个贝类和藻类新品种中，25个是近4年完成的。(2) 是部分品种优势明显。鲤鱼、鲫鱼、罗非鱼和海带新品种较多。鲤鱼登记引进种4个，自主培育了24个；鲫鱼无登记引进种，自主培育了12个；罗非鱼登记引进种3个，自主培育了8个；海带无登记引进种，自主培育了10个。有5个及以上水产新品种的还有南美白对虾。另外，系列品种不断推出，如吉富罗非鱼—新吉富罗非鱼—吉丽罗非鱼，建鲤—津新鲤—津新鲤2号，"中科红"海湾扇贝—海湾扇贝"中科2号"，"蓬莱红"扇贝—栉孔扇贝"蓬莱红2号"，海带"东方1号"—"东方2号"—"东方3号"—"东方6号"—"东方7号",坛紫菜"申福1号"—"申福2号";条斑紫菜"苏通1号"—"苏通2号";裙带菜"海宝1号"—"海宝2号"等。(3)是研发主体多元化。统计发现,新品种(包括引进种)第一培育单位基本涵盖了我国水产行业的主要研究机构,包括科研院所、水产类大专院校、水产推广机构、水产类企业等,分别为80个、40个、7个、29个。研发主体主要以科研院所与大专院校为主,近年来企业参与程度增加。1家单位独立培育的品种有93个,占60%;2家及以上单位联合培育的品种有63个,占40%;3家及以上单位联合培育的品种14个,不足10%。

(四) 成果推广

目前我国的水产新品种推广主体基本以培育单位为核心,以市场需求为导向。培育单位直接或间接将新品种苗种销售给养殖户。养殖效果良好的品种会受到养殖户的青睐,进而刺激苗种销量增加,推广效应逐步扩大。建鲤、彭泽鲫、团头鲂"浦江1号"、中国对虾"黄海1号"等一批品种经养殖效果检验,推广效果良好,其知名度不断提高,占据一定的市场,生产贡献不断提高。农业部每年会发布主推品种技术,引导各地从政策层面进一步支持推广可能具备一定潜力的品种。据调研了解,仍然存在一部分品种,因为培育单位自身推广能力限制或者品种特异性等因素,获得新品种证书后即被束之高阁,无法发挥生产效应。

二、澳洲龙纹斑引种与开发研究现状

优良的养殖品种是养殖者所期盼的,它不仅给养殖者带来好的经济效益,更为人们提供了优质美味的动物蛋白。过去十年里,有关水产养殖单位先后自澳大利亚引进了养殖新品种,其中有的养殖种类已取得了阶段性的成果,如人们熟知的澳洲宝石斑,有的仍处于试养阶段,有的养殖效果不太好。曹凯德于2002年4月下旬至5月上旬应澳大利亚水产养殖公司的邀请赴维多利亚省的Euroa镇对该公司主养澳洲龙纹斑进行了考察(曹凯德,2001)。

通过实地考察,总的印象是澳洲龙纹斑对水温的耐受范围大,对那里所有的内陆水域均适应,饲料转化率为1.05,比较适宜集约化养殖。国内已有几家引养了澳洲龙纹斑,如青岛晟华种苗公司,江苏省淡水水产研究所等。实践说明此鱼可以在我国开展规模养殖,应特别注意早期投喂人工饲料的驯化及光照的控制。澳洲龙纹斑的营养价值、肉质胜过乌鳢,其经济价值远远超出乌鳢,在有些方面也超过了鳜鱼。如果在引养成功的基础上尽快解决其在国内的人工繁殖,做到苗种自给,毋庸置疑澳洲龙纹斑在我国是一种有发展前景的优良养殖新品种，目前已初见端倪（韩茂森，2003）。

澳洲龙纹斑是澳洲迄今最大的淡水鱼类，在世界淡水鱼类行列中也是最大的。1990年以来，有关水产研究所开展了澳洲龙纹斑人工繁殖的研究，目的在于恢复因受水利工程设施及环境污染等影响而衰退的鱼类资源。目前已能够在人工控制的环境中产卵孵化，一部分苗放流于自然水域，另一部分苗作为人工养殖的苗源，也有部分苗出口国外。澳洲龙纹斑向港澳台地区及东南亚出口（韩茂森，2003）。

1999年，我国台湾首次从澳洲引进澳洲龙纹斑鱼苗养殖，首批引进台湾的一寸鱼苗为数4500尾，当时向澳洲龙纹斑繁殖业者购买的价格是NT＄150元/尾。2001年3月梁忠宝再度引进1500尾寸苗，这次鱼苗价格是每尾12美元。1~2kg的商品鱼在澳洲方面开出的池边价是每尾台币3~5万元，3~5kg的成鱼每对要价更高达50万台币。大陆自2001年从澳洲引进，经过多年的研究，近几年取得了澳洲龙纹斑规模化繁养殖技术的突破，其养殖得以蓬勃发展（郑石勤，2011）。

澳洲龙纹斑为白肉鱼，在澳洲排名四大经济鱼类之首，是世界上最大的淡水鱼之一，与宝石鲈、黄金鲈、银鲈等原产于澳洲的淡水鱼并列为澳洲鱼。由于澳洲龙纹斑有特殊的外型以及鲜美口感，在澳大利亚素有"国宝鱼"美称。澳洲龙纹斑于1999首次由梁忠宝等从澳洲引进台湾，2001年由青岛现代农业开发中心与澳大利亚水产公司合作引入2300条苗种，随后引进浙江、江苏、广东、福建等地，并人工驯化，繁殖成功。2015年，福建省农业科学院中心实验室科技人员在邵武和福清基地，成功培育亲鱼50对，繁育水花25万尾，鱼苗成活率达89.6%，为澳洲龙纹斑种苗培育积累了丰富的经验，为其推广奠定了坚实的基础（翁伯琦，2016）。

澳洲龙纹斑在烹饪过程中皮易剥，刺易去，在澳大利亚及其周边国家和地区常以澳洲龙纹斑烤制佳肴或制成鱼排招待宾客。澳洲龙纹斑肉质白嫩，去皮可加工成鱼排和鱼片。主要市场有日本、台湾、新加坡、香港等国家和地区。其价格在墨尔本为20美元/kg。澳大利亚鲜活澳洲龙纹斑的出塘价在20澳元/kg左右，折合人民币约130元/kg，1~3kg的澳洲龙纹斑在市场上最受欢迎，价格也最高。澳大利亚产地5~6公分规格澳洲龙纹斑鱼苗价格约人民币7元/尾，但是东莞生产的同样规格鱼苗仅需人民币5元/尾。澳洲龙纹斑塘头价约为40元/斤。澳洲龙纹斑目前市场批发价格在60元/斤以上。澳洲龙纹斑商品鱼养殖阶段，饲料转化率为1.2~1.3，利润空间较大（曹凯德，2001；韩茂森，2003；李娴，2013；吕华当，2017；欧小华，2017）。

澳洲龙纹斑是澳洲迄今为止最大的淡水鱼类，在世界淡水鱼类行列中也是最大的。因其具有生长速度快、抗病能力强、对水温的耐受范围较大、人工配合饲料转化率高等特点，适合不同养殖区域集约化养殖。同时，研究结果表明，澳洲龙纹斑具有营养丰富（高蛋白、低脂肪、富含微量元素和人体必须的氨基酸和不饱和脂肪酸）、味道鲜美等优点，其经济价值远超我国常规养殖品种，比较迎合人们日益增长的饮食需求，澳洲龙纹斑具有成为我国淡水养殖名贵经济鱼类的开发潜力。如果在国内进一步完善和推广澳洲龙纹斑的人工繁殖技术，作为一种新的名贵鱼养殖品种的澳洲龙纹斑在引种鱼类中将有较大的发展空间（郭松，2012；宋理平，2013）。

三、澳洲龙纹斑养殖开发研究现状

澳洲龙纹斑经济价值高，对水温适应范围较广，7℃仍有摄食，对水质要求不高。驯化后可摄食人工配合饲料，生长速度快。2年即可上市。通过养殖实验证明可以开展规模化养殖目前国内已有部分单位开始引进养殖，养殖技术成熟后有很好的推广前景（李娴，2013）。

1998年时梁忠宝看到美国CNN报导澳洲龙纹斑，决定请做水产贸易的朋友韩先生引进种苗。韩先生找到澳洲的朋友，透过管道成功的取得了出口执照。澳洲龙纹斑在1999年1月底引进台湾，首批引进台湾的澳洲龙纹斑一寸鱼苗为数4500尾，当时向澳洲澳洲龙纹斑繁殖业者购买的价格是NT＄150元/尾。第一批鱼苗由5个人均分，但是因为不熟悉这条鱼的生态习性，大部分都已经折损掉了。梁忠宝当时分得约1000尾，养殖三年也只剩80多尾。后来这些鱼在2004年移到南投，当年又不幸遇到七二水灾，被冲走大半，最后只剩28尾。在出口执照期限到期前，2001年3月梁忠宝再度引进1500尾寸苗，这次鱼苗价格是每尾12美元。梁忠宝说澳洲龙纹斑属政府列管鱼种，必须取得澳洲政府官方核准，业者才能出口。由于非常看好澳洲龙纹斑的未来市场，梁忠宝将手中的鱼苗小心翼翼的进行饲养和选种，因为还不熟悉养殖的要件，期间折损了许多鱼。从1999年至今，梁忠宝只有出售过很少量的鱼，大部分都留在自场里，进行严格的筛选，只把体型好的、成长快的留下来进行培育，对质量相当坚持的梁忠宝，为了建立种原，把差的都淘汰掉了。只有体质硬、线条顺畅、成长速度快的才挑选出来留下当种鱼，为的是建立澳洲龙纹斑的优良种原场。还好突破了重重难关，亲代繁殖的一代与二代已达15000尾之多。十多年来，梁忠宝一直只有投入没有回收，现在他已经建构足够的种鱼数量，预计在2013年终于可以开始出苗了（郑石勤，2011）。

曾经与梁忠宝是合作伙伴的江天赐，目前也在进行澳洲龙纹斑的繁殖与养殖，养殖场位在宜兰壮围乡滨海路旁。江天赐接受采访时表示，对于澳洲龙纹斑的体型、肉质以及未来的发展性也相当看好，可以媲美龙胆石斑，只不过澳洲龙纹斑的养殖和繁殖的难度更高。江天赐说进入澳洲龙纹斑养殖行列的门坎很高，因为它有很特殊的生态习性。江天赐已经开始出售澳洲龙纹斑鱼苗了，每年有数万尾的量可以供应，寸苗售价台币100元/尾。大陆市场也已开始拓展，他说这条鱼的成长快及肉质细，与龙胆石斑不相上下（郑石勤，2011）。

数据显示，大陆也在2001年间自澳洲引进澳洲龙纹斑，但至今尚未有大批成鱼上市或进行人工繁殖的报告。根据最近的大陆报载，厦门东晟水族研发有限公司在2011年再度自澳洲引种，并首度养殖成功，于2011年6月12日参加第五届海峡两岸（厦门）农渔业交流暨产业对接会，开始在厦门推广养殖。东晟水族负责人林东晟表示，在厦门餐厅要花人民币2000多元才能吃到澳洲龙纹斑，不过目前想花钱也未必能吃到。因为澳洲龙纹斑要7年才会转性，也就是要经过7年才能辨别出种鱼的雌雄，然后才能做进一步的繁育研究。当时好不容易从国外引回来的种鱼，在研究过程中死了很多，后来不得不多次再进再养。从成本上来说，引进一条澳洲龙纹斑要好几万，所以基本都是来研究的，很少拿来吃。目前澳洲龙纹斑已经由养殖试验阶段进入了人工繁殖育种、生态养殖阶段。澳洲龙纹斑若能全面推广养殖，3年内预计可带动农户1500~2000户（郑石勤，2011）。

张善发（2009）等人将澳洲龙纹斑引进乐清市港龙渔业有限公司进行养殖，养殖时间为2008年4月12日到10月12日，投放鱼苗27000尾，平均重1.15g/尾，总重31kg，6个月后剩余鱼苗24354尾，成活率90.2%，总重600kg，平均每尾重24.6g。其中45g左右的鱼有7568尾、占总数的31.07%，20g左右的10358尾、占总数的42.53%，20g以下的占26.4%。本次试验鱼的生长速度和成活率相对于澳大利亚工厂化养殖模式下偏低，主要是由于对澳洲鳕鲈的认识还有一定的局限性，包括对其前期饵料的优化、后期人工颗粒饲料的选择，以及光线的控制和疾病的预防都有待于进一步的探索。本试验经过引种、运输、饵料的选择、驯养等6个月时间，对澳洲龙纹斑的生活习性、日常管理、疾病预防及摄食生长状况积累了一定的经验。在澳洲龙纹斑鲜美的味道和昂贵的市场价格诱惑，澳洲龙纹斑的养殖前景比较广阔（张善发，2009）。

记者近日获悉，广东东莞安业水族养殖有限公司（以下简称安业水族）全人工繁育养殖澳洲龙纹斑获得成功。据安业水族董事长罗汝安介绍，公司早在2011年就引进澳洲龙纹斑，并开展了全人工繁育及养殖试验。2013年春季，培育的一百多尾亲鱼便逐渐成熟并少量产卵，今年春季亲鱼开始大量繁殖。截至今年4月底，安业水族已成功繁育出体长4~10cm的种苗六万多尾，罗汝安估计今年一共能培育出8万~10万尾的澳洲龙纹斑鱼苗。澳洲龙纹斑全人工繁育养殖在该市首获成功。这标志东莞市在研究推广应用名优特稀新品种上又取得了新成果,这对于东莞调整优化养殖品种结构,提升产业科技创新水平,加快都市现代渔业发展具有重要的意义(刘怡,2014;中国农业新闻网,2014)。

青岛现代农业开放中心于2001年6月从澳大利亚引进澳洲龙纹斑养殖,并特聘青岛市渔业技术推广站曹金祥研究员、烟台大学海洋学院韩茂森教授为技术顾问,经6个月的饲养,鱼的长势良好,两位专家在中心领导及年轻助手的通力合作下,结合生产潜心研究,始终坚持走健康养殖可持续发展之路,让人们及早品尝到名贵新品种(农中侃,2002)。

澳洲龙纹斑是澳大利亚南部墨瑞河水域具有很高营养价值和经济价值的游钓、商业性生产和资源增殖的地方名贵品种,有澳大利亚"国宝鱼"的美称。澳洲龙纹斑生长速度相对较快,比较适合在我国进行养殖,因此浙江省乐清市水产科学研究所于2008年4月份将澳洲龙纹斑引进并驯化养殖。通过1年多的养殖时间,初步掌握了该引进品种在我市淡水水域的养殖生产技术,为将来澳洲龙纹斑在国内推广养殖奠定了基础。从2008年4月13日开始养殖算起,共投入规格为1.15g/尾的澳洲龙纹斑鱼苗总数27000尾。养至2009年4月17日的计算结果为,收获规格为62.9g/尾的澳洲鳕鲈鱼苗总数17516尾,养殖成活率为64.87%。本实验通过引种、运输、饵料的选择、驯养等1年的养殖时间,对澳洲龙纹斑的生活习性,日常管理、疾病预防及摄食生长状况积累了一定的经验,独创性地解决了一些养殖中所存在的关键性技术难题,基本掌握了澳洲龙纹斑在我国水泥池养殖模式技术,为将来澳洲龙纹斑在国内推广养殖奠定了基础。本次驯养,在鱼生长速度和成活率上,相对于澳大利亚工厂化下养殖模式的偏低,主要由于对于澳洲龙纹斑饲养技术的掌握还不够,包括对其前期的饵料的优化,后期人工颗粒饲料的选择,以及光线的控制和疾病的预防都有待于进一步的探索(蔡乘成,2012)。

据吕华当(2017)报道,以后中国养殖户要养殖澳洲龙纹斑,再也不用漂洋过海进口鱼苗了,在广东省东莞市就能买到本地繁育的澳洲龙纹斑鱼苗!"鱼苗100%中国繁育,更加适合中国环境养殖。"广东省东莞市安业水族养殖有限公司总经理罗汝安自豪地告诉《海洋与渔业》记者。澳大利亚名优鱼种澳洲龙纹斑闻名于世,中国也有一些企业引进养殖,但是由于进口鱼苗成本高,路途中也会有所损伤。因此,国内也有一些企业在探索人工苗种繁育研究。去年,罗汝安花费数百万元从澳大利亚引进成熟的苗种繁育技术,按照现有亲鱼数量,未来年产鱼苗数量可达200万尾(吕华当,2017)。

(罗钦、李冬梅、张斌)

第二节 研究论文摘编

141. The Australian Murray cod-a new candidate for intensive production systems①.

［英文摘要］：More recently there has been considerable interest (from both producers and markets) in the grow-out of Murray cod to satisfy a significant domestic and export demand for human consumption (Ingram et al. 2005a). Farming Murray cod to produce high-quality table fish for domestic and export markets is a rapidly emerging agri-business sector in south eastern Australia.

［译文］：

澳洲墨瑞鳕鱼-集约化养殖系统的新品种

最近，对墨瑞鳕鱼养殖出现了相当大的兴趣（来自生产者和市场），为满足国内外对墨瑞鳕鱼消费的巨大需求。为国内外市场的高品质餐桌鱼类生产墨瑞鳕鱼是澳洲东南部迅速兴起的农业产业。

参考文献：Ingram, B. A., De Silva, S. S., Gooley, G. J. 2005. The australian murray cod-a new candidate for intensive production systems [J]. World Aquaculture (3), 37-43, 69.

（罗钦、李冬梅、黄惠珍 译）

142. Finfish Mariculture in Inland Australia: A Review of Potential Water Sources, Species, and Production Systems

［英文摘要］：Secondary salinization has rendered over 100 million hectares of arable land throughout the world, and over 5 million hectares in Australia, unsuitable for conventional agriculture. The use of this land and associated water for mariculture is an adaptive approach to this environmental problem with many potential economic, social, and environmental benefits. In this article, we review three key technical aspects for the development of a finfish mariculture industry in inland Australia, namely the potential sources of water, the species suitable for culture, and the production systems available to produce them. Based on factors such as their quality, quantity, and proximity to infrastructure, the most appropriate water sources are

---

① 研究论文摘编序号续上一章，全书同。

groundwater obtained from interception schemes and waters from operational or disused mines. Pond-based culture methods using these water sources have many specific advantages; however, few species can thrive in the wide range of seasonal water temperatures experienced within the temperate regions where secondary salinity is most abundant. Recirculating aquaculture systems (RAS) would enable production of more species in a greater number of water sources; however, the benefits typically associated with RAS are not as apparent in rural areas. Mulloway, Argyrosomus japonicus, are a temperate species that appear to have many positive attributes for inland mariculture; however, more data are required on their rate of growth across a wider range of temperatures. Seasonal production of barramundi, Lates calcarifer, in ponds has been demonstrated in the temperate climatic zones but may be more appropriate for those salinized water sources located in the warmer parts of the country. Freshwater species, such as silver perch, Bidyanus bidyanus, and Murray cod, Maccullochella peelii peelii, are likely to be suitable for low-salinity waters. Rainbow trout, Oncorhynchus mykiss, also have excellent potential provided water temperature can be maintained below the upper lethal limit.

[译文]：

澳洲内陆长须鲸海水养殖：潜在的水源，物种和生产系统的评估

次生盐碱化已经渲染了全球超过1亿公顷的耕地，澳洲就有500多万公顷，不适合常规农业生产。这片土地的利用和相应的水用于海水养殖是一种适合的解决这种环境问题的方法，有许多潜在的经济、社会和环境效益。本文中，我们评估了澳洲内陆长须鲸海水养殖业发展的三个关键技术点，即潜在的水源，适合的养殖品种，以及可用来生产它们的生产系统。根据其质量、数量和临近基础设施等因素，最适合的水源是横断层的地下水和从运营或废弃矿井中获得的水。使用这些水源进行池塘养殖的方法有很多特别的优点；然而，在次生盐碱化最丰富的温带地区，季节性的水温变化范围经常很广，很少鱼类可以繁荣发展。循环水养殖系统（RAS）能在更多种水源中养殖更多种的鱼类；然而，RAS的收益在农村并不明显。石首鱼，日本银身鰔，是一种温带的鱼类，似乎对内陆海水养殖有很多积极的属性；然而，在更广的温度范围内，需要收集更多关于其生长速率的数据。在温带气候地区已经证明了可以在池塘中季节性养殖尖吻鲈鱼，但可能更适合那些位于该国较热地区的盐化水源。淡水鱼类，如澳洲银鲈鱼和墨瑞鳕鱼，都有可能适用于养殖在低盐度的水域。只要水温能维持在致死上限以下，虹鳟鱼也有很好的潜力。

参考文献：Partridge, G. J., Lymbery, A. J., George, R. J. 2008. Finfish mariculture in inland australia: a review of potential water sources, species, and production systems [J]. Journal of the World Aquaculture Society, 39 (3), 291-310.

（罗钦、李冬梅、黄惠珍　译）

143. Movements and habitat use by the endangered Australian freshwater Mary River cod, Maccullochella peelii mariensis

[英文摘要]：The decline of the endangered Mary River cod, Maccullochella peelii mariensis, of Queensland, Australia, has been attributed to anthropomorphic habitat alterations, however, the relationships between this subspecies and its physical environment are poorly understood. We used radiotelemetry to investigate the movements and use of habitats by nine Mary River cod (total length 420~760mm; weight 1.3~5.5kg) in the Mary River system over a 20 month period. The mean distance moved each month was positively correlated with monthly stream discharge, but the direction of movements was unpredictable. Patterns of movement varied considerably among individuals and appeared to be unrelated to size. Cod established home ranges of between 70 and 820min length. Five of the tagged cod returned to a previous home range after moving more than 10 km, whereas three did not move more than 2 km from their capture location for the duration of the study. Cod used large woody debris complexes more thanany other type of habitat, and rarely used areas of open water. Managers should give high priority to maintenance of fish passage and protection/rehabilitation of large woody debris habitats if cod populations are to recover.

[译文]：

澳大利亚濒临灭绝的淡水墨瑞鳕鱼的迁徙和栖息地使用

澳大利亚昆士兰州濒临灭绝的墨瑞鳕鱼的衰退归因于拟人栖息地的变化，然而，这个亚种与其物理环境之间的关系目前知之甚少。我们使用无线电遥测技术，20个月内在玛丽河系统中调查了九条玛丽河鳕鱼（总长度420~760mm；体重1.3~5.5kg）的栖息地活动和使用情况。每月移动的平均距离与河流月流量正相关，但移动方向是不可预测的。个体之间的移动方式差异很大，似乎与体型无关。鳕鱼的移动范围在70~820m。在移动超过10km后，五条标记鳕鱼返回之前的生活范围，而在研究期间，三条没有从其捕获位置移动超过2km。相比其他类型的栖息地，鳕鱼更多地活动在有大型木质碎屑混合物的栖息地，而且很少活动在露天水域。如果要复原鳕鱼种群，管理者应高度重视维护鱼道和保护/恢复大型木质碎片栖息地。

参考文献：Simpson, R. R., Mapleston, A. J. 2002. Movements and habitat use by the endangered australian freshwater mary river cod, maccullochella peelii mariensis [J]. Environmental Biology of Fishes, 65 (4), 401-410.

144. A piecewise regression approach for determining biologically relevant hydraulic thresholds for the protection of fishes at river infrastructure

[英文摘要]：A piecewise regression approach was used to objectively quantify barotrauma injury thresholds in two physoclistous species, Murray cod Maccullochella peelii and silver perch Bidyanus bidyanus, following simulated infrastructure passage in a barometric chamber. The probability of injuries such as swimbladder rupture, exophthalmia and haemorrhage, and emphysema in various organs increased as the ratio between the lowest exposure pressure and the acclimation pressure (ratio of pressure change, $R_{(NE:A)}$) reduced. The relationship was typically non-linear and piecewise regression was able to quantify thresholds in $R_{(NE:A)}$ that once exceeded resulted in a substantial increase in barotrauma injury. Thresholds differed among injury types and between species but by applying a multispecies precautionary principle, the maintenance of exposure pressures at river infrastructure above 70% of acclimation pressure ($R_{(NE:A)}$ of 0.7) should protect downstream migrating juveniles of these two physoclistous species sufficiently. These findings have important implications for determining the risk posed by current infrastructures and informing the design and operation of new ones.

[译文]：

用确定生物相关水力阈值的分段回归方法，来保护河流基础设施中的鱼类

采用分段回归的方法，客观量化了气压室内模拟基础设施通道后，两种闭鳔物种墨瑞鳕鱼和银鲈的气压损伤阈值。随着最低暴露压力和驯化压力（压力变化率 $R_{(NE:A)}$）之间的比率降低，诸如鱼鳔破裂、突眼和出血以及各器官中的肺气肿等伤害的概率增加。这种关系通常是非线性的，分段回归能够量化 $R_{(NE:A)}$ 的阈值，一旦超过，就会导致气压伤损伤的显着增加。临界值在伤害类型和物种之间有所不同，但通过应用多种预防原则，河流基础设施暴露压力维持在适应压力70%以上（$R_{(NE:A)}$ 为0.7）应该充分保护下游这两种闭鳔物种的迁徙幼体。这些发现对于确定当前基础设施造成的风险并为新设计和运行提供信息具有重要的意义。

参考文献：Boys, C. A., Robinson, W., Miller, B., et al. 2016. A piecewise regression approach for determining biologically relevant hydraulic thresholds for the protection of fishes at river infrastructure [J]. Journal of Fish Biology, 88, 1677-1692.

(罗钦、李冬梅、黄惠珍 译)

145. Comparative movements of four large fish species in a lowland river

[英文摘要]: A multi-year radio-telemetry data set was used to comparatively examine the concurrent movements of the adults of three large-bodied Australian native freshwater fishes (Murray cod Maccullochella peelii, trout cod Maccullochella macquariensis and golden perch Macquaria ambigua) and the introduced carp Cyprinus carpio. The study was conducted over a reach scale in the regulated Murray River in south-eastern Australia. Differences were identified in the movements among these species. The predominant behaviour was the use of small movements (1 km) did occur, the frequency varied considerably among species. Large-scale movements were least evident for M. macquariensis and more common for M. ambigua and C. carpio with these two species also having a greater propensity to change locations. Macquaria ambigua displayed the largest movements and more M. ambigua moved on a 'continual' basis. Although a degree of site fidelity was evident for all species, the highest levels were exhibited by M. macquariensis and M. peelii. Homing was also evident to some degree in all species, but was greatest for M. peelii.

[译文]:

比较四种大型鱼类在低地河流中的活动

多年无线电遥测数据集被用于比较研究三种体型较大的成年澳大利亚本土淡水鱼（墨瑞鳕鱼，圆尾麦氏鲈和黄金鲈鱼）和引进鲤鱼的并发运动。这项研究在澳大利亚东南部受调节的墨累河上达到了一个达标范围。在这些物种间的运动中发现差异。主要行为是在使用小的运动（1km）中确定发生，频率在物种间差异很大。大型活动对圆尾麦氏鲈而言最不明显，对黄金鲈鱼和引进鲤鱼而言更为常见，因为这两物种也更倾向于改变地点。圆尾麦氏鲈表现出最大的运动，更多圆尾麦氏鲈在"持续"的基础上移动。尽管所有物种的遗传保真度都很明显，但最高水平由圆尾麦氏鲈和墨瑞鳕鱼展现。所有物种中归航在一定程度上也是显而易见的，但是墨瑞鳕鱼的程度最大。

参考文献：Koehn, J. D., Nicol, S. J. 2016. Comparative movements of four large fish species in a lowland river [J]. Journal of Fish Biology, 88 (4), 1350-1368.

(罗钦、李冬梅、黄惠珍 译)

146. Efficiency of electrofishing in turbid lowland rivers: implications for measuring temporal change in fish populations

[英文摘要]: To quantify how electrofishing capture probability varies over time and across physiochemical and disturbance gradients in a turbid lowland river, we tagged between 68 and 95 fish·year^{-1} with radio transmitters and up to 424 fish·year^{-1} with external and passive integrated transponder (PIT) tags. We surveyed the site noninvasively using radiotelemetry to determine which of the radio-tagged fish were present (effectively closing the radio-tagged population to emigration) and then electrofished to estimate the proportion of available fish that were captured based on both this and standard mark-recapture methods. We replicated the electrofishing surveys three times over a minimum of 12 days each year, for 7 years. Electrofishing capture probability varied between 0.020 and 0.310 over the 7 years and between four different large-bodied species (Murray cod (Maccullochella peelii), trout cod (Maccullochella macquariensis), golden perch (Macquaria ambigua ambigua), and silver perch (Bidyanus bidyanus)). River turbidity associated with increased river discharge negatively influenced capture probability. Increasing fish length increased detection of fish up to 500 mm for Murray cod, after which capture probability decreased. Variation in capture probability in large lowland rivers results in additional uncertainty when estimating population size or relative abundance. Research and monitoring programs using fish as an indicator should incorporate strategies to lessen potential error that might result from changes in capture probabilities.

[译文]：

在浑浊的河流下游中电捕鱼的效率：衡量鱼类种群时间变化的影响

为了量化电捕获概率随着时间的推移以及浑浊的低地河流的物理化学和干扰梯度的变化，我们用无线电发射器每年标记了68~95条鱼，最高达到一年424条鱼具有外部和无源集成转发器标签。我们使用无线电遥测技术非侵入性地对现场进行了无线电遥测，以确定哪些放射性标记的鱼存在（有效地关闭带有放射性标记鱼群以便进行移出），然后进行电捕捞，最终估计出那些可捕获鱼（基于这一标准和标准标记-重新捕获方法）的比例。我们在每年至少12d内复制了三次电捕捞调查，为期7年。这7年中，在四种不同的大体型物种（墨瑞鳕鱼、鳟鱼鳕鱼、黄金鲈鱼和澳洲银鲈）之间，电捕鱼捕获概率在0.020和0.310之间变化。随着河流排放而相应增加的河流浊度，负面影响着捕获概率。鱼类长度的增加使墨瑞鳕鱼的鱼类检测量增加到500mm，之后捕获概率降低。在估算种群规模或相对丰富度时，大型低地河流中捕获概率的变化会导致额外的不确定性。使用鱼类作为指标的研究和监测计划应纳入战略，以减少捕获概率变化可能导致的潜在错误。

参考文献：Lyon, J. P., Bird, T., Nicol, S., et al. 2014. Efficiency of electrofishing in turbid lowland rivers: implications for measuring temporal change in fish populations [J].

Canadian Journal of Fisheries & Aquatic Sciences, 71 (6), 878-886 (9).

(罗钦、李冬梅、黄惠珍 译)

147. Recovery of the endangered trout cod, Maccullochella macquariensis: what have we achieved in more than 25 years?

[英文摘要]: Recovery of threatened species is often necessarily a long-term process. The present paper details the progress towards the recovery of trout cod, Maccullochella macquariensis, an iconic, long-lived fish species first listed as threatened in the 1980s. The objectives, actions and progress over three successive national recovery plans (spanning 18 years) are assessed, documenting changes to population distribution and abundance and updating ecological knowledge. Increased knowledge (especially breeding biology and hatchery techniques, movements, habitats and genetics) has greatly influenced recovery actions and the use of a population model was developed to assist with management options and stocking regimes. Key recovery actions include stocking of hatchery-produced fish to establish new populations, regulations on angling (including closures), education (particularly identification from the closely related Murray cod, M. peelii) and habitat rehabilitation (especially re-instatement of structural woody habitats). In particular, the establishment of new populations using hatchery stocking has been a successful action. The importance of a coordinated long-term approach is emphasised and, although there is uncertainty in ongoing resourcing of the recovery program, much has been achieved and there is cautious optimism for the future of this species.

[译文]:

濒危的鳟鱼鳕鱼恢复：在 25 年内我们已经取得了什么成就？

濒危物种的恢复往往是一个长期过程。本文详细介绍了鳟鱼鳕鱼的种群复苏进程。鳟鱼鳕鱼是一种在 20 世纪 80 年代首次被列为濒危物种的具有标志性长寿鱼类。评估了连续三次国家复兴计划（跨越 18 年）的目标，行动和进展，记录了种群分布和丰富度的变化，更新了生态知识。增加知识（尤其是繁殖生物学和孵化技术，种群活动，栖息地和遗传学）极大地影响了复苏行动，并开发利用种群模型来协助管理种群选择和储备制度。关键的复苏行动包括孵化场生产的鱼储备以建立新种群，捕鱼规定（包括休渔），教学（特别是有密切关系的墨瑞鳕鱼的鉴别）和栖息地恢复（尤其是结构木质栖息地的重新安置）。特别是使用孵化场储备建立新种群是一项成功的行动。它强调了长期协调一致方法的重要性，虽然复苏计划的资源不断存在不确定性，但已取得很多成就，并对该物种的未来持谨慎乐观的态度。

参考文献：Koehn, J. D., Lintermans, M., Lyon, J. P., *et al.* 2013. Recovery of the endangered trout cod, maccullochella macquariensis: what have we achieved in more than 25 years? [J]. Marine & Freshwater Research, 64 (9), 822-837.

（罗钦、李冬梅、黄惠珍　译）

148. Refining the activity component of a juvenile fish bioenergetics model to account for swimming costs

[英文摘要]：We develop a swimming costs model that accounts for the influence of flow velocity and body weight on the net active metabolic rate of Murray cod (Maccullochella peelii). Laboratory trials indicated that swimming costs increased with flow velocity (exponent = 2.36) and declined allometrically with body weight (exponent = -0.27). The newly derived swimming costs model provided a more dynamic estimate of Murray cod energy consumption, which explained 74% of variation in the swimming costs. This new model was compared to traditional bioenergetics models (fixed proportion and optimal swimming speed) to determine swimming costs in a variable temperature (6.4~26.1℃) and flow velocity (0.06~0.46 ms^{-1}) regime downstream of a large hypolimnetic-releasing impoundment on a major Australian river. Incorporating species-specific swimming cost models, such as the one developed here, into bioenergetics modelling allows the exploration of the impact of flow velocity in lotic systems on the growth responses of freshwater fish.

[译文]：

优化幼鱼生物能源模型的活动成份，以考虑游泳成本

我们开发了一个游泳成本模型，来描述流速和体重对墨瑞鳕鱼净活性代谢率的影响。实验室试验表明，游泳成本随着流速（指数=2.36）而增加，并随体重（指数=-0.27）不等式下降。新推出的游泳成本模型提供了对墨瑞鳕鱼能量消耗的更加动态的估计，其解释了74%的游泳成本变化。这个新模型与传统的生物能量模型（固定比例和最佳游泳速度）相比较，以确定在澳大利亚主要河流上大型深层释放蓄水区的具有可变温度（6.4~26.1℃）和流速（0.06~0.46m/s）的下游区内产生的游泳成本。将物种特定的游泳成本模型（如本文开发的模型）纳入生物能量模型，可以探索激流群落体系中流速对淡水鱼生长反应的影响。

参考文献：Nick S. Whiterod, Shaun N. Meredith, Paul Humphries. 2013. Refining the activity component of a juvenile fish bioenergetics model to account for swimming costs [J]. Marine & Freshwater Behaviour & Physiology, 46 (4), 201-210.

(罗钦、李冬梅、黄惠珍 译)

149. Post-release mortality of angled golden perch Macquaria ambigua and Murray cod Maccullochella peelii

[英文摘要]：Short-term (≤4 days) post-release mortalities of two large, culturally and ecologically important Australian freshwater teleosts golden perch Macquaria ambigua (Richardson) and Murray cod Maccullochella peelii (Mitchell) were investigated. There was no angler-induced mortality among golden perch that were immediately released in winter and spring, but 24% of this species and 15% of Murray cod died after delayed release in summer. Significant predictors of mortality were limited to times caught and total length for golden perch, and restraint methods and recovery times for Murray cod, but other parameters were also implicated as cumulative influences. The estimated mortalities may be sufficient to produce population-level impacts for these two long-lived species but could be considerably reduced through revisions of tournament regulations.

[译文]：

被钓的黄金鲈鱼和墨瑞鳕鱼释放后的死亡率

对两种大型有文化和生态重要性的澳大利亚淡水硬骨鱼圆尾麦氏鲈和墨瑞鳕鱼进行了短期（≤4d）释放后死亡率调查。在冬季和春季，立即释放的黄金鲈鱼中没有一种由垂钓引起的死亡，但在夏季延迟释放后，该物种的24%和墨瑞鳕鱼的15%出现死亡。死亡率的显著预测因子仅限于黄金鲈鱼的捕获时间和总长度，以及墨瑞鳕鱼的限制方法和复苏时间，但其他参数也与渐增的影响密切关联。估计的死亡率可能足以对这两类长寿鱼种产生种群水平的影响，但可以通过修改休闲渔业捕捞条例大大减少这种影响。

参考文献：Hall, K. C., Broadhurst, M. K., Butcher, P. A. 2012. Post-release mortality of angled golden perch macquaria ambigua and murray cod maccullochella peelii [J]. Fisheries Management & Ecology, 19 (1), 10-21.

(罗钦、李冬梅、黄惠珍 译)

150. A simulation model to explore the relative value of stock enhancement versus harvest regulations for fishery sustainability

[英文摘要]：Harvest restrictions and stock enhancement are commonly proposed management responses for sustaining degraded fisheries, but comparisons of their relative effectiveness have seldom been considered prior to making policy choices. We built a population model that incorporated both size-dependent harvest restrictions and stock enhancement contributions to explore trade-offs between minimum length limits and stock enhancement for improving population sustainability and fishery metrics (e. g., catch). We used a Murray cod Maccullochella peelii peelii population as a test case, and the model incorporated density-dependent recruitment processes for both hatchery and wild fish. We estimated the spawning potential ratio (SPR) and fishery metrics (e. g., angler catch) across a range of minimum length limits and stocking rates. Model estimates showed that increased minimum length limits were much more effective than stock enhancement for increasing SPR and angler catches in exploited populations, but length limits resulted in reduced harvest. Stocking was predicted to significantly increase total recruitment, population sustainability, and fishery metrics only in systems where natural reproduction had been greatly reduced via habitat loss, fishing mortality was high, or both. If angler fishing effort increased with increased fish abundance from stocking efforts, fishing mortality was predicted to increase and reduce the benefits realized from stocking. The model also indicated that benefits from stock enhancement would be reduced if reproductive efficiency of hatchery-origin fish was compromised. The simulations indicated that stock enhancement was a less effective method to improve fishery sustainability than measures designed to reduce fishing mortality (e. g., length limits).

[译文]：

一种模拟模型，用于探讨可持续性渔业的库存增强与捕获法则的相对价值

捕获限制和库存增加通常是为维持退化渔业提出的管理措施，但在作出政策选择之前，很少考虑比较其相对有效性。我们建立了一个种群模型，包含了捕获大小的限制和增加库存的贡献，以探索最小长度限制与增加种群之间的权衡以改善种群可持续性繁育和渔业指标（例如捕获量）。我们使用墨瑞鳕鱼群体作为测试样本，并且该模型高度依赖孵化场和野生鱼的补充进程。我们在最小长度限制和放养率范围内估计了产卵潜力比（SPR）和渔业指标（如钓鱼者捕获量）。模型估计表明，对于增加SPR和垂钓者捕获量，增加最小长度限制比增加种群更有效，但长度限制会导致捕获减少。预计只有在因丧失栖息地而大大减少自然繁殖，捕捞死亡率很高或两者情况兼有的系统中；放养才能显著提高总补给，种群可持续性和渔业指标。如果垂钓者的捕鱼力随着来自放养形成鱼类丰富度增加而增加，捕捞死亡率预计会增加，并减少从放养中获得的收益。该模型还表明，如果孵化场的鱼的繁殖效率受到影响，则增加放养带来的利益将会减少。模拟结果表明，为提高渔业可持续性，增加放养不是太有效的方法，主要还是降低捕捞死亡率（如长度限制）。

参考文献：Rogers, M. W., Allen, M. S., Brown, P., et al. 2010. A simulation model to explore the relative value of stock enhancement versus harvest regulations for fishery sustainability [J]. Ecological Modelling, 221 (6), 919-926.

（罗钦、李冬梅、黄惠珍　译）

151. Evaluating relative impacts of recreational fishing harvest and discard mortality on Murray cod (Maccullochella peelii peelii)

[英文摘要]：Murray cod Maccullochella peelii peelii support popular recreational fisheries in Australia. Catch and release of Murray cod is common due to regulations and a developing trend of voluntary angler release of harvestable fish, but no previous studies have investigated discard mortality of released fish. We estimated discard mortality by measuring post-angling survival of angler-caught wild Murray cod. Angled fish were monitored for five days after hooking, and overall survival rates of 98% were observed. We explored implications of catch and release as a fishing mortality source by comparing our results to roving creel survey estimates of harvest from six fisheries. We applied the maximum likelihood mortality estimate and upper 95% confidence interval to creel survey estimates of the number of Murray cod that were released in the fishery to estimate the total deaths resulting from catch and release mortality. Estimated ratios of deaths from discard mortality to harvest indicated that high numbers of released fish could contribute as much or more to fishing mortality as harvest in some systems. Future Murray cod research should aim to estimate annual exploitation rates to determine the population-level impacts of fishing mortality, and thus, allow effects of hooking and harvest mortality to be considered in future regulation decisions.

[译文]：

评估对墨瑞鳕鱼的休闲捕捞和丢弃死亡率的相对影响

墨瑞鳕鱼支持澳大利亚流行的休闲渔业。捕捞和释放墨瑞鳕鱼是常见的，由于规则和垂钓者志愿放生的发展趋势，但以前的研究没有调查释放鱼的丢弃死亡率。我们通过测量垂钓者捕获野生墨瑞鳕鱼后的存活率，来估计丢弃物死亡率。在挂钩后钓出的鱼监测5d，并且观察到总体存活率为98%。我们通过将我们的结果与来自六个渔厂的流动鱼笼调查估计出的收成进行比较，探讨了捕捞量和释放量作为捕捞死亡率来源的影响。我们应用最大似然死亡率估计值和上限95%置信区间，对鱼类释放的墨瑞鳕鱼数量进行调查估计，以估计捕捞和释放死亡率造成的总死亡率。从丢弃物死亡率到捕获死亡率的估计比例表明，在某些系统中，大量释放的鱼类可能对捕捞死亡率造成或更大的影响。

未来的墨瑞鳕鱼研究应旨在估计年开采率,以确定捕捞死亡率的种群水平影响,从而在未来的管理决策中考虑挂钩和捕获死亡率的影响。

参考文献：Douglas, J., Brown, P., Hunt, T., et al. 2010. Evaluating relative impacts of recreational fishing harvest and discard mortality on murray cod (maccullochella peelii peelii) [J]. Fisheries Research, 106 (1), 18-21.

（罗钦、李冬梅、黄惠珍 译）

152. An assessment of recreational fishery harvest policies for Murray cod in southeast Australia

[英文摘要]：Murray cod Maccullochella peelii peelii is one of the world's largest freshwater fish and supports popular fisheries in southeast Australia, but no previous modelling efforts have evaluated the effects of fisheries regulations or attempted to develop sustainable harvest policies. We compiled existing population metrics and constructed an age-structured model to evaluate the effects of minimum length limits (MLLs) and fishing mortality rates on Murray cod fisheries. The model incorporated a Beverton and Holt stock recruit curve, age-specific survivorship and vulnerability schedules, and discard (catch and release) mortality for fish caught and released. Output metrics included yield (kg), spawning potential ratio (SPR), total angler catch, total harvest, and the proportion of angler trips that would be influenced by each regulation based on recent creel survey data. The model suggested that annual exploitation (U) should be held to less than 0.15 under the current MLL of 500 mm total length to achieve an SPR>0.3, a target usually considered to prevent recruitment overfishing. Exploitation rates at or exceeding 0.3 would cause SPR values to drop below typical management targets unless the MLL was set at or above 700 mm. Regulations that protected Murray cod from overfishing created higher angler catches and higher catch of trophy fish, but at a cost of reducing the proportion of angler trips resulting in a harvested fish. Expressing model output on a per-angler trip basis may help fishery managers explain regulation trade offs to anglers.

[译文]：

对澳大利亚东南部墨累-鳕鱼休闲渔业收获政策的评估

墨瑞鳕鱼是世界上最大的淡水鱼之一，支持澳大利亚东南部最受欢迎和最支持的渔业，但之前的模拟工作没有评估渔业法规的影响或试图制定可持续捕获政策。我们编制了现有的种群指标，并构建了一个年龄结构模型来评估最小长度限制（MLLs）和捕捞死亡率对墨瑞鳕鱼渔业的影响。该模型为捕获和释放的鱼类引进贝弗顿-霍尔特新增曲线，特定年龄组的存活率和发育危险期，以及丢弃（捕获和释放）的死亡率。产量指标包括产量（kg），产卵潜力比（SPR），钓鱼总捕获量，总收获量以及基于最近鱼篓捕捞调查数据且受调控影响的钓鱼者出游比例。该模型表明，在目前500mm总长MLL下，年度开采量（U）应保持在0.15以下，以达到SPR>0.3，这是通常被认为可以防止过度捕捞的标准。如果开采率等于或大于0.3，将导致SPR值下降到低于典型管理目标值，除非MLL设定在700mm或以上。保护墨瑞鳕鱼免受过度捕捞的法规，创造了更高的垂钓者捕获量和更高的捕捞量，但是以降低捕捞鱼类的垂钓者出游比例为代价。基于每一位钓鱼者之旅的表达模型的输出可能有助于渔业管理者向钓鱼者解释监管制度。

参考文献：Allen, M. S., Brown, P., Douglas, J., *et al.* 2009. An assessment of recreational fishery harvest policies for murray cod in southeast australia [J]. Fisheries Research, 95 (2), 260-267.

（罗钦、李冬梅、黄惠珍　译）

153. Behavioral responses of Murray cod Maccullochella peelii peelii to pulse frequency and pulse width from electric fishing machines

[英文摘要]：The effective electrical conductivity (C_f) of Murray cod Maccullochella peelii peelii was evaluated as part of an investigation into effective electric fishing settings for this important species. We describe an evaluation of minimum thresholds of power applied to an electrical field required to elicit four responses (escape, forced swimming, immobilization, and narcosis) from Murray cod. Estimates of C_f varied from 46 to 80 μS/cm depending on the response threshold of interest; these results support hypotheses that a global standard for C_f could be applied to power transfer calculations. Results indicate that at 60 Hz, pulse widths (PWs) of 6.7-10 ms are most efficient; however, the electrical dose required for all responses is minimized using higher frequencies resulting in PWs of less than 2 ms. Our observations support other work suggesting that the complexity and range of responses are determined jointly by the nature of the stimulus and the orientation to the field. We suggest that a measure of electrical dose required for a given stimulus provides insight into settings for maximizing electric fishing effectiveness.

[译文]：
墨瑞鳕鱼对电动钓鱼机的脉冲频率和脉冲宽度的行为反应
墨瑞鳕鱼的有效电导率（C_f）被作为对这种重要物种的有效电捕鱼设置的调查的一

部分进行评估。我们描述了从墨瑞鳕鱼引发4个反应（逃跑，强迫游泳，固定和麻醉）所需的电场最低阈值的评估。根据感兴趣的响应阈值，C_f 的估计值从46到80μS/cm不等；这些结果支持关于 C_f 的全球标准可用于功率传输计算的假设。结果表明，在60Hz时，6.7~10ms 的脉冲宽度（PW_s）是最有效的；然而，使用更高的频率所有响应所需的电剂量最小化，导致 PW_s 小于2ms。我们的观察支持了其他一些工作，这些工作表明响应的复杂性和范围是由刺激的性质和领域的方向共同决定的。我们建议衡量给定刺激所需的电剂量可以提供对电捕鱼效果最大化设置的了解。

参考文献：Andrew R. Bearlin, Simon J. Nicol, Terry Glenane. 2008. Behavioral responses of murray cod maccullochella peelii peelii to pulse frequency and pulse width from electric fishing machines [J]. Transactions of the American Fisheries Society, 137 (1), 107-113.

（罗钦、李冬梅、黄惠珍 译）

154. Factorial aerobic scope is independent of temperature and primarily modulated by heart rate in exercising Murray cod (Maccullochella peelii peelii)

[英文摘要]: Several previous reports, often from studies utilising heavily instrumented animals, have indicated that for teleosts, the increase in cardiac output (V_b) during exercise is mainly the result of an increase in cardiac stroke volume (V_S) rather than in heart rate (f_H). More recently, this contention has been questioned following studies on animals carrying less instrumentation, though the debate continues. In an attempt to shed more light on the situation, we examined the heart rates and oxygen consumption rates (Mo_2; normalised to a mass of 1 kg, given as Mo_2) of six Murray cod (Maccullochella peelii peelii; kg mean mass±SE = 1.81±0.14 kg) equipped with implanted fH and body temperature data loggers. Data were determined during exposure to varying temperatures and swimming speeds to encompass the majority of the biological scope of this species. An increase in body temperature (T_b) from 14℃ to 29℃ resulted in linear increases in (26.67~41.78 μmol min^{-1} kg^{-1}) and f_H (22.3~60.8 beats min^{-1}) during routine exercise but a decrease in the oxygen pulse (the amount of oxygen extracted per heartbeat; 1.28~0.74 μmol beat^{-1} kg^{-1}). During maximum exercise, the factorial increase in was Mo_2 kg calculated to be 3.7 at all temperatures and was the result of temperature-independent 2.2-and 1.7-fold increases in f_H and oxygen pulse, respectively. The constant factorial increases in f_H and oxygen pulse suggest that the cardiovascular variables of the Murray cod have temperature-independent maximum gains that contribute to maximal oxygen transport during exercise. At the expense of a larger factorial aerobic scope at an optimal temperature, as has been reported for species of salmon and trout, it is possible that the Murray cod has evolved a lower, but temperature-independent, factorial aerobic scope as an adaptation to the largely fluctuating and unpredictable thermal climate of southeastern Australia.

[译文]：

阶乘有氧范围（相对代谢空间）与温度无关，主要通过锻炼墨瑞鳕鱼的心率来调节

以前的一些报道，往往利用大量仪器对动物进行的研究表明，对于硬骨鱼类，运动时心输出量（V_b）的增加主要是心搏量（V_S）增加而不是心率（f_H）增加的结果。最近，尽管争论仍在继续，但随着对携带较少仪器的动物进行研究这个事件发生之后，这一争论一直受到质疑。为了更好地了解这种情况，我们检查了6只墨瑞鳕鱼（kg平均质量±SE＝1.81±0.14kg）的心率和耗氧率（Mo_2；标准化的质量为1kg，作为Mo_2），配备植入的fH和体温数据记录仪。为涵盖该物种的大部分生物学范围，在不同温度和游泳速度环境中测定数据。常规运动中，体温（T_b）从14℃升高到29℃时，引起（26.67～41.78μmol/min/kg）和f_H（22.3～60.8beats/min）线性增加，但氧脉搏下降（氧气量/心跳；1.28～0.74μmol/beat/kg）。在最大活动期间，所有温度下摩尔比计算因子增加为2.0mg，计算结果为3.7，分别是在温度无关的情况下心率和氧脉搏的2.2倍和1.7倍。f_H和氧脉搏冲的恒定因子增加表明，在活动中，墨瑞鳕鱼的心血管变量具有与温度无关的最大氧输送的最佳增益效果。以最佳温度下的较大相对代谢空间为代价，正如已经报道的鲑鱼和鳟鱼物种一样，墨瑞鳕鱼有可能演变出较低但与温度无关的相对代谢空间，以适应澳大利亚东南部波动大且不可预测的气候。

参考文献：Clark, T. D., Ryan, T., Ingram, B. A., *et al.* 2005. Factorial aerobic scope is independent of temperature and primarily modulated by heart rate in exercising murray cod (maccullochella peelii peelii) [J]. Physiological & Biochemical Zoology, 78 (3), 347-355.

（罗钦、李冬梅、黄惠珍　译）

155. Collection and distribution of the early life stages of the Murray cod (Maccullochella peelii peelii) in a regulated river

[英文摘要]：The Murray cod (M. peelii peelii) is a large fish species keenly sought by anglers. However, this species has declined in distribution and abundance and is now listed nationally as vulnerable. This study was undertaken in the Ovens and Murray rivers, to collect larvae and age-o Murray cod and determine the distribution of larval Murray cod around the mid-Murray River irrigation storage of Lake Mulwala in Victoria, Australia. Murray cod larvae were collected from 17 of 18 sites: main channels and flowing anabranch channels of regulated and unregulated rivers, sites upstream and downstream of the lake, in the upper and lower reaches of the lake and in the outflowing Yarrawonga irrigation channel during 1993 - 1994. Larval Murray cod were collected only by methods that sampled drift in flowing waters. Age-o Murray cod were collected by electrofishing in the main river, but not in off-channel waters, suggesting that cod are likely to settle into habitats in the main channel at a postlarval stage. The widespread occurrence of drifting larvae suggests that this species may be subject to previously unrecognized threats as they pass through hydroelectric power stations or become stranded in anabranch and irrigation channels. Results of this study are likely to be applicable to other species with drifting larval stages and are relevant to other locations in the Murray-Darling Basin.

[译文]：

监管河流中墨瑞鳕鱼早期生命阶段的收集和分布

墨瑞鳕鱼是垂钓者热衷追求的大型鱼类。然而，这个物种的分布和丰度下降，现在被列为国家濒危物种。这项研究在Ovens和Murray河中进行，收集幼体和0龄的墨瑞鳕鱼，来确定在澳大利亚维多利亚州Mulwala湖中Murray河中部灌区里幼年墨瑞鳕鱼的分布。从18个地点中的17个位置收集了墨瑞鳕鱼幼体：1993年至1994年期间，收集的地点有受管制和不受管制的主要渠道和河流的支流、湖泊的上下游流域、湖泊的上下游地点和亚勒旺加流动灌溉渠中。只有在流水中漂移中通过采样的方法才能收集墨瑞鳕鱼幼苗。0岁的墨瑞鳕鱼主要是在河流中通过电捕鱼收集的，而不是在非航道水域中收集的，这表明鳕鱼在幼年后期更喜欢进入主航道的栖息地。漂流幼体的广泛出现表明，当这些物种经过水电站或滞留在分支和灌溉渠道时，它们可能会受到之前未被认识的威胁。这项研究的结果可能适用于其他有漂流幼体阶段的物种，并且与Murray-Darling盆地的其他地点有关。

参考文献：Koehn, J. D., Harrington, D. J. 2005. Collection and distribution of the early life stages of the murray cod (maccullochella peelii peelii) in a regulated river [J]. Australian Journal of Zoology, 53 (3), 137-144.

（罗钦、李冬梅、黄惠珍　译）

156. Microsatellite loci for studies of wild and hatchery Australian Murray cod Maccullochella peelii peelii (Percichthyidae)

[英文摘要]：Murray cod is a species of conservation concern that is subject both to wild harvest and to aquaculture production. Six polymorphic microsatellite loci have been developed for this species, which will facilitate studies of wild-stock structure, will help the management of hatchery diversity, and will aid genetic improvement programmes. The markers exhibited nonindependence of genotypes between loci and deviations from Hardy-Weinberg expectations, although these most probably reflect practices at the hatchery from which the genotyped individuals were sourced.

[译文]：

野生和养殖的澳大利亚墨瑞鳕鱼（真鲈科）微卫星点位的研究

墨瑞鳕鱼是一种既野生生长又有水产养殖生产的受保护物种。已经为该物种开发了六个多态微卫星基因座，这将有助于野生种群结构的研究，多样性孵化场的管理，并将有助于遗传改良计划。这些标记显示基因型与Hardy-Weinberg期望偏差之间的遗传型非独立性，尽管这些很可能反映了在孵化场基因型个体来源的实践。

参考文献：Loughnan, S.R., Baranski, M.D., Robinson, N.A., et al. 2004. Microsatellite loci for studies of wild and hatchery australian murray cod maccullochella peelii peelii (percichthyidae) [J]. Molecular Ecology Resources, 4 (3), 382-384.

（罗钦、李冬梅、黄惠珍 译）

157. Observations on the distribution and abundance of carp and native fish, and their responses to a habitat restoration trial in the Murray River, Australia

[英文摘要]：A native fish strategy has been initiated to rehabilitate native fish populations in the Murray-Darling Basin, Australia. The reintroduction of large woody debris (LWD) into the Basins large lowland rivers is one of the restoration activities in the% strategy. The results from three separate studies undertaken on the Murray River between Yarrawonga and Tocumwal are presented on the relationship between carp (Cyprinus carpio), native species, and LWD to examine whether native species and carp compete for LWD habitat. The first study reports on the relative abundance of carp and native fish in a river reach. Since 1995 carp abundance has declined, whereas the abundance of native fish populations has remained relatively constant providing little support for the hypothesis that competition for LWD habitat is having effects at the population level effects. The second study reports on the relationship between LWD, river channel position, and its use as habitat by carp and native species. A statistically significant relationship was observed between native fish, LWD, the location within a meander, and curvature of the meander. There was no statistically significant relation between carp and any of these parameters indicating that carp utilise a variety of riverine habitats, whereas native species were strongly associated with LWD. The third study reports on an experiment that tested the response of carp to the placement of new LWD habitat. The response from carp was statistically inconclusive. The combination of these studies suggest that it is unlikely that carp and native species are directly competing for LWD habitat and it is unlikely that carp will inundate restored LWD habitats and preclude native species.

[译文]：

关于鲤鱼和本地鱼的分布和丰度的观察及其对澳大利亚墨累河栖息地恢复试验的反应

澳大利亚 Murray-Darling 盆地开始采用天然鱼类策略来恢复本地鱼类种群。将大型木质碎片（LWD）重新引入盆地大型低地河流是该战略中的复原活动之一。在 Yarrawonga 和 Tocumwal 之间对墨累河进行的三项独立研究结果，提出了鲤鱼（Cyprinus carpio），本地物种和 LWD 之间的关系，以检查本地物种和鲤鱼是否存在 LWD 栖息地竞争。第一份研究报告报道了河段中鲤鱼和本地鱼的相对丰度。自 1995 年以来，鲤鱼丰度下降，而本地鱼类种群数量保持相对稳定，这一点几乎不支持"竞争 LWD 栖息地对种群水平具有影响"的假设。第二项研究报告了 LWD，河道位置，河道位置作为鲤鱼和本地物种栖息地这三者间的关系。在本地鱼类中，观察到 LWD，河曲内的位置和河曲的曲率之间统计学上显著的关系。鲤鱼和任何这些关于表明鲤鱼利用多种河流栖息地的参数之间没有统计学上的显著关系，而本地物种与 LWD 密切相关。第三项研究报告了一项实验，该实验测试了鲤鱼对放置新 LWD 栖息地的反应。对鲤鱼的研究在统计学上没有明显结果。这些研究的总体表明，鲤鱼和本地物种不可能直接竞争 LWD 栖息地，鲤鱼不太可能在泛滥成灾的 LWD 栖息地并排除本地物种。

参考文献：Simon J. Nicol, Jason A. Lieschke, Jarod P. Lyon, *et al.* 2004. Observations on the distribution and abundance of carp and native fish, and their responses to a habitat restoration trial in the murray river, australia [J]. New Zealand Journal of Marine & Freshwater Research, 38 (3), 541-551.

（罗钦、李冬梅、黄惠珍　译）

158. Variable stocking effect and endemic population genetic structure in Murray cod Maccullochella peelii

[英文摘要]：Microsatellite markers were utilized to examine the genetic structure of Murray cod Maccullochella peelii throughout its distribution in the Murray-Darling Basin (MDB) of eastern Australia, and to assess the genetic effects of over three decades of stocking hatchery-reared fingerlings. Bayesian analysis using the programme Structure indicated that the species is largely genetically panmictic throughout much of its extensive range, most probably due to the high level of connectivity between catchments. Three catchments with terminal wetlands (the Lachlan, Macquarie and Gwydir), however, contained genetically distinct populations. No stocking effects were detected in the catchments that were genetically panmictic (either because of low genetic power or lack of effects), but the genetically differentiated Gwydir and Macquarie catchment populations were clearly affected by stocking. Conversely, there was no genetic evidence for survival and reproduction of stocked fish in the Lachlan catchment. Therefore, stocking of M. peelii throughout the MDB has resulted in a range of genetic effects ranging from minimal detectable effect, to substantial change in wild population genetic structure.

[译文]：

墨瑞鳕鱼多变的放流效果和地方种群遗传结构

微卫星标记物被用于检测澳洲东部墨瑞-达令盆地（MDB）分布的墨瑞鳕鱼的遗传结构，并评估放流了三十多年的人工培育的鱼种的遗传效应。使用程序结构的贝叶斯分析表明，在大部分地区该鱼类主要是遗传性泛型，最可能是由于流域之间高度的连通性。但是，终端湿地的3个流域（拉克兰河，麦格理和圭迪尔）包有基因上有区别的种群。在遗传性泛型的流域（由于遗传力低或缺乏效力）没有检测到放流效应，但遗传分化的圭迪尔和麦格理流域种群显然受到放流的影响。相反，拉克兰河流域没有放流鱼的存活和繁殖的遗传证据。因此，在整个MDB中放流墨瑞鳕鱼已经导致了一系列遗传效应，从最小的可检测的效果到野生种群遗传结构的实质性变化。

参考文献：Rourke, M. L., H. C. McPartlan †, Ingram, B. A., et al. 2011. Variable stocking effect and endemic population genetic structure in murray cod maccullochella peelii [J]. Journal of Fish Biology, 79 (1), 155-177.

（罗钦、李冬梅、黄惠珍　译）

159. Flow magnitude and variability influence growth of two freshwater fish species in a large regulated floodplain river

[英文摘要]：Fish are often targets for environmental watering outcomes under the premise that aspects of the flow regime are linked to key components of their life-history. This study examined the conceptual link between variability in river discharge and fish productivity by measuring annual growth patterns (generated using sclerochronology over a 22-year period) of two native freshwater cod Maccullochella spp. species over a range of flow conditions in a regulated Australian floodplain River. We found a positive relationship between fish growth, flow variability and river discharge. Flow variability during spring and summer-autumn, as well as their antecedent values, was particularly important in explaining annual growth of the nationally endangered Maccullochella macquariensis. Growth of Maccullochella peelii displayed similar patterns, though were more closely aligned with spring discharge. These results are consistent with the general view that increased river regulation, due to its suppression of flow magnitude and variability, has been a major contributing factor in the decline of native fish populations throughout the world. Our results provide support and guidance for the use of environmental water delivery, and have broad application to rivers worldwide for which any quantification of ecological impacts of regulation, and responses to water management remain scarce.

[译文]：

流量和可变性对在某个受管制的洪泛区大河中的2种淡水鱼类生长的影响

鱼类通常是以环境的有水的出口为目标，在流态的各部分与它们生活史的关键部分有关的条件下。本研究调查了河水流量的变化与通过年度增长模式（用22年以上的年轮生成）测量的2种本土淡水鳕鱼类生产力之间的理论联系的通过测量两种天然淡水鳕鱼，在1条受一系列的水流条件管控的澳洲的洪泛区大河中。我们发现鱼类生长，流量变化和河流排放之间是正相关的。春季和夏秋季的流量变化（和它们的原先值一样）对国家濒危的吻麦鳕鲈年生长量的解释特别重要。墨瑞鳕鱼的生长显示出相似的模式，尽管与春季的排放量更看齐。这些结果与一般观点一致，即加强河道整治是造成全世界本土鱼类种群下降的主要原因，由于它可控制水流强度和流量变动。我们的结果为使用环境供水提供了支持和指导，并且广泛应用于世界各地以任何法规量化生态影响的河流中，以及对水资源管理仍然很匮乏。

参考文献：Tonkin, Z., Kitchingman, A., Lyon, J., et al. 2017. Flow magnitude and variability influence growth of two freshwater fish species in a large regulated floodplain river [J]. Hydrobiologia, 797 (1), 289-301.

（罗钦、李冬梅、黄惠珍　译）

160. Accounting for false mortality in telemetry tag applications

[英文摘要]：Deaths of animals in the wild are rarely observed directly, which often limits understanding of survival rates. Telemetry transmitters offer field ecologists the opportunity to observe mortality events in cases as the absence of animal movement. When observations of mortality are based on factors such as the absence of animal movement, live individuals can be mistaken for dead, resulting in biased estimates of survival. Additionally, tag failure or emigration might also influence estimates of survival in telemetry studies. Failing to account for mis-classification, tag failure, and emigration rates can result in overestimates of mortality rates by up two-fold, even when the data are corrected for obviously mistaken entries. We use a multi-state capture-recapture model with a misclassification parameter in estimating both the rate of permanent emigration and/or tag failure and the rate at which individuals are mistakenly identified as dead. We use this method on an annual telemetry survey of three species of native fish in the Murray river, Australia: Murray cod (Maccullochella peelii), trout cod (Maccullochella macquariensis) and golden perch (Macquaria ambigua). Evidence for higher mortality rates in the first year post-implantation occurred for Murray cod and golden perch, which is likely an effect of tagging and/or the transmitter, or transmitters shedding. Using simulations, we confirm that our model approach is robust to a broad range of misclassification and transmitter failure rates. With these simulations we also demonstrate that misclassification models that do not account for emigration will likely be erroneous if live and dead animals have different probabilities of detection. These findings will have a broad interest to ecologists wishing to account for multiple sources of misclassification error in capture-mark-recapture studies, with the caveat that the specifics of the approach are dependent on species, transmitter types and other aspects of experimental design which may or may not be amenable to the misclassification framework.

[译文]：

对遥测标签应用程序中假死亡率的解释

野生动物的死亡很少被直接观察到，这往往限制了对其存活率的认识。遥测发射机为野外生态学家提供观察到死亡事件的机会，假如在动物不动的情况下。当死亡率观察是基于诸如动物不动等因素时，活体可能被误认为死亡，导致存活数的估算有偏差。此外，标签失效或动物迁移也可能影响遥测研究中存活数的估算。无法解释错误分类，标签失效，迁移率可能导致死亡率高估了两倍，即使数据对明显错误的项目进行了修正。我们使用了一个多状态捕获-再捕获模型，其用了一个估算两个不变的迁移和（或）标签失效的比率以及活体被误认为死亡的比率的误分类参数。我们使用这种方法对澳洲墨瑞河的3种本地鱼进行一年度的遥控测量：墨瑞鳕鱼，吻麦鳕鲈和黄金鲈鱼。墨瑞鳕鱼和黄金鲈鱼植入后第一年出现较高死亡率的迹象，这可能是受标签和（或）发射器的影响，或发射器脱落。通过模拟，我们确认本模型方法对广泛范围的误分类和发射机故障率是稳健的。通过这些模拟，我们还证明不考虑迁移的误分类模型可能是错误的，如果活的和死亡的动物有不同的检测概率。这些发现将对生物学家有广泛的兴趣，他们希望对在捕获–标记–再捕获研究中分类错误率的多个来源进行解释，同时要注意的是，该方法的细节取决于物种，发射器类型和其他关于可能或者不可能适合误分类框架的实验设计的部分。

参考文献：Bird, T., Lyon, J., Wotherspoon, S., et al. 2017. Accounting for false mortality in telemetry tag applications. Ecological Modelling, 355, 116-125.

（罗钦、李冬梅、黄惠珍　译）

161. Recovery from a fish kill in a semi-arid Australian river: Can stocking augment natural recruitment processes?

[英文摘要]：Localized catastrophic events can dramatically affect fish populations. Management interventions, such as stocking, are often undertaken to re-establish populations that have experienced such events. Evaluations of the effectiveness of these interventions are required to inform future management actions. Multiple hypoxic blackwater events in 2010-2011 substantially reduced fish communities in the Edward-Wakool river system in the southern Murray-Darling Basin, New South Wales, Australia. These events led to extensive fish kills across large sections of the entire system following a period of prolonged drought. To expedite recovery efforts, 119 661 golden perch Macquaria ambigua and 59 088 Murray cod Maccullochella peelii fingerlings were stocked at five locations over 3 years. All fish stocked were chemically marked with calcein to enable retrospective evaluation of wild or hatchery origin. Targeted collections were undertaken 3 years post-stocking to investigate the relative contribution of stocking efforts and recovery via natural recruitment in the system. Of the golden perch retained for annual ageing (n=93) only nine were of an age that could have coincided with stocking activities. Of those, six were stocked. The dominant year-class of golden perch were spawned in 2009; before the stocking programme began and prior to blackwater events. All Murray cod retained (n=136) were of an age that coincided with stocking activities, although only eight were stocked. Among the Murray cod captured, the dominant year-class was spawned in 2011, after the blackwater events occurred. The results from this study provide first evidence that natural spawning and recruitment, and possibly immigration, were the main drivers of golden perch and Murray cod recovery following catastrophic fish kills. Interpreted in the context of other recent examples, the collective results indicate limited benefit of stocking to existing connected populations already naturally recruiting in riverine systems.

[译文]：

在澳洲半干旱河流中灭亡的一种鱼的恢复：放流能否提升自然补充进程？

局部的灾难性事件可能会严重影响鱼类种群。管理干预，如放流，经常用来重建遭受该事件的种群。评估这些干预措施的有效性，以便为今后的管理行动提供信息。2010—2011年多个缺氧黑水事件大大减少了爱德华-瓦科尔河系的鱼类群落，位于澳洲新南威尔士州墨瑞-达令盆地的南部。在随后的持续性干旱时期，这些事件导致整个系统的大部分鱼类大面积死亡。为加快恢复速度，超过3年时间，在五个地点放流了119 661尾黄金鲈鱼和59 088墨瑞鳕鱼苗。所有鱼都用钙黄绿素进行化学标记，以便对野生的鱼苗或人工孵化的鱼苗能进行回溯性评价。3年针对性地收样，用以调查在该系统中放流力度和经由自然补充恢复的相对贡献。每年保存的黄金鲈鱼（n=93）中，只有9尾的年龄可能与放流活动时间相吻合。其中有6尾被标记。孵化出最多黄金鲈鱼苗的是2009年；在放养计划开始之前和在黑水事件之前。所有保存的墨瑞鳕鱼（n=136）与放流活动时间相吻合，尽管只有8尾被标记。在捕获的墨瑞鳕鱼中，孵化出最多墨瑞鳕鱼苗的是2011年，是在黑水事件发生后。本研究首次证明了自然的产卵和补充（可能是迁移），是灾难性鱼类死亡后黄金鲈鱼和墨瑞鳕鱼恢复的主要驱动力。解读其他近期的实例后，所有结果表明，放流对现存相关种群的好处有限，其已经在河流系统中自然地补充了。

参考文献：Thiem, J. D., Wooden, I. J., Baumgartner, L. J., *et al*. 2017. Recovery from a fish kill in a semi-arid australian river: can stocking augment natural recruitment processes? [J]. Austral Ecology. 42, 218-226.

（罗钦、李冬梅、黄惠珍　译）

162. Assessment of stocking effectiveness for Murray cod (Maccullochella peelii) and golden perch (Macquaria ambigua) in rivers and impoundments of south-eastern Australia

[英文摘要]：Stock enhancement is a management tool used for fishery recovery worldwide, yet the success of many stocking programs remains unquantified. Murray cod (Maccullochella peelii) and golden perch (Macquaria ambigua) are important Australian recreational target species that have experienced widespread decline. Stocking of these species has been undertaken for decades, with limited assessment of effectiveness. A batch marking and recapture approach was applied to assess stocked Murray cod and golden perch survival, contributions to wild fisheries, and condition in rivers and impoundments. Stocked fish were marked with calcein. Marked fish were detected during surveys undertaken 3 years and 10 months from initial marking, and it is probable that marks will persist beyond this time. The proportion of calcein marked fish in the population sub-sample whose age was equal to, or less than, the number of years since release, varied by 7% ~ 94% for Murray cod, and 9% ~ 98% for golden perch. Higher proportions of marked fish were found in impoundments than rivers. Marked Murray cod had significantly steeper length-weight relationships (i.e. higher weight at a given length) to unmarked fish. Our results show that application of methods for discriminating stocked and wild fish provides critical information for the development of adaptive, location-specific stocking strategies.

[译文]：

在澳洲东南部的河流和湖泊中墨瑞鳕鱼和黄金鲈鱼放流效果的评估

放流增加是用于全球渔业恢复的管理工具，但许多放养计划的成功仍然无法量化。墨瑞鳕和金鲈是重要的澳洲休闲渔业指标鱼类，其曾经急剧下降。这些鱼类的放流已经进行了几十年，对其效果的评估有限。应用一批标记物和再捕获方法来评估放流的墨瑞鳕鱼和黄金鲈鱼的存活率，对河流和湖泊的野生渔业的贡献。放流的鱼被钙黄绿素标记。从刚开始标记算起，在3年零10个月的调查中发现了标记鱼，而且在这个时间之后那个标记鱼很可能会持续发现。在年龄等于或小于放流年数的子样品中钙黄绿素标记鱼的比例，墨瑞鳕鱼的变化范围为7%~94%，金鲈的变化范围为9%~98%。湖泊中标记鱼的比例高于在河流中的比例。标记的墨瑞鳕鱼相对于未标记的鱼具有明显更陡的长度-重量关系（即给定长度的较高重量）。我们的研究结果表明，鉴别放流鱼与野生鱼方法的应用，为适应性的和定位性的放流策略的发展提供了关键信息。

参考文献：Forbes, J., Watts, R. J., Robinson, W. A., et al. 2016. Assessment of stocking effectiveness for murray cod (maccullochella peelii) and golden perch (macquaria ambigua) in rivers and impoundments of south-eastern australia. Marine & Freshwater Research, 67: 1410-1419.

（罗钦、李冬梅、黄惠珍 译）

163. Impacts of stock enhancement strategies on the effective population size of Murray cod, Maccullochella peelii, a threatened Australian fish

[英文摘要]：Murray cod, Maccullochella peelii (Mitchell) is an iconic Australian species endemic to the Murray-Darling Basin (MDB) of inland south-eastern Australia. Murray cod has been a valuable food source and supported a large commercial fishery throughout much of the 20th century. Over-fishing and habitat destruction have resulted in significant declines in Murray cod populations throughout much of its range. Since the early 1980s, large numbers of Murray cod have been stocked into waterways to support both recreational fishing and conservation efforts. In this study, the likely impacts of past and current stocking practices on genetic diversity of Murray cod were modelled and new strategies to maximise genetic diversity in stocked populations are explored. The results suggest that a large, well-managed breeding and stocking programme could help maintain genetic diversity of Murray cod across the MDB. In catchments within the MDB where the effective population size is very small, a well-designed stocking programme, following strict guidelines for numbers of families reared and number of individuals maintained per family, could increase genetic diversity in a few generations.

［译文］：

放流增大策略对有效的濒危澳洲墨瑞鳕鱼群种数量的影响

墨瑞鳕鱼是澳洲内陆东南部墨瑞-达令盆地（MDB）的标志性鱼类。墨瑞鳕鱼一直是重要的食物来源，并在20世纪的大部分时间里支撑着大型渔业商业活动。过度捕捞和栖息地破坏导致了大部分地区的墨瑞鳕鱼种群显著下降。自20世纪80年代初以来，大量的墨瑞鳕鱼已被放流在水路中，用以支持休闲捕鱼和保护工作。本研究中，对过去和现在放流的墨瑞鳕鱼的遗传多样性可能的影响进行了模拟，并开发了放流种群遗传多样性最大化的新策略。结果表明，一个大型的，管理良好的育种和放流计划可以帮助保持游经MDB的墨瑞鳕鱼的遗传多样性。在MDB流域内有效的种群规模是非常小的，设计良好的放流计划，接着对家庭养殖户数和每户保持养殖数量进行严格的指导，可以在几代种群中增加遗传多样性。

参考文献：Ingram, B. A., Hayes, B., & †, M. L. R. 2011. Impacts of stock enhancement strategies on the effective population size of murray cod, maccullochella peelii, a threatened australian fish. Fisheries Management & Ecology, 18 (6), 467-481.

（罗钦、李冬梅、黄惠珍 译）

164. Evaluating mark-recapture sampling designs for fish in an open riverine system

［英文摘要］：Sampling designs for effective monitoring programs are often specific to individual systems and management needs. Failure to carefully evaluate sampling designs of monitoring programs can lead to data that are ineffective for informing management objectives. We demonstrated the use of an individual-based model to evaluate closed-population mark-recapture sampling designs for monitoring fish abundance in open systems, using Murray cod (Maccullochella peelii (Mitchell, 1838)) in the Murray-Darling River basin, Australia, as an example. The model used home-range, capture-probability and abundance estimates to evaluate the influence of the size of the sampling area and the number of sampling events on bias and precision of mark-recapture abundance estimates. Simulation results indicated a trade-off between the number of sampling events and the size of the sampling reach such that investigators could employ large sampling areas with relatively few sampling events, or smaller sampling areas with more sampling events to produce acceptably accurate and precise abundance estimates. The current paper presents a framework for evaluating parameter bias resulting from migration when applying closed-population mark-recapture models to open populations and demonstrates the use of simulation approaches for informing efficient and effective monitoring-program design.

［译文］：

在开放河流系统中评估鱼的标记-捕回抽样设计

有效监控程序的抽样设计通常针对个别的系统和管理的需要。未能准确评估监控程序的抽样设计可能会导致报告管理目标的数据是无效的。我们用澳洲墨瑞-达令河流域的墨瑞鳕鱼举个例子，论证用基于个体的模型来评估封闭种群的标记-捕回抽样设计，用于监测开放系统中的鱼类丰度。该模型使用活动区，捕获概率和丰度概数来评估抽样面积的大小和采样数量对标记-捕回丰度概数的偏差和精度的影响。模拟结果表明采样数量与采样范围大小之间可以权衡，使得调查人员可以使用大范围的采样和相对小样品数量，或者使用较多样品数量和相对小采样范围，都可以产生令人满意的精确的丰度估概数。本文提出了一个框架，用于评估由迁移导致的参数偏差的产生，当对开放种群应用闭合种群标记-捕回模型时，而且论证使用模拟方法报告高效和有效的监控程序设计。

参考文献：Tetzlaff, J. C. 2011. Evaluating mark-recapture sampling designs for fish in an open riverine system [J]. Marine & Freshwater Research, 62 (7), 835-840.

（罗钦、李冬梅、黄惠珍　译）

165. Balancing conservation and recreational fishery objectives for a threatened fish species, the Murray cod, Maccullochella peelii

［英文摘要］：Murray cod Maccullochella peelii (Mitchell) is a large, iconic Australian fish species targeted by anglers but also listed as nationally threatened. A consultative process that included conservation and fishery interests helped to develop a population model for this species and agree on management scenarios to be tested. The modelled scenarios illustrated that threats to populations (risk of decline) can be substantially reduced and catch rates increased through harvest slot length limits (HSLL) rather than minimum legal limits (MLL). A 600-to 1000-mm HSLL provided lower risk of decline and greater catch rates than the existing 500-mm MLL, but better results were achieved with a 400-to 600-mm HSLL. Importantly, a range of other impacts (fish kills, stocking, thermal impacts, larval mortalities, habitat changes) were recognised and incorporated. This study provides an example of the utility of a population model to improve management decision-making for both conservation and fishery objectives.

[译文]：

濒危鱼类墨瑞鳕鱼保护与休闲渔业的平衡目标

墨瑞鳕鱼是钓鱼者垂钓大型标志性澳洲鱼类的目标，也被列为国家濒危级。一个包括保护和渔业利益在内的协商进程有助于为这一物种建立种群模型，并同意管理方案进行测试。模拟情景说明，通过收获槽长度限制（HSLL）而不是最低法定限值（MLL）可以使种群（下降风险）威胁大大降低并且提高捕捞率。600~1000mm 的 HSLL 比现有 500mm 的 MLL 具有更低的下降风险和更大的捕获率，但是 400~600mmHSLL 的使用效果更好。重要的是，一系列其他影响（鱼类杀戮，放养，热影响，幼虫死亡率，栖息地变化）被确认和合并。本研究提供了一个利用种群模型来提升保护和渔业生产的管理决策的实例。

参考文献：Koehn, J. D., Todd, C. R. 2012. Balancing conservation and recreational fishery objectives for a threatened fish species, the murray cod, maccullochella peelii [J]. Fisheries Management & Ecology, 19 (5)：410-425.

（罗钦、李冬梅、黄惠珍　译）

166. Management and Ecological Note Field detection of calcein in marked fish-a cautionary note

[英文摘要]：The article presents study on field detection of calcein, fluorochrome dye in Murray cod, Maccullochella peelii (Mitchell), species. It mentions that the study revealed that that non-destructive field detection methods for calcein during large fishery surveys can be unreliable. It adds the need for review of non-destructive methods for discriminating hatchery-bred fish from wild fish.

[译文]：

管理和生态，鱼中标记物钙黄绿素的现场检测-一个注意事项

本文介绍了在墨瑞鳕鱼类中荧光染料钙黄绿素的现场检测。研究显示在大型渔业调查中钙黄绿素的非破坏性现场检测方法是不可靠的。它增加了对从野生鱼类中区分出人工孵化养殖鱼的非破坏性方法的需求。

参考文献：Ingram, B. A., Hunt, T. L., Lieschke, J., et al. 2015. Field detection of calcein in marked fish-a cautionary note [J]. Fisheries Management & Ecology, 22（3）：265-268.

（罗钦、李冬梅、黄惠珍　译）

167. 澳洲虫纹鳕鲈的生物学特性及引养前景

主要内容：优良的养殖品种是养殖者所期盼的，它不仅给养殖者带来好的经济效益，更为人们提供了优质美味的动物蛋白。过去十年里，有关水产养殖单位先后自澳大利亚引进了养殖新品种，其中有的养殖种类已取得了阶段性的成果，如人们熟知的澳洲宝石斑（Scortum barcoo），有的仍处于试养阶段，有的养殖效果不太好。笔者于2002年4月下旬至5月上旬应澳大利亚水产养殖公司的邀请赴维多利亚省的Euma镇对该公司主养的宝石斑及虫纹鳕鲈进行了考察。当我们到达Euma时，映人眼帘的尽是郁郁葱葱的草地，树木成荫，鲜花盛开，湾塘、河溪随处可见，是一个花园式的休闲垂钓的旅游圣地。澳大利亚水产养殖公司是该镇颇有名望的企业，我们与当地人交谈时，提及宝石斑及虫纹鳕鲈时人尽皆知。如果来自远方的朋友能品尝到虫纹鳕鲈，招待规格就比较高了。澳大利亚是一个地广人稀的发达国家，淡水养殖主要分布于东南部的维多利亚省和新南威尔省以及中北部的昆士兰省，现将号称澳大利亚的国宝—虫纹鳕鲈（见图1）扼要介绍于后，供国内养殖者参考。

参考文献：韩茂森.2003.澳洲虫纹鳕鲈的生物学特性及引养前景[J].淡水渔业，33（4）：50-52。

注：无中文和英文摘要。

168. 着力农业科技创新 服务新兴产业发展——写在《澳洲龙纹斑种苗繁育与养殖技术研究》完稿之际

摘要：概述澳洲龙纹斑的特征与开发意义，以及"澳洲龙纹斑设施种苗工厂化繁育技术产业化工程"项目的实施成效与体会；并介绍《澳洲龙纹斑种苗繁育与养殖技术研究》一书的主要内容及其出版的目的与意义。

Focus on the innovation of agricultural science and service to emerging industrial development-Writing at the occasion when completing the manuscript of 《Research on seedling breedingand cultivation techniques of Maccullochella peelii》

Abstract：In this paper, the features and development significance of Australia Maccullochella peelii, and the implement achievements and experience on the project of 44 Industrialization projects of factory breeding technology on Australia Maccullochella peelii seedling" were summarized, also the main content and published purpose and significance of the book 《Research on seedling breeding and cultivation technology of Australia Maccullochella peeliiyweie introduced.

参考文献：翁伯琦，罗土炎.2016.着力农业科技创新服务新兴产业发展——写在《澳洲龙纹斑种苗繁育与养殖技术研究》完稿之际[J].福建农业科技（10）：59-61。

169. 墨瑞鳕在广东全人工繁育成功

摘要：墨瑞鳕是澳洲的"国宝鱼"，记者近日获悉，广东东莞安业水族养殖有限公司（以下简称安业水族）全人工繁育养殖墨瑞鳕获得成功。据安业水族董事长罗汝安介绍，公司早在2011年就引进墨瑞鳕，并开展了全人工繁育及养殖试验。2013年春季，培育的一百多尾亲鱼便逐渐成熟并少量产卵，今年春季亲鱼开始大量繁殖。截至今年4月底，安业水族已成功繁育出体长4~10cm的种苗六万多尾，罗汝安估计今年一共能培育出8万~10万尾的墨瑞鳕鱼苗。

参考文献：刘怡.2014.墨瑞鳕在广东全人工繁育成功[J].海洋与渔业（5）：28-30。

注：无英文摘要。

170. 墨瑞鳕成功养殖"五步曲"

摘要：墨瑞鳕鱼原产于澳洲，为大型淡水鱼类，分布于澳洲墨瑞（Murray-Darling）水系，是澳洲极为古老鱼类之一。由于味道鲜美，素有澳洲"国宝鱼"美誉。广东省东莞安业水族养殖公司于2013年成功人工繁育出鱼苗并掌握了全套的养殖技术。现将墨瑞鳕从幼苗到成鱼养殖所需设施条件、养殖关键技术、病害防治等简述如下。

参考文献：赵志玉.2014.墨瑞鳕成功养殖"五步曲"[J].海洋与渔业（12）。

注：无英文摘要。

171. 澳洲鳕鲈养殖试验初探

摘要：鳕鲈是澳洲个体最大的淡水鱼，其外形美丽，肉质细嫩鲜美，价格昂贵，有澳洲国宝鱼的美称。鳕鲈生长速度相对较快，成活率高，病害较少。

参考文献：张善发，林昌华，李琼文，等.2009.澳洲鳕鲈养殖试验初探[J].科学养鱼，2009（3）：21-21。

注：无英文摘要。

172. 澳洲淡水鳕鱼及其人工繁殖

主要内容：鳕鱼 Murray cod（Maccullochella peeli）为肉食性鱼类，其食料包括了种类繁多的双壳类、软体动物、甲壳类和其他小型鱼类等。鳕鱼生长迅速，一龄鱼可长至 23cm，以后 2~5 龄分别是 34cm、46cm、56cm 和 64cm。检查 23kg 个体的耳石，其年龄约为 15 龄。鳕鱼 4 龄以上成熟，通常雌雄性征不明显，近性成熟的雌鱼，生殖孔较大、突出、紫红色，腹部膨胀。每年春季，当水温上升至 20℃ 时开始产卵，受精卵较大，直径 3~4mm，具多个油球，粘性。受精卵黏附在浅水处的树洞、水管孔内或其他固体如树木、礁石及泥土的表面。孵化期间雄鱼护卵，产卵量视鱼体大小而定，一般在 2 万~4 万粒之间。相对产卵量是 3200~7600 粒/kg。70~22℃ 水温时，第 5d 开始孵化，第 8d 全部孵完。由于强度捕捞和环境污染及产卵场被破坏，鳕鱼资源量日益减少，而市场的需求量日增加。进行鳕鱼的人工繁殖，并增加其鱼类资源已摆上人们的议事日程。随着外源激素（HcG、CPC-）诱导鱼类产卵技术的应用和完善，在澳大利亚 HCG 和 CPe 已成功应用于鳕鱼的人工繁殖，孵化率可高达 70%~90%。

参考文献：陈永乐，朱新平，刘毅辉，等 . 199. 澳洲淡水鳕鱼及其人工繁殖[J] . 淡水渔业，29（7）：6-7。

注：无中文和英文摘要。

173. 又一条养殖亩产值达 20 万的鱼问世 墨瑞鳕鱼苗东莞生产–更适合中国养殖环境

主要内容：以后中国养殖户要养殖墨瑞鳕鱼，再也不用漂洋过海进口鱼苗了，在广东省东莞市就能买到本地繁育的墨瑞鳕鱼苗！"鱼苗 100% 中国繁育，更加适合中国环境养殖。"广东省东莞市安业水族养殖有限公司（以下简称安业养殖）总经理罗汝安自豪地告诉《海洋与渔业》记者。澳大利亚名优鱼种墨瑞鳕鱼闻名于世，中国也有一些企业引进养殖，但是由于进口鱼苗成本高，路途中也会有所损伤。因此，国内也有一些企业在探索人工苗种繁育研究。去年，罗汝安花费数百万元从澳大利亚引进成熟的苗种繁育技术，按照现有亲鱼数量，未来年产鱼苗数量可达 200 万尾。"这是一条非常优质的鱼，争取用两三年时间在国内推广养殖，为养殖户带来良好效益，为百姓餐桌增加一道佳肴。"罗汝安说。

参考文献：吕华当，罗茵 . 2017. 又一条养殖亩产值达 20 万的鱼问世 墨瑞鳕鱼苗东莞生产 更适合中国养殖环境[J] . 海洋与渔业，5（277）：36-37。

注：无中文和英文摘要。

174. 澳大利亚"国宝鱼"在广东东莞全人工繁育成功

主要内容：记者从广东东莞市海洋与渔业局获悉，澳大利亚墨瑞鳕鱼全人工繁育在东莞首获成功。据东莞市海洋与渔业局相关负责人介绍，早在2011年，东莞市安业水族有限公司引进新品种澳大利亚墨瑞鳕鱼（Murray cod）开展全人工繁育试验。截至目前，该公司育有亲本100多尾，个体质量达8~10kg，已成功培育出体长4~10cm的苗种6万多尾，可供应市场。中央电视台对此进行了专题采访。据了解，墨瑞鳕鱼是原产于澳大利亚的大型淡水鱼类，分布于澳洲最大的最重要的墨瑞（Marray-Darling）水系，并以墨瑞河命名。墨瑞鳕鱼是大洋洲极为古老的鱼类之一，同大多数淡水鱼类类似，其祖先来自海洋。有研究调查显示，墨瑞鳕鱼肉质结实、白而细嫩、味道鲜美、无腥味，且有一股淡淡的独特香味，肉嫩刺少，含有丰富的4中氨基酸及EPA和DHA，营养价值相当高，故澳洲对该鱼的经济价值评价极高。由于墨瑞鳕鱼有特殊的外型以及鲜美的口感，故在澳大利亚素有"国宝鱼"美称。

参考文献：杨宝琴．2014．澳大利亚"国宝鱼"在广东东莞全人工繁育成功[J]．水产科技情报，41（3）：162-163。

注：无中文和英文摘要。

175. 来自澳大利亚的独特娇客-墨瑞鳕

主要内容：墨瑞鳕是澳大利亚的原生鱼种，为最顶级的白肉鱼，在澳洲排名四大经济鱼种之首。墨瑞鳕在1999年引进台湾，目前业者已突破了重重难关，亲代繁殖的一代与二代已达15000尾之多。墨瑞鳕原产地在澳大利亚东南部墨瑞达令河流域，为澳洲四种主要人工繁殖的暖水性淡水鱼类之一，也是当地著名的淡水养殖品种。墨瑞鳕的学名：Maccullochella peelii peelii，中文名：墨瑞河真鲈，英文名：Murray cod，俗称：墨瑞鳕、河鳕、东洋鳕、鳕鲈、澳洲淡水鳕鲈等，大陆称为澳洲龙纹斑、虫纹鳕鲈或虫纹石斑。此鱼是世界上最大的淡水鱼之一，与宝石鲈、黄金鲈、银鲈等原产于澳洲的淡水鱼并列为澳洲鱼。由于墨瑞鳕有特殊的外型以及鲜美口感，在澳大利亚素有"国宝鱼"美称。

参考文献：郑石勤．2011．来自澳大利亚的独特娇客-墨瑞鳕[J]．养鱼世界，7，73。

注：无中文和英文摘要。

176. 澳大利亚"国宝鱼"在东莞全人工繁育养殖成功

主要内容：日前，笔者从广东东莞市海洋与渔业局获悉，澳大利亚墨瑞鳕鱼全人工繁育养殖在该市首获成功。这标志东莞市在研究推广应用名优特稀新品种上又取得了新成果，这对于东莞调整优化养殖品种结构，提升产业科技创新水平，加快都市现代渔业发展具有重要的意义。据东莞市海洋与渔业局相关负责人介绍，早在2011年东莞市安业水族有限公司引进新品种澳大利亚墨瑞鳕鱼，开展全人工繁养殖试验。该公司经过近4年的努力，取得全人工繁养殖试验成功。截至目前，该公司育有亲本100多尾，个体重达8~10kg，已成功繁育出体长4~10cm苗种6万多尾，可供应市场。据了解，墨瑞鳕鱼是原产于澳大利亚的大型淡水鱼类，分布于澳洲最大和最重要的墨瑞水系，并以墨瑞河命名。墨瑞鳕鱼是大洋洲极为古老的鱼类之一，同大多数淡水鱼类类似，其祖先来自海洋。有研究调查显示，墨瑞鳕鱼肉质结实、白而细嫩、味道鲜美、无腥味，且有一股淡淡的独特香味，肉嫩刺少，含有丰富的四种氨基酸及EPA与DNA，营养价值相当高，故澳洲对该鱼经济性评价极高。由于墨瑞鳕鱼有特殊的外型以及鲜美的口感，故在澳大利亚素有"国宝鱼"美称。

参考文献：中国农业新闻网.2014.澳大利亚"国宝鱼"在东莞全人工繁育养殖成功[J].江西饲料，3：47-48。

注：无中文和英文摘要。

177. 澳大利亚养殖鱼类落户青岛

主要内容：宝石鲈 Scortum barcoo、鳕鲈 Maceullochella peeli 是澳大利亚久负盛名的养殖鱼类，具有生长周期短、抗病力强，其鱼肉租盖率均达52%、57%。这两种鱼既可以工厂化养殖，也可在池塘内养殖，鱼的营养丰富，肉质细嫩，很少肌间刺，青岛现代农业开放中心于2001年6月从澳大利亚引进养殖，并特聘青岛市渔业技术推广站曹金祥研究员、烟台大学海洋学院韩茂森教授为技术顾问，经6个月的饲养，鱼的长势良好，两位专家在中心领导及年轻助手的通力合作下，结合生产潜心研究，始终坚持走健康养殖可持续发展之路，让人们及早品尝到名贵新品种。

参考文献：农中侃.2002.澳大利亚养殖鱼类落户青岛[J].现代渔业信息，17（2）：17。

注：无中文和英文摘要。

178. 澳洲淡水养殖珍品——鳕鲈

主要内容：鳕鲈是澳洲个体最大的淡水鱼，一般情况下个体长度在55~65厘米，重量为2~5kg。鳕鲈外形美丽，肉质细嫩鲜美，被誉为澳洲淡水珍品，其价格相当昂贵。由于人工繁殖难度大，所以影响了鳕鲈养殖业的发展。近几年来，澳洲个别单位经过精心研究，突破了人工繁育难关，成功地进行了工厂化生产养殖，获得很高的经济效益。鳕鲈在中下层水域生活，肉食性，以水中的软体动物、甲壳动物、小鱼等为食。经驯化，可摄食配合饲料。鳕鲈适应的温度范围为4~40℃。最适生长温度为23~32℃。

在温度适宜条件下，养殖1周年的个体重达0.5~0.6kg，此后生长速度逐渐加。

参考文献：顾志敏，刘启文．2005．澳洲淡水养殖珍品——鳕鲈[J]．农村百事通，5。

注：无中文和英文摘要。

179. 澳洲鳕鲈引种驯化养殖的报告

主要内容：虫纹鳕鲈属鲈形目，暖鲈科，鳕鲈属，是澳大利亚南部墨累河水域具有很高营养价值和经济价值的游钓、商业性生产和资源增殖的地方名贵品种，有澳大利亚国宝鱼的美称。澳洲鳕鲈生长速度相对较快，比较适合在我国进行养殖，因此于2008年4月将澳洲鳕鲈引进并驯化养殖。通过1年多的养殖时间，初步掌握了该引进品种在我市淡水水域的养殖生产技术，为将来澳洲鳕鲈在国内推广养殖奠定了基础。现将养殖报告简介如下。

参考文献：蔡乘成，陈鹏，张祖兴．2012．澳洲鳕鲈引种驯化养殖的报告[J]．养殖技术顾问，1。

注：无中文和英文摘要。

180. 东莞成功繁育养殖澳大利亚"国宝鱼"

主要内容：日前，记者从广东省东莞市海洋与渔业局获悉，澳大利亚墨瑞鳕鱼全人工繁育养殖在该市首获成功。这对于该市加快都市现代渔业发展具有重要的意义。墨瑞鳕鱼是原产于澳大利亚的大型淡水鱼类，因分布于澳洲的墨瑞水系得名。墨瑞鳕鱼肉质结实、白而细嫩、味道鲜美、无腥味，且有一股淡淡的独特香味，肉嫩刺少，营养价值、经济性评价俱高，故在澳大利亚素有"国宝鱼"美称。

参考文献：钟云．东莞成功繁育养殖澳大利亚"国宝鱼"[J]．千里眼·顺风耳，79。

注：无中文和英文摘要。

181. 连城县水产养殖现状及发展对策

摘要：对连城县水产养殖现状进行分析，并建议发展休闲旅游渔业，采用封闭式循环水养殖模式，同时引进澳洲龙纹斑、香鱼、鲟鱼等优质鱼类开展水产养殖。

Present situation of aquatic breeding and development countermeasure in Liancheng County

Abstract: Liancheng County breed mainly in Ctenopharyngodon idellus, Cyprinus carpio, Hypophthalmichthys molitrix and Aristichthys nobilis, aquaculture production for four consecutive years of more than 15 thousand tons, production for many years in Longyan city the first. The Development Countermeasures is tourism and leisure fishery, using a closed recirculating aquaculture system, while the introduction of Maccullochella peelii peelii, Plecoglossus altivelis, Sturgeon.

参考文献：罗钦，罗土炎，饶秋华. 2016. 连城县水产养殖现状及发展对策[J]. 福建农业科技，10：45-48。

182. 名优新品种　墨瑞鳕

主要内容：墨瑞河真鲈（Maccullochella peelii peelii），英文名 Murray cod，俗称墨瑞鳕、河鳕、东洋鳕、鳕鲈、澳洲淡水鳕鲈等，中国大陆称为澳洲龙纹斑、虫纹鳕鲈或虫纹石斑。墨瑞鳕原产于澳大利亚东南部墨瑞达令河流域，为澳洲四种主要人工繁殖的暖水性淡水鱼类之一，也是当地著名的淡水养殖品种。墨瑞河与宝石鲈、黄金鲈、银鲈等原产于澳洲的淡水鱼并列为澳洲鱼。由于墨瑞鳕有特殊的外型以及鲜美口感，在澳大利亚素有"国宝鱼"的美称。

参考文献：2014. 名优新品种 墨瑞鳕[J]. 海洋与渔业，5。

注：无中文和英文摘要。

183. 墨瑞鳕在广东全人工繁育成功

主要内容：墨瑞鳕是澳大利亚的"国宝鱼"，记者近日获悉，广东东莞安业水族养殖有限公司（以下简称安业水族）全人工繁育墨瑞鳕获得成功。据安业水族董事长罗汝安介绍，公司早在2011年就引进墨瑞鳕，并开展了全人工繁育及养殖试验。2013年春季，培育了一百多尾亲鱼便逐渐成熟并少量产卵，今年春季亲鱼开始大量繁殖。截至今年4月底，安业水族已成功繁育出体长4~10cm的种苗六万多尾，罗汝安估计今年一共能培育出8万~10万尾的墨瑞鳕鱼苗。据了解，我国最早于2001年从澳大利亚引进墨瑞鳕，但一直以来没有见到大批墨瑞鳕成鱼上市或人工繁殖成功的报道，业界没能攻克墨瑞鳕全人工繁殖技术。时隔十多年，安业水族终于打破了这一僵局。

参考文献：刘怡. 2014. 墨瑞鳕在广东全人工繁育成功[J]. 海洋与渔业，28-30。

注：无中文和英文摘要。

184. 一种鱼类自动称重装置（专利）

摘要：本实用新型提供一种鱼类自动称重装置，该称重装置包括：鱼类包裹布、支架、秤，所述的鱼类包裹布挂于支架上，所述的支架置于秤上；所述的鱼类包裹布包括第一挂杆、第二挂杆和称重布，第一挂杆、第二挂杆分别与称重布的一相对侧的两端连接，称重布的另一相对侧的两端处于自由可折叠状态，使得整个鱼类包裹布在称鱼状态下呈现"长条状U型"的样式。本实用新型称量在鱼类不仅不会对鱼类的构成损害，有利于鱼类鲜活保鲜与运输；而且还避免了大型鱼类在称量过程中的"晃动"影响称量的准确性。

参考文献：罗钦，李巍，罗土炎，等. 2017. 一种鱼类自动称重装置. CN201720131608.7。

注：无英文摘要。

185. 一种墨瑞鳕礼品袋（专利）

摘要：一种墨瑞鳕礼品袋，包括外套壳，所述外套壳为左右两端敞口的弹性壳体，所述外套壳内装有内袋，所述内袋的底面设有凹腔，所述凹腔内设有循环管、电源和控制开关，所述循环管的两端分别延伸至内袋内腔的左右两侧，且循环管的中部依次安装有过滤器和微型增压泵，本墨瑞鳕礼品袋，结构简单，包装方便，通过对气囊袋充气对内袋进行封闭，礼品袋可以进行循环利用，保证利用率，内袋可以装水，延长了礼品袋使用过程中鱼的生命，保证新鲜度，同时通过循环管和微型增压泵使水体进行循环流动，过滤器可以过滤掉鱼产生的排泄物，使水保持澄清，内袋使用透明材料制成，通过外套壳的展示窗可以看到鱼，保证美观度。

参考文献：张栋国. 2017. 一种墨瑞鳕礼品袋. CN201621432488.6。

注：无英文摘要。

186. 一种低水分澳洲龙纹斑鱼籽酱的制作方法（专利）

摘要：本发明属于食品加工技术领域，具体涉及一种低水分澳洲龙纹斑鱼籽酱的制作方法。具体通过选鱼暂养、杀鱼沥血、取卵搓卵、漂洗沥干、腌制装盒、低温脱水、冷藏保存、品质检测这八个步骤来制备低水分澳洲龙纹斑鱼籽酱。本发明制得的澳洲龙纹斑鱼籽酱水分含量低，只有50.0%，富含氨基酸、高度不饱和脂肪酸（EPA、DHA）和铁锌硒等微量元素，具有口感好、口味纯正、有弹性、安全可靠的特点。

参考文献：罗钦，陈红珊，罗土炎，等. 2018. 一种低水分澳洲龙纹斑鱼籽酱的制作方法. CN201710939962.7。

注：无英文摘要。

第七章 澳洲龙纹斑产业与展望

第一节 产业发展对策与建议

我国是世界上水产引种最多的国家之一。据不完全统计，我国共引进水产养殖种类140多种，其中鱼类90余种，虾类10余种，贝类10余种，藻类10余种，其他种类10余种，登记为水产新品种的有30种。另外，直接利用引进种研发获得的新品种有23个，可以说我国1/3的水产新品种源自于引进种。水产苗种的引进对促进我国水产养殖业的发展起到了重要作用，其中罗非鱼、大菱鲆、鳗鱼、南美白对虾、扇贝（海湾扇贝和虾夷扇贝）等已成为我国重要水产品，创造了巨大的产业价值。2013年我国罗非鱼的养殖产量达到165.77万吨，南美白对虾的养殖产量达到142.98万吨。从2009年开始，国家未再对引进种进行认定。

引种促进了养殖业的发展，但不能解决我国水产苗种发展的根本问题。目前引种工作非法引进多，合法引进少；盲目引进多，科学论证少；未经检疫多，防病处理少；急于推广多，认真评估少；重复引进多，保种选育少。引种不可避免存在以下风险：一是引种受制于人。不但价格被垄断控制而且所引种一般从第二代开始就明显表现出退化，无法继续使用。二是面临外来物种入侵的威胁。目前已显现出危害的水生入侵动植物有小龙虾、鳄鱼、福寿螺、水葫芦等。三是引进种携带寄生虫及病害的问题。有报道称，造成我国水产养殖重大损失的病害很多都是由被动引进的病毒造成，如：对虾白斑杆状和桃拉病毒、牙鲆虹彩病毒等。

与种植业和畜牧业比较，我国水产养殖良种化还较落后，与美国斑点叉尾鲴、南美白对虾以及挪威的大西洋鲑产业比较，差距还非常大。55%的水产良种覆盖率和25%的遗传改良率仍有较大提升空间。现代水产种业发展对水产新品种的需求非常迫切。然而，现阶段我国水产新品种研发面临一些挑战。一是水产种质资源状况堪忧。水域生态环境恶化与自然渔业资源过度捕捞导致大多数水生生物资源量急剧下降，遗传多样性减少。保种资源有种质退化迹象。二是良种研发基础仍较薄弱。良种研发需要时间、人力、财力的持续支持，品种太多，削弱了单一品种研发投资的实力。另外，仍存在如鳗鱼等养殖品种苗种尚未实现人工繁育。三是依赖引种的隐患。四是培育主体间缺乏合力。某一品种的研发单位众多，多边合作明显欠缺，科技资源没有良好整合，无法形成合力，造成资源浪费。五是研发成果推广助力不足。新品种的制种或制苗成本往往较普通苗种高，且质量优势往往不易表现，一般老百姓只认价格造成良种质优价难优的问题，研发成果的初期推广步履艰难。

随着澳洲龙纹斑技术研究与生产实践不断深入，给人们深刻启示：要发展澳洲龙纹斑养殖产业，必须把握5个重要环节，一是种质资源征集与有效保存，建立区域性种质资源库与基地；二是鱼苗有效繁殖与基地建设，研发基本设施与人才队伍建设；三是优化设计设施与合理布局，构建生产性养殖基地群与经营；四是有效防控疾病与建立体系，不断提高健康养殖水平与成效；五是研究饲料配方与生产工艺，构建产业化饲料生产线与开发。同时在国内进一步完善和培育澳洲龙纹斑的经营市场则是迫在眉睫。让优质鱼可以获得优质价格，不仅要丰富市场供应，同时可以提升饮食消费水平（翁伯琦，2016）。

综述所述，针对我国目前引进优质水产新品质的问题与困难，我们提出一下几点建议。

一、进一步完善澳洲龙纹斑人工繁育技术

由于强度捕捞和环境污染及产卵场被破坏，澳洲龙纹斑资源量日益减少，而市场的需求量日增加。进行澳洲龙纹斑的人工繁殖，并增加其鱼类资源已摆上人们的议事日程。我国幅员辽阔，适合养殖澳洲龙纹斑的地区很多，但是澳洲龙纹斑品种资源较稀少，因此选择一个较适合或最适合澳洲龙纹斑繁育养的地区，先建立其规模化繁育基地，然后再进行全国产业化推广。澳洲龙纹斑是一种淡水鱼类，生存水温5～35℃，最适水温21～24℃，福建省具有丰富的淡水、海水及山泉水资源，大部分地区水温能常年保持在18～25℃，非常适合澳洲龙纹斑的生长需求。同时福建省于2014年起，依托福建省种业创新与产业化工程项目"澳洲龙纹斑设施种苗工厂化繁育技术产业化工程（2014S1477-2）"率先开展澳洲龙纹斑繁育养及加工产业化开发。建议应用水产学、动物营养学及畜牧兽医学理论与技术方法，进一步系统研究澳洲龙纹斑人工繁、育、养及疾病防控等技术并开展集成推广应用，自主培育亲鱼10万尾，后备亲鱼100万尾，孵

化鱼苗1亿尾。力求为促进福建乃至于国内适宜区域淡水渔业的持续发展提供新品种与新技术支撑。

二、进一步研发出一套高效澳洲龙纹斑"繁、育、养"设备

我国的工厂化养殖起步较晚，20世纪70年代才从国外引进该项技术，经过10多年的消化吸收，于20世纪90年代初才进入产业化发展阶段，以工厂化养殖、网箱养殖、封闭式循环水养殖为代表的设施渔业发展迅速，目前国内已形成了一个因地制宜、多种模式并举、立体开发应用的海水养殖新局面，具有科技含量高、投入高、产出高的特点，被视为现代渔业的代表。建议加工制作高效供氧系统、曝气系统、产卵器、孵化器、循环水养殖和水库水箱养殖系统等设备，并重点突破关键技术。降低养殖企业成本、提高规模养殖效益、推进澳洲龙纹斑规模养殖的产业化进程。

三、进一步研发出一系列高效澳洲龙纹斑人工配合饲料

我国是世界水产养殖大国之一，水产养殖总产量已连续多年居世界第一。水产饲料是水产养殖的重要物质基础，被称为水产养殖业的粮草，在水产养殖业中具有举足轻重的地位。水产动物对饲料蛋白质水平要求较高，一般为畜禽的2~4倍，通常占配方的25%~50%，甚至更多。饲料蛋白质含量过高或过低，均不利于澳洲龙纹斑生长。因此，合理的饲料配方及蛋白来源是水产饲料能否得到高效利用的关键。同时作为水产饲料的主要蛋白源的鱼粉，由于随着水产品需求量的日益增加，养殖规模的逐渐扩大，鱼粉的需求量呈现快速增长之势，然而全球渔业自然资源的衰退导致世界鱼粉产量逐年下降，鱼粉供应矛盾日益突出、价格不断攀升。因此寻找价格低廉、来源丰富的饲料蛋白源来替代鱼粉具有重要的意义。建议研发出高效环保的并且使用动物性、植物性和单细胞3类蛋白源最优比例替代部分鱼粉的澳洲龙纹斑开口饵料、稚鱼配合饲料、幼鱼配合饲料、中成鱼配合饲料和亲鱼软颗粒饲料，并制定稚鱼、幼鱼和中成鱼饲料配方及企业标准，为澳洲龙纹斑高效的规模化"繁、育、养"提供产业化技术支持。

四、进一步完善澳洲龙纹斑疾病防控体系

随着水产养殖规模扩大、品种增多、产量增高，生态环境也遭受到了破坏。有害气体增多，有害细菌和条件致病菌的大量滋生，造成养殖动物的体质下降，抗应激能力差，极易引发鱼病。在养殖过程中，鱼病发生后使用药物控制不但造成养养殖成本上升，而且还会导致鱼的品质下降。随着人们生活水平的不断提高，广大消费者对水产

质量也提出了更高的要求。因此，水产养殖者在养殖生产中，应采用生态防病技术，控制养殖鱼类不发病或少发病，确保养殖鱼类的健康生长和水产品的质量安全。建议及时发现、快速诊断出澳洲龙纹斑寄生虫病、细菌性疾病、霉菌性疾病和病毒性疾病，同时研发出一套高治愈率的澳洲龙纹斑疾病防治技术，为澳洲龙纹斑规模化开发提供有效的技术保障。

五、积极挖掘水产养殖面积的潜力确保澳洲龙纹斑发展空间

改革开放以来，我国水产养殖业高速发展取得了举世瞩目的成就，成为世界第一水产养殖大国、世界第一渔业大国、世界第一水产品出口大国，是世界上唯一渔业养殖产量超过捕捞产量的国家，即中国是通过人类活动干预水域自然生态系统提升食物供给功能获得极大成功的一个国家。几十年来，我国的水产养殖作为大农业的重要产业，不仅在保障市场供应、解决吃鱼难、促进农村产业结构调整、增加农民收入、提高农产品出口竞争力、优化国民膳食结构和保障食物安全等方面作出了重要贡献。到2030年，随着我国人口增加接近峰值和城镇化发展对水产品需求的增加，我国水产品总产量还需要增加近2000万吨，并将主要通过水产养殖的方式获得。2010年水产养殖面积约1.15亿亩的基础上，至少还要再增加约2000万亩养殖水域才能保证产量的增加。因此，在加快转变水产养殖发展方式，坚持推进产业结构、产品结构、区域结构调整，不断提高养殖生产科技贡献率的同时，"要像重视耕地一样重视水域的治理和开发利用"，参照基本农田保护制度，建立重点养殖水域、养殖基地保护制度，确保水产养殖最小使用面积，防止随意挤压水产养殖发展空间。最小养殖水域使用面积保障线应设置在约1.35亿亩，以便保障我国水产品的基本产出，为国家食物安全作出新的贡献（唐启升，2014）。

六、提升澳洲龙纹斑养殖装备推进渔业现代化

全面推进中低产养殖池塘标准化改造工程，全国有200多万公顷养殖池塘，针对投入不足的问题，建议中央财政尽快启动标准化池塘改造财政专项；加强中央财政引导性补贴支持，组织和调动地方各级财政、社会力量和群众自筹，加大对老旧池塘的改造投入，进一步推进养殖池塘标准化改造，以稳定池塘养殖面积，提高养殖单产，增强综合生产能力，保证水产品的有效持续供给；改造的内容不仅包括清淤、护坡、平整道路、疏通渠道，还要根据当地实际情况，配套建设进水净化、排水处理装备，提高水质调控能力，减少废水排放；结合养殖池塘标准化改造工程，完善承包责任制，建立养殖池塘

维护和改造的长效机制。同时，大力促进粗放型、简易型水产养殖向现代养殖设施工程化方向转变，建议设立研发专项，针对我国陆基池塘养殖、网箱养殖和工厂化养殖等主要养殖方式设施装备简陋、生产方式粗放、节水减排问题突出、产能较低等现状，加大水产养殖设施机械化、自动化和信息化研发的科技投入，加快养殖环境精准化调控以及节水、循环、减排养殖模式的研究，发展一批池塘循环水养殖、工厂化循环水养殖、规模化筏式养殖等系统模式，建立一批具有工程化养殖水平、较高经济效益、符合可持续发展要求的现代养殖示范园；通过政策支持和财政补贴，重点发展养殖筏架作业设备、机械化采捕设备和抗台风养殖设施，推广规范化循环水养殖设施、各类增氧机械、水质净化设备、水质监测与精准化调控装备，以及自动化投喂设备等。

七、加强澳洲龙纹斑产业支撑体系建设

进一步明确澳洲龙纹斑种业发展布局、重点任务和区域良种繁育体系，配套国家级良种场设施设备，增强良种保种和亲本培育能力，强化内部管理，提高良种亲本质量，加快水产遗传育种中心建设步伐，提高遗传育种中心的技术装备水平，加快生长速度快、抗病力强的优良新品种培育；探索新型运行机制，建立符合我国澳洲龙纹斑养殖生产实际的良种繁育体系，提高品种创新能力和供应能力；鼓励大型苗种生产企业开展新品种选育，探索繁育推一体化的商业化良种繁育运作模式，提高水产苗种质量和良种覆盖率。以县级水生动物防疫站建设为基础，主要水生动物疫病参考实验室、重点实验室及省、部级水生动物疫病预防控制中心建设为重点，加快澳洲龙纹斑防疫体系建设，逐步构建预防监测能力强、范围广、反应快、经费有保障的澳洲龙纹斑防疫体系。加大实施补贴政策的力度，参照种植业和畜牧业发展的有关政策，加大中央财政资金投入，提高建设标准，在澳洲龙纹斑良种繁育与推广、装备购置、疫病防控等方面实施补贴政策，促进澳洲龙纹斑支撑体系建设。同时加强领导，强化县级和乡镇推广体系能力建设，提高技术推广服务能力。构建层次分明、双向互动的水产养殖科技推广机制，建立技术交易市场，加强水产养殖技术创新知识的转化与交流，强化高校和科研院所"三农"职责，探索科技成果转化、推广的直通模式。

八、加强澳洲龙纹斑养殖业管理与执法能力建设

进一步完善水域滩涂养殖权、种苗管理、水生动物防检疫、水产品质量安全、养殖水域生态环境保护以及养殖业执法等方面的法律法规。加快澳洲龙纹斑健康养殖、重大疫病防控、药物安全使用、有害物质残留及检测方法等方面的标准和技术规范的制定和

修订。稳定水面承包经营关系,延长承包期,规范承包费的使用。借鉴土地、林地、草地等的做法,承包期稳定在 30 年左右,促使经营者加强对生产设施的维护和管理,稳定和提升渔业综合生产能力。要求用于养殖设施维护和更新改造的费用不少于水面承包费的 50%。完善澳洲龙纹斑养殖证制度,保障养殖者权利。根据《物权法》规定,修改或增加有关从事养殖权利取得、登记、期限和保护等方面的条款,保护渔业生产者的水域滩涂使用权。进一步明确养殖证登记管理办法,明确发放范围、发放程序和有效期限,对养殖证的流转做出明确规定。加快实施内陆水域和浅海滩涂资源分类管理,科学划定可围区、限围区和禁围区。坚持科学围垦、生态围垦,有序推进围垦工程建设。开展水域使用权与土地使用权转换试点,提高围垦土地集约利用水平。加强涉海、涉水项目的区域规划论证和生态评价,规范海洋和内陆产业、滩涂水域围垦、海洋工程、水利工程等的规划审批、建设监管和跟踪评估。积极探索在市场经济条件下的多元化投入机制,保障水生生物资源和养殖环境养护事业资金投入。加快建立以渔政机构为主,技术推广、质量检验检测和环境监测等机构协作配合的水产养殖业执法工作机制。重点针对养殖证、水产苗种生产许可证、养殖投入品和企业各项管理记录档案建立等,加大执法检查力度。根据新修订的《动物防疫法》和《兽药管理条例》,建立渔业官方兽医制度、渔用兽药处方制度和渔业执业兽医制度,从制度上规范养殖用药。同时,要加强养殖执法机构和队伍建设,提高执法技术装备,建立执法监督检查机制和绩效考核制度。加强澳洲龙纹斑作为水产外来物种的科学管理刻不容缓。要完善水生外来物种立法,强化国家管理职能,建立完善的管理体系,加强水生外来入侵生物预防和控制研究。

九、推行澳洲龙纹斑质量可追溯制度完善质检体系建设

根据《农产品产地安全管理办法》《农产品包装和标识管理办法》和《农产品产地证明管理规定》等规章要求,加强澳洲龙纹斑产地安全环境调查、监测与评价。按照法定程序,建立澳洲龙纹斑养殖区环境质量预警机制,逐步推行澳洲龙纹斑养殖区调整或临时性关闭措施;加强澳洲龙纹斑产地保护和环境修复,积极开展无公害水产品产地认定;加强鲜活澳洲龙纹斑生产质量安全可追溯制度建设,建立企业、政府、消费者等全社会可追溯查询平台,实施产地准出、市场准入制度,建立并完善鲜活澳洲龙纹斑质量安全监测网络体系,并据此开展风险分析和预警发布。同时,按照"预防与善后并重"原则,建立并完善澳洲龙纹斑质量安全重大突发事件预警应急处置预案。要严格执行澳洲龙纹斑质量安全重大事件报告制度,不得瞒报、迟报。加强疫情监测,发挥科研、推广、质检、环境监测和行业协会等方面的作用,及时报告所发现的问题,深入分析评

估，提出预警和处置意见，将事件控制在萌芽状态。一旦水产品质量安全事件发生，各级渔业主管部门和有关单位要立即启动预案，快速应对，密切配合，科学处置，妥善解决，保护消费者健康和生产者的合法权益。为了推行澳洲龙纹斑质量可追溯制度、提高应对突发事件的预警处置能力，必须尽快建设并完善国家、区域、地方多层次的养殖水产品质量检测体系，提高对养殖水产品质量的检测和监管能力。

第二节 产业发展趋势与展望

我国适合水产养殖的水域辽阔，适合养殖的品种众多，但由于消费能力和习惯的差异，加上缺少及时的市场信息，所以在各地区的主要养殖品种和养殖模式相对较固定，进而导致了市场供需矛盾，引发了终端产品价格的波动。唯有持续跟踪、关注市场信息，结合市场需求合理调整养殖品种并采用新型养殖模式，才能从根本上解决市场供需矛盾，将终端产品价格波动对养殖效益的影响降低至最低。作为水产品生产、贸易和消费大国，我国水产品产量居世界首位，且是世界上唯一一个水产养殖产量超过捕捞产量的国家。从20世纪开始，渔业行业开始在中国兴起，大量渔场的开设，带动了一批饲料、加工、养殖企业的萌生。近十余年来，我国渔业经济保持快速增长，在不到16年的时间内，我国水产总量从1998年的3906万吨已发展到2014年的6300万吨左右。随着渔业行业的快速兴起，越来越多的人开始关注淡水养殖。从最开始的普通池塘养殖到现在的多元化养殖模式，工厂化养鱼池、家庭庭院水池、河渠水库等方式被广泛运用，我国已有将近1838万平方米的淡水面积被大幅度开发用于水产养殖。据相关数据分析，2014—2017年我国淡水鱼产量将持续增长。据统计，2013年我国淡水养殖面积同比增长1.67%，此外，2009—2013年我国淡水鱼产量年均增长率为5.5%，预计2014—2018年中国淡水鱼产量的年增长率不会低于5%。通过一系列数据可以看出，我国淡水养殖依旧存在发展潜力。同时，目前我国淡水养殖整体水平不高，通过加速整合养殖资源、加快技术推广和应用，行业将会有更大的空间。此外，我国淡水鱼类在出口、加工、食品等领域亦有广阔的发展前景。

我国淡水鱼种主要以草鱼、鲢鱼、鲤鱼、鳙鱼、鲫鱼和罗非鱼为主，这六大类淡水鱼产量占我国养殖淡水鱼总产量的78%，其他品种则产量较小。与其他国家相比，我国养殖鱼种尤为单一，这一问题普遍存在于各省市地区。品种的单一，不仅造成市场竞争激烈，同时极易破坏水产养殖生态环境。同时，消费者面临着从"吃不起"到"不敢吃"食品安全困扰，此问题将进一步推动淡水养殖进行"由量到质"的转变。为顺应渔业市场的不断扩张，养殖技术同样在不断寻找新方向，从而提高产量与质量。目前，国家大力推行渔业专业合作社与渔业园区，养殖鱼种的单一性问题得到了一定程度的解决。渔业园区与渔业专业合作社的兴起，也让越来越多的养殖户加入其中，从原先的"单打独斗"中走出来，共享资源，同时共享利润。这表明我国淡水养殖已经转为向规模要效益的路子，养殖观念的革新、养殖模式的改变、养殖规模化的形成将会是淡水养殖发展的必然方向。

在建设现代水产种业，迈向世界水产种业强国的道路上，我们必须吸收国内外一切先进的技术和经验，在加强种质资源保护与合理开发资源的前提下，建立符合中国水产养殖业特色的品种开发和管理的理念，对于重点的养殖品种制定长期研发发展规划，加强引进种开发、利用和管理，整合社会资源逐步完善新品种系统性研发与推广，有计划、有重点地推进主要养殖种类的良种化，最终实现水产养殖业良种化，提高我国水产养殖产品国际市场竞争力。未来我们应该在现有工作基础上，重点打造完善以下五大体系。

一、系统完善的澳洲龙纹斑种质资源保存体系

水产育种需要以丰富的品种资源为基础。广泛收集和保存种质资源对水产新品种研发具有重要价值。虽然目前我国已建设种质资源保护区、遗传育种中心和原良种场等工程，从种群、个体、细胞、基因等多水平进行资源保护与保存；但是整体未形成完整体系，个体内部资源数据信息化水平低，相互间缺乏横向、纵向联系，各种资源暂无法共享。未来应由政府扶持建立大数据保障平台，完成体系联动与资源调配。体系重点收集保存三大类资源：一是现有养殖主导品种和优质野生种的种质材料；二是抗病、抗寒、耐盐碱以及高蛋白等优良种质材料；三是保持生物多样性可供基础研究的科研种质材料。

二、完善澳洲龙纹斑引种开发利用和管理体系

建立健全全国性水产养殖品种引进管理体系与信息系统。加强引进种的风险评估，严格执行检疫法规，将外来物种对我国种质资源的危害降到最低限度。对我国已经引进和待引进的水产养殖新品种进行全过程跟踪管理。根据产业发展，制定出今后引进的中长期发展规划。因地制宜将引进种消化吸收、再创新和推广，重视和加强对当地生态环境的保护。

三、提升澳洲龙纹斑研发新体系

打破单兵作战，相互竞争的格局，采取强强联合和多学科作战策略。从国家和市场对种质利用和水产养殖品种需求出发，明确主要方向是培育高产、优质、抗病、抗寒、抗盐碱和适应设施渔业高密度养殖的品种。同时进行育种的应用基础和应用技术研究。每个研发团队因至少包括科研院所、企业和推广部门。各单位间各司其职，科研院所负责技术路线制定与难点攻关，企业负责路线执行并反馈问题，推广部门负责研发测试并反馈市场需求。由全国统一平台对各育种团队进行信息登记与公开，保证各团队见相互取长补短，避免造成资源冲突。

四、配套完善的澳洲龙纹斑推广体系

国家应对新品种推广配套相应的资金支持，保证新品种公告后有后续推广经费。建立渠道完善的宣传体系，利用网络、电视、书刊、报纸、杂志等媒体，广泛开展新品种宣传，提高市场认识度。建立逐级示范的推广基地，以国家、省、地（市）级推广站示范基地为主体，进行新品种层层示范推广，保证推广产业全国一盘棋。建立技术指导服务中心，广泛开展新技术、新品种培训与并对产品售后的问题进行咨询。另外积极引导社会力量进行良种推广合作。

纵观中国澳洲龙纹斑产业，近年来，养殖规模不断扩大，技术水平和基础设施建设也得到很大提高，但集约化水平低、技术集成不足等问题依旧十分突出。开展澳洲龙纹斑优质苗种生产、从养殖环境控制、营养饲料、疾病防控等方面提升养殖技术、亟待进行，从而保证中国的澳洲龙纹斑能够高品质、高产量、高标准的满足国内外市场需求。随着澳洲龙纹斑养殖业的快速发展，产业缺乏系统的种质保护和选种育种，种质退化，生长速度降低，抗病力下降，严重影响了澳洲龙纹斑养殖产业的发展。优化澳洲龙纹斑规模化繁育技术，着力提高苗种质量，建立苗种培育的技术规范，形成优质苗种质量评价体系；建立澳洲龙纹斑良种选育技术体系，运用传统育种技术和分子育种技术，培育出优质、高产、抗逆、抗病的澳洲龙纹斑优良品种，逐步实现澳洲龙纹斑养殖良种化。同时，着力优化养殖模式，提升养殖技术水平。加强澳州龙纹斑养殖生物学基础研究，更深入系统地了解澳州龙纹斑的生殖、生长特点、营养需求、疾病发病机理，研制高效、优质、安全的澳州龙纹斑配合饲料；建立疾病防控技术体系等。澳洲龙纹斑作为具有殖潜力的水产品种，整个产业方兴未艾，我们相信，通过大家的共同努力，福建澳洲龙纹斑产业化开发将走在全国前列，起到良好的示范引领

作用，中国澳洲龙纹斑产业会进入一个持续稳定的发展阶段。在新的发展时期，着力开发与优化发展澳洲龙纹斑新兴产业，必将为淡水养殖转型升级与乡村农民增收致富作出新的更大的贡献。

（罗钦、黄惠珍、徐庆贤）

附录　研究进展摘编索引

1. 澳洲龙纹斑种质资源及生物学特性分析 ……………………………………（1）

［1］Harrisson, K. A., Yen, J. D., Pavlova, A., *et al.* 2016. Identifying environmental correlates of intraspecific genetic variation［J］. Heredity.

［2］Austin, C. M., Tan, M. H., Lee, Y. P., *et al.* 2016. The complete mitogenome of the murray cod, maccullochella peelii (mitchell, 1838) (teleostei: percichthyidae)［J］. Mitochondrial Dna Part A Dna Mapping Sequencing & Analysis.

［3］Koehn, J. D., Nicol, S. J. 2013. Comparative habitat use by large riverine fishes［J］. Marine & Freshwater Research.

［4］Llewellyn, L. C. 2014. Movements of golden perch macquaria ambigua (richardson) in the mid murray and lower murrumbidgee rivers (new south wales) with notes on other species［J］. Australian Zoologist.

［5］Whiterod, N. S. 2013. The swimming capacity of juvenile murray cod (maccullochellapeelii): an ambush predator endemic to the murray-darling basin, australia. Ecology of Freshwater Fish.

［6］Disspain, M. C. F., Wilson, C. J., Gillanders, B. M. 2013. Morphological and chemical analysis of archaeological fish otoliths from the lower murray river, south australia［J］. Archaeology in Oceania.

［7］Koehn, J. D. 2009. Multi-scale habitat selection by murray cod maccullochella peelii peelii, in two lowland rivers［J］. Journal of Fish Biology.

［8］Nock, C. J., Baverstock, P. R. 2008. Polymorphic microsatellite loci for species of australian freshwater cod, maccullochella［J］. Conservation Genetics.

［9］Meaghan, R., Jenny, N., Hayley, M., *et al.* 2010. Isolation and characterization of 102 new microsatellite loci in murray cod, maccullochella peelii peelii (percichthyidae), and assessment of cross-amplification in 13 australian native and six introduced freshwater species [J]. Molecular Ecology Resources.

[10] Tonkin, Z. D., Humphries, P., Pridmore, P. A. 2006. Ontogeny of feeding in two native and one alien fish species from the murray-darling basin, australia [J]. Environmental Biology of Fishes.

[11] A. J. King †. 2010. Ontogenetic patterns of habitat use by fishes within the main channel of an australian floodplain river [J]. Journal of Fish Biology.

[12] Boys, C. A., Thoms, M. C. 2006. A large-scale, hierarchical approach for assessing habitat associations of fish assemblages in large dryland rivers [J]. Hydrobiologia.

[13] Leigh, S. J., Zampatti, B. P. 2013. Movement and mortality of murray cod, maccullochella peelii, during overbank flows in the lower river murray, australia [J]. Australian Journal of Zoology.

[14] Ellis, I. M. 2004. An independent review of the february 2004 lower darling river fish deaths: guidelines for future release effects on lower darling river fish populations [J]. Open Access This Report Has Been Reproduce with the Publishers Permission.

[15] Baumgartner, L. J. 2007. Diet and feeding habits of predatory fishes upstream and downstream of a low-level weir [J]. Journal of Fish Biology.

[16] Kopf, S. M., Humphries, P., Watts, R. J. 2014. Ontogeny of critical and prolonged swimming performance for the larvae of six australian freshwater fish species [J]. Journal of Fish Biology.

[17] Murphy, M. J. 2012. The vertebrate fauna of currawananna state forest and adjacent agricultural and aquatic habitats in the new south wales south western slopes bioregion [J]. Australian Zoologist.

[18] Skye H. Woodcock, Bronwyn M. Gillanders, Andrew R. Munro, et al. 2011. Determining mark success of 15 combinations of enriched stable isotopes for the batch marking of larval otoliths [J]. North American Journal of Fisheries Management.

[19] Nock, C. J., Elphinstone, M. S., Rowland, S. J., et al. 2010. Phylogenetics and revised taxonomy of the Australian freshwater cod genus, Maccullochella [J]. Marine and Freshwater Research.

[20] Rourke, M. L., Mcpartlan, H. C., Ingram, B. A., et al. 2010. Biogeography and life history ameliorate the potentially negative genetic effects of stocking on murray cod

(maccullochella peelii peelii) [J]. Marine & Freshwater Research.

[21] Whiterod, N. S. 2010. Calibration of a rapid non-lethal method to measure energetic status of a freshwater fish (murray cod, maccullochella peelii peelii) [J]. Marine & Freshwater Research.

[22] Koehn, J. D., Mckenzie, J. A., O'Mahony, D. J., *et al.* 2009. Movements of murray cod (maccullochella peelii peelii) in a large australian lowland river [J]. Ecology of Freshwater Fish.

[23] Turchini, G. M., Quinn, G. P., Jones, P. L., *et al.* 2009. Traceability and discrimination among differently farmed fish: a case study on australian murray cod [J]. Journal of Agricultural & Food Chemistry.

[24] Koehn, J. D. 2009. Using radio telemetry to evaluate the depths inhabited by murray cod (maccullochella peelii peelii) [J]. Marine & Freshwater Research.

[25] 翁伯琦, 罗土炎, 刘洋, 等. 2016. 澳洲龙纹斑生物学特征及其繁养殖技术研究进展[J]. 福建农业学报.

[26] 郭松, 王广军, 方彰胜, 等. 2012. 澳洲鳕鲈的生物学特征及人工繁养技术[J]. 江苏农业科学.

[27] 李娴, 朱永安, 钟君伟, 等. 2013. 虫纹鳕鲈的生物学特性及人工养殖技术研究[J]. 湖北农业科学.

[28] 张龙岗, 杨玲, 李娴, 等. 2013. 利用 mtdna coi 基因序列分析引进的澳洲虫纹鳕鲈群体遗传多样性[J]. 水产学杂志.

[29] 罗钦, 罗土炎, 林旋, 等. 2016. 澳洲龙纹斑鱼种生长特性研究[J]. 福建农业学报.

[30] 王波, 张艳华, 韩茂森. 2003. 虫纹麦鳕鲈的形态和生物学性状[J]. 水产科技情报.

[31] 张龙岗, 杨玲, 张延华, 等. 2012. 虫纹鳕鲈线粒体 COI 基因片段的克隆与序列分析[J]. 长江大学学报(自然科学版).

2. 澳洲龙纹斑繁育与养殖 ……………………………………………………(38)

[32] Forbes, J. P., Watts, R. J., Robinson, W. A., *et al.* 2015. Recreational fishing effort, catch, and harvest for murray cod and golden perch in the murrumbidgee river, australia [J]. North American Journal of Fisheries Management.

[33] Danem, N., Paull, J., Bretta, I. 2008. Sexing accuracy and indicators of matu-

ration status in captive murray cod maccullochella peelii peelii using non-invasive ultrasonic imagery [J]. Aquaculture.

[34] Tonkin, Z. D., Humphries, P., Pridmore, P. A. 2006. Ontogeny of feeding in two native and one alien fish species from the murray-darling basin, australia [J]. Environmental Biology of Fishes.

[35] Austasia Aquaculture.

[36] Danem, N., Paull, J., Bretta, I. 2010. Advanced ovarian development of murray cod maccullochella peelii peelii via phase-shifted photoperiod and two temperature regimes [J]. Aquaculture.

[37] Newman, D. M., Jones, P. L., Ingram, B. A. 2008. Age-related changes in ovarian characteristics, plasma sex steroids and fertility during pubertal development in captive female murray cod maccullochella peelii peelii [J]. Comparative Biochemistry & Physiology Part A.

[38] Jonathan, D., David, G., William, B., et al. 2008. Cryopreservation of sperm from murray cod, maccullochella peelii peelii [J]. Aquaculture.

[39] Daly, J., Galloway, D., Bravington, W., et al. 2008. Cryopreservation of sperm from Murray cod (Maccullochella peelii peelii) and silver perch (Bidyanus bidyanus) [J]. Skretting Australasian Aquaculture. Innovation in a Global Market (3-6 August 2008, Brisbane).

[40] Todd, C. R., Ryan, T., Nicol, S. J., et al. 2010. The impact of cold water releases on the critical period of post-spawning survival and its implications for murray cod (maccullochella peelii peelii): a case study of the mitta mitta river, southeastern australia [J]. River Research & Applications.

[41] Newman D M, Jones P L, Ingram B A. 2007. Temporal dynamics of oocyte development, plasma sex steroids and somatic energy reserves during seasonal ovarian maturation in captive murray cod maccullochella peelii peelii [J]. Comparative Biochemistry and Physiology.

[42] Koehn, J. D., Harrington, D. J. 2006. Environmental conditions and timing for the spawning of murray cod (maccullochella peelii peelii) and the endangered trout cod (m. macquariensis) in southeastern australian rivers [J]. River Research & Applications.

[43] Humphries, P. 2005. Spawning time and early life history of murray cod, maccullochella peelii, (mitchell) in an australian river. Environmental Biology of Fishes.

[44] Ingram, B. A. 2009. Culture of juvenile murray cod, trout cod and macquarie perch (percichthyidae) in fertilised earthen ponds [J]. Aquaculture.

[45] Mellor, P., Fotedar, R. (2011). Physiological responses of murray cod (maccullochella peelii peelii) (mitchell 1839) larvae and juveniles when cultured in inland saline water [J]. Indian Journal of Fisheries.

[46] Rourke, M. L., Mcpartlan, H. C., Ingram, B. A., et al. 2009. Polygamy and low effective population size in a captive murray cod (maccullochella peelii peelii) population: genetic implications for wild restocking programs [J]. Marine & Freshwater Research.

[47] Lennard, W. A., Leonard, B. V. 2006. A comparison of three different hydroponic sub-systems (gravel bed, floating and nutrient film technique) in an aquaponic test system [J]. Aquaculture International.

[48] Abery, N. W., De Silva, S. S. 2015. Performance of murray cod, maccullochella peelii peelii (mitchell) in response to different feeding schedules [J]. Aquaculture Research.

[49] Ryan, S. G., Smith, B. K., Collins, R. O., et al. 2010. Evaluation of weaning strategies for intensively reared australian freshwater fish, murray cod, maccullochella peelii peelii [J]. Journal of the World Aquaculture Society.

[50] Allen-Ankins, S., Stoffels, R. J., Pridmore, P. A., et al. 2012. The effects of turbidity, prey density and environmental complexity on the feeding of juvenile murray cod maccullochella peelii [J]. Journal of Fish Biology.

[51] Couch, A. J., Unmack, P. J., Dyer, F. J., et al. 2016. Who's your mama? riverine hybridisation of threatened freshwater trout cod and murray cod [J]. Peerj.

[52] Baumgartner, L. J., Mclellan, M., Robinson, W., et al. 2012. The influence of saline concentration on the success of calcein marking techniques for hatchery-produced murray cod (maccullochella peelii) [J]. Journal and Proceedings-Royal Society of New South Wales.

[53] 欧小华,罗汝安.2017.澳洲"国宝鱼"墨瑞鳕养殖关键技术[J].海洋与渔业.

[54] 张志山,朱树人,朱永安.2016.澳洲虫纹鳕鲈工厂化养殖技术[J].齐鲁渔业.

[55] 杨小玉,郭正龙.2013.澳洲龙纹斑工厂化养殖技术[J].水产养殖.

[56] 郭正龙，杨小玉，孟庆宇．2012．澳洲龙纹斑工厂化养殖技术［J］．科学养鱼．

[57] 曹凯德．2001．澳洲墨累河鳕鱼养殖技术［J］．水生态学杂志．

[58] 左瑞华，汪学军．2001．虫纹鳕鲈苗种培育影响因素初步分析［J］．安徽农学通报．

[59] 曹凯德．2001．澳大利亚墨累河鳕鱼养殖技术［J］．水利渔业．

[60] 陈杰，杨兴丽，申秀英．2002．墨累鳕的养殖技术［J］．河南水产．

[61] 罗钦，罗土炎，饶秋华．2016．澳洲龙纹斑的繁殖方法．CN105660479A．

[62] 罗钦，罗土炎，饶秋华．2016．澳洲龙纹斑鱼苗的培养方法．CN105706971A．

[63] 罗钦，罗土炎，饶秋华．2015．澳洲龙纹斑受精卵孵化器．CN205093374U．

[64] 蔡星，秦巍仑，周忠良．2014．一种新品种龙纹斑温室规模化养殖方法．CN103947585A．

[65] 罗土炎，张志灯，饶秋华，等．2017．一种澳洲龙纹斑受精卵的孵化方法．CN201710667649.2．

[66] 钱晓明，朱永祥，刘大勇，等．2015．一种龙纹斑幼鱼的仿自然生态投喂模式．CN201510125321.9．

[67] 蔡星，秦巍仑，周忠良．2014．一种新品种龙纹斑池塘生态养殖技术．CN201410146331.6．

[68] 蔡星，周忠良，秦巍仑．2014．一种新品种龙纹斑的自然繁殖方法．CN201410146971.7．

[69] 蔡星，周忠良，郭正龙．2015．一种新品种龙纹斑的人工繁殖方法．CN201410146334.X．

[70] 钱晓明，金加余，温松来．2017．一种通用型集约化循环养殖池．CN201710612275.4．

[71] 钱晓明，温松来，凌诚，等．2016．一种用于养殖池水过滤回用处理方法．CN201610345392.4．

[72] KAUSE ANTTI；MAENTYSAARI ESA；JAERVISALO OTSO．2009．METHOD FOR THE PRODUCTION OF FISH PROGENY．FI：2007000301：W．

[73] MERRYFULL ALBERT．2016．THERMAL CONTROLLED AQUACULTURE TANKS FOR FISH PRODUCTION．AU：2004000152：W．

3. 澳洲龙纹斑疾病与防控 ……………………………………………（87）

[74] Liu, Y., Zeng, R., Weng, B., et al. 2016. Draft genome sequence of strepto-

coccus sp. x13sy08, isolated from murray cod (maccullochella peelii peelii). Genome Announcements.

[75] Gunasekera, R. M., Gooley, G. J., Silva, S. S. D. 1998. Characterisation of 'swollen yolk-sac syndrome' in the australian freshwater fish murray cod, maccullochella peelii peelii, and associated nutritional implications for large scale aquaculture [J]. Aquaculture.

[76] Boys, C. A., Rowland, S. J., Gabor, M., et al. (2012). Emergence of epizootic ulcerative syndrome in native fish of the murray-darling river system, australia: hosts, distribution and possible vectors [J]. Plos One.

[77] Baily, J. E., Bretherton, M. J., Gavine, F. M., et al. 2005. The pathology of chronic erosive dermatopathy in murray cod, maccullochella peelii peelii (mitchell) [J]. Journal of Fish Diseases.

[78] Aarong, S., Sarahl, S., Paull, J., et al. 2011. Groundwater pre-treatment prevents the onset of chronic ulcerative dermatopathy in juvenile murray cod, maccullochella peelii peelii (mitchell) [J]. Aquaculture.

[79] Schultz, A G, Healy, J M, Jones, P L, et al. 2008. Osmoregulatory balance in Murray cod, Maccullochella peelii peelii (Mitchell), affected with chronic ulcerative dermatopathy [J]. Aquaculture.

[80] A G Schultz, P L Jones, T Toop. 2014. Rodlet cells in murray cod, maccullochella peelii peelii, (mitchell), affected with chronic ulcerative dermatopathy [J]. Journal of Fish Diseases.

[81] Rona Barugahare, Joy A. Becker, Matt Landos, et al. 2011. Corrigendum to "gastric cryptosporidiosis in farmed australian murray cod, maccullochella peelii peelii" [J]. Aquaculture.

[82] Gomes, G. B., Miller, T. L., Vaughan, D. B., et al. 2017. Evidence of multiple species of chilodonella, (protozoa, ciliophora) infecting australian farmed freshwater fishes [J]. Veterinary Parasitology.

[83] Becker, J. A., Tweedie, A., Gilligan, D., et al. 2013. Experimental infection of australian freshwater fish with epizootic haematopoietic necrosis virus (ehnv). Journal of Aquatic Animal Health.

[84] Rimmer, A. E., Whittington, R. J., Tweedie, A., et al. 2017. Susceptibility of a number of australian freshwater fishes to dwarf gourami iridovirus (infectious spleen and kid-

ney necrosis virus)[J]. Journal of Fish Diseases.

[85] Go, J., Whittington, R. 2006. Experimental transmission and virulence of a megalocytivirus (family iridoviridae) of dwarf gourami (colisa lalia) from asia in murray cod (maccullochella peelii peelii) in australia[J]. Aquaculture.

[86] Go, J., Lancaster, M., Deece, K., *et al*. 2006. The molecular epidemiology of iridovirus in murray cod (maccullochella peelii peelii) and dwarf gourami (colisa lalia) from distant biogeographical regions suggests a link between trade in ornamental fish and emerging iridoviral diseases[J]. Molecular & Cellular Probes.

[87] Harford, A. J., O'Halloran, K., Wright, P. E. 2010. Effect of in vitro and in vivo organotin exposures on the immune functions of murray cod (maccullochella peelii peelii)[J]. Environmental Toxicology & Chemistry.

[88] Harford, A. J., O'Halloran, K., Wright, P. F. 2005. The effects of in vitro pesticide exposures on the phagocytic function of four native australian freshwater fish[J]. Aquatic Toxicology.

[89] 罗钦, 饶秋华, 李巍, 等. 2017. 小瓜虫对澳洲龙纹斑不同生长阶段苗种的感染与致病性研究[J]. 中国预防兽医学报.

[90] 罗土炎, 罗钦, 饶秋华, 等. 2016. 不同药物对澳洲龙纹斑幼鱼小瓜虫病的治疗效果[J]. 福建农业学报.

[91] 罗土炎, 罗钦, 涂杰峰, 等. 2015. 澳洲龙纹斑养殖过程中主要疾病诊断及其防治[J]. 福建农业学报.

[92] 罗土炎, 罗钦, 饶秋华, 等. 2017. 重金属离子与消毒药物对澳洲龙纹斑幼鱼的急性毒性[J]. 福建农业学报.

[93] 蔡星, 朱永祥, 刘大勇, 等. 2013. 一种生态防治澳洲龙纹斑小瓜虫的技术. CN103404467A.

[94] 罗钦, 罗土炎, 张志灯, 等. 2017. 澳洲龙纹斑鱼防治小瓜虫的方法. CN201510111037.6.

[95] 罗钦, 罗土炎, 陈华, 等. 2017. 一种鱼类盾纤虫感染症的治疗方法. CN201710078170.5.

4. 澳洲龙纹斑营养与品质 ………………………………………… (138)

[96] Palmeri, G., Turchini, G. M., Caprino, F., *et al*. 2008. Biometric, nutritional and sensory changes in intensively farmed murray cod (maccullochella peelii peelii,

mitchell) following different purging times [J] . Food Chemistry.

[97] Rasanthi M. Gunasekera, Sena S. De Silva, Brett A. Ingram. 1999. The amino acid profiles in developing eggs and larvae of the freshwater percichthyid fishes, trout cod, maccullochella macquariensis, and murray cod, maccullochella peelii peelii [J] . Aquatic Living Resources.

[98] Gunasekera R. M, De Silva S. S, Ingram B. A. 1999. Early ontogeny-related changes of the fatty acid composition in the percichthyid fishes trout cod, maccullochella macquariensis and murray cod, m. peelii peelii. Aquatic Living Resources.

[99] Palmeri, G., Turchini, G. M., De Silva, S. S. 2007. Lipid characterisation and distribution in the fillet of the farmed australian native fish, murray cod (maccullochella peelii peelii) [J]. Food Chemistry.

[100] Palmeri, G., Turchini, G. M., Marriott, P. J., et al. 2009. Biometric, nutritional and sensory characteristic modifications in farmed murray cod (maccullochella peelii peelii) during the purging process [J] . Aquaculture.

[101] Palmeri, G., Turchini, G. M., Silva, S. S. D. 2010. Short-term food deprivation does not improve the efficacy of a fish oil finishing strategy in murray cod [J] . Aquaculture Nutrition.

[102] Small, K., Kopf, R. K., Watts, R. J., et al. 2014. Hypoxia, blackwater and fish kills: experimental lethal oxygen thresholds in juvenile predatory lowland river fishes [J] . Plos One.

[103] Lee J. Baumgartner, Nathan Reynoldson, Dean M. Gilligan. 2006. Mortality of larval murray cod (maccullochella peelii peelii) and golden perch (macquaria ambigua) associated with passage through two types of low-head weirs [J] . Marine and Freshwater Research.

[104] Forbes, J. P., Watts, R. J., Robinson, W. A., et al. 2015. System-specific variability in murray cod and golden perch maturation and growth influences fisheries management options [J] . North American Journal of Fisheries Management.

[105] Shigdar, S., Harford, A., Ward, A. C. 2009. Cytochemical characterisation of the leucocytes and thrombocytes from murray cod (maccullochella peelii peelii, mitchell) [J]. Fish Shellfish Immunol.

[106] Shigdar, S., Cook, D., Jones, P., et al. 2007. Blood cells of murray cod

maccullochella peelii peelii (mitchell) [J]. Journal of Fish Biology.

[107] 罗钦, 罗土炎, 颜孙安, 等. 2015. 澳洲龙纹斑肌肉氨基酸的分析研究[J]. 福建农业学报.

[108] 宋理平, 冒树泉, 胡斌, 等. 2013. 虫纹鳕鲈肌肉营养成分分析与品质评价[J]. 饲料工业.

5. 澳洲龙纹斑饲料与制备 ………………………………………………… (167)

[109] De Silva, S. S., Gunasekera, R. M., Collins, R. A., et al. 2002. Performance of juvenile murray cod, maccullochella peelii peelii (mitchell), fed with diets of different protein to energy ratio [J]. Aquaculture Nutrition.

[110] Gunasekera, R. M., Silva, S. S. D., Collins, R. A., et al. 2000. Effect of dietary protein level on growth and food utilization in juvenile murray cod maccullochella peelii peelii (mitchell) [J]. Aquaculture Research.

[111] Abery, N. W., Gunasekera, R. M., De Silva, S. S. 2002. Growth and nutrient utilization of murray cod maccullochella peelii peelii (mitchell) fingerlings fed diets with varying levels of soybean meal and blood meal [J]. Aquaculture Research.

[112] Silva, S. S. D., Gunasekera, R. M., Gooley, G. 2000. Digestibility and amino acid availability of three protein-rich ingredient-incorporated diets by murray cod maccullochella peelii peelii, (mitchell) and the australian shortfin eel anguilla australis, richardson [J]. Aquaculture Research.

[113] De Silva S S, Gunasekera R M, Ingram B A. 2015. Performance of intensively farmed murray cod maccullochella peelii peelii (mitchell) fed newly formulated vs. currently used commercial diets, and a comparison of fillet composition of farmed and wild fish [J]. Aquaculture Research.

[114] Varricchio Ettore, Russo Finizia, Coccia Elena, et al. 2012. Immunohistochemical and immunological detection of ghrelin and leptin in rainbow trout oncorhynchus mykiss, and murray cod maccullochella peelii peelii, as affected by different dietary fatty acids [J]. Microscopy Research and Technique.

[115] Senadheera, S. D., Turchini, G. M., Thanuthong, T., et al. 2011. Effects of dietary α-linolenic acid (18: 3n-3) /linoleic acid (18: 2n-6) ratio on fatty acid metabolism in murray cod (maccullochella peelii peelii) [J]. J Agric Food Chem.

[116] Francis, D. S., Turchini, G. M., Jones, P. L., et al. 2007. Growth perform-

ance, feed efficiency and fatty acid composition of juvenile murray cod, maccullochella peelii peelii, fed graded levels of canola and linseed oil [J]. Aquaculture Nutrition.

[117] Francis, D. S., Turchini, G. M., And, P. L. J., et al. 2007. Dietary lipid source modulates in vivo fatty acid metabolism in the freshwater fish, murray cod (maccullochella peelii peelii) [J]. J Agric Food Chem.

[118] Francis, D. S., Turchini, G. M., Jones, P. L., et al. 2006. Effects of dietary oil source on growth and fillet fatty acid composition of murray cod, maccullochella peelii [J]. Aquaculture.

[119] Turchini, G. M., Francis, D. S., De Silva, S. S. 2010. Modification of tissue fatty acid composition in murray cod (maccullochella peelii peelii, mitchell) resulting from a shift from vegetable oil diets to a fish oil diet [J]. Aquaculture Research.

[120] Francis, D. S., Turchini, G. M., Smith, B. K., et al. 2010. Effects of alternate phases of fish oil and vegetable oil-based diets in murray cod [J]. Aquaculture Research.

[121] Turchini, G., Gunasekera, R. M., De Silva, S. 2003. Effect of crude oil extracts from trout offal as a replacement for fish oil in the diets of the Australian native fish Murray cod Maccullochella peelii peelii [J]. Aquaculture Research.

[122] Turchini G M, Francis D S, De Silva S S. 2006. Fatty acid metabolism in the freshwater fish murray cod (maccullochella peelii peelii) deduced by the whole-body fatty acid balance method [J]. Comp Biochem Physiol B Biochem Mol Biol.

[123] Francis, D. S., Peters, D. J., Turchini, G. M. 2009. Apparent in vivo Δ-6 desaturase activity, efficiency, and affinity are affected by total dietary c18 pufa in the freshwater fish murray cod [J]. Journal of Agricultural & Food Chemistry.

[124] TURCHINI, G. M, FRANCIS, D. S, DE SILVA, S. S. 2010. Finishing diets stimulate compensatory growth: results of a study on murray cod, maccullochella peelii peelii [J]. Aquaculture Nutrition.

[125] Lennard, W. A., Leonard, B. V. 2005. A comparison of reciprocating flow versus constant flow in an integrated, gravel bed, aquaponic test system [J]. Aquaculture International.

[126] Sierp, M. T., Qin, J. G., Recknagel, F. 2009. Biomanipulation: a review of biological control measures in eutrophic waters and the potential for murray cod maccullochella

peelii peelii, to promote water quality in temperate australia [J]. Reviews in Fish Biology & Fisheries.

[127] Kaminskas, S., Humphries, P. 2009. Diet of murray cod (maccullochella peelii peelii) (mitchell) larvae in an australian lowland river in low flow and high flow years [J]. Hydrobiologia.

[128] 钱晓明, 叶建华, 董志航, 等. 2015. 龙纹斑仔稚鱼配合饲料及其制备方法. CN104642817A.

[129] 王芬, 张伟, 唐佳, 等. 2015. 一种虫纹鳕鲈配合饲料及其制备方法. CN105211500A.

[130] 王芬, 张伟, 唐佳, 等. 2015. 一种预防虫纹鳕鲈体表溃疡的饲料及其制备方法. CN105211565A.

[131] 张蕉南, 胡兵, 李惠, 等. 2014. 澳洲龙纹斑幼鱼膨化颗粒配合饲料. CN104304800A.

[132] 胡兵, 张蕉南, 李惠, 等. 2014. 澳洲龙纹斑中成鱼膨化颗粒配合饲料. CN104286581A.

[133] 林虬, 张蕉南, 胡兵, 等. 2014. 澳洲龙纹斑稚鱼膨化颗粒配合饲料. CN104286574A.

[134] 文远红, 张璐, 米海峰, 等. 2014. 澳洲虫纹鳕鲈育成期配合饲料及其制备方法. CN103783268A.

[135] 文远红, 米海峰, 张璐, 等. 2014. 一种澳洲虫纹鳕鲈鱼苗开口配合饲料及其制备方法. CN103798549A.

[136] 刘大勇, 蔡星, 秦桂祥, 等. 2013. 澳洲龙纹斑成鱼膨化配合饲料及其制备方法. CN103250912A.

[137] 钱晓明, 叶建华, 董志航, 等. 2015. 一种改善龙纹斑成鱼风味品质的配合饲料及其制备方法. CN201510125509.3.

[138] 刘洋, 曾润颖, 罗土炎, 等. 2016. 一种海藻寡糖鱼饲料添加剂. CN201510894666.0.

[139] 詹森·A·布茨马, 威廉姆·R·吉本斯, 迈克尔·L·布朗. 2016. 用于生产高质量蛋白浓缩物和脂类的固态发酵系统和方法. CN201480055185.4.

[140] 朱永祥, 刘大勇, 蔡星, 等. 2013. 一种生态防治澳洲龙纹斑水霉病的中草药制备方法及应用. CN103272005A.

6. 澳洲龙纹斑引种与开发 ………………………………………………… (221)

[141] Ingram, B. A., De Silva, S. S., Gooley, G. J. 2005. The australian murray cod--a new candidate for intensive production systems [J]. World Aquaculture.

[142] Partridge, G. J., Lymbery, A. J., George, R. J. 2008. Finfish mariculture in inland australia: a review of potential water sources, species, and production systems [J]. Journal of the World Aquaculture Society.

[143] Simpson, R. R., & Mapleston, A. J. 2002. Movements and habitat use by the endangered australian freshwater mary river cod, maccullochella peelii mariensis [J]. Environmental Biology of Fishes.

[144] Boys, C. A., Robinson, W., Miller, B., et al. 2016. A piecewise regression approach for determining biologically relevant hydraulic thresholds for the protection of fishes at river infrastructure [J]. Journal of Fish Biology.

[145] Koehn, J. D., Nicol, S. J. 2016. Comparative movements of four large fish species in a lowland river [J]. Journal of Fish Biology.

[146] Lyon, J. P., Bird, T., Nicol, S., et al. 2014. Efficiency of electrofishing in turbid lowland rivers: implications for measuring temporal change in fish populations [J]. Canadian Journal of Fisheries & Aquatic Sciences.

[147] Koehn, J. D., Lintermans, M., Lyon, J. P., et al. 2013. Recovery of the endangered trout cod, maccullochella macquariensis: what have we achieved in more than 25 years? [J]. Marine & Freshwater Research.

[148] Nick S. Whiterod, Shaun N. Meredith, Paul Humphries. 2013. Refining the activity component of a juvenile fish bioenergetics model to account for swimming costs [J]. Marine & Freshwater Behaviour & Physiology.

[149] Hall, K. C., Broadhurst, M. K., Butcher, P. A. 2012. Post-release mortality of angled golden perch macquaria ambigua and murray cod maccullochella peelii [J]. Fisheries Management & Ecology.

[150] Rogers, M. W., Allen, M. S., Brown, P., et al. 2010. A simulation model to explore the relative value of stock enhancement versus harvest regulations for fishery sustainability [J]. Ecological Modelling.

[151] Douglas, J., Brown, P., Hunt, T., et al. 2010. Evaluating relative impacts of recreational fishing harvest and discard mortality on murray cod (maccullochella peelii peelii)

[J]. Fisheries Research.

[152] Allen, M. S., Brown, P., Douglas, J., et al. 2009. An assessment of recreational fishery harvest policies for murray cod in southeast australia [J]. Fisheries Research.

[153] Andrew R. Bearlin, Simon J. Nicol, Terry Glenane. 2008. Behavioral responses of murray cod maccullochella peelii peelii to pulse frequency and pulse width from electric fishing machines [J]. Transactions of the American Fisheries Society.

[154] Clark, T. D., Ryan, T., Ingram, B. A., et al. 2005. Factorial aerobic scope is independent of temperature and primarily modulated by heart rate in exercising murray cod (maccullochella peelii peelii) [J]. Physiological & Biochemical Zoology.

[155] Koehn, J. D., Harrington, D. J. 2005. Collection and distribution of the early life stages of the murray cod (maccullochella peelii peelii) in a regulated river [J]. Australian Journal of Zoology.

[156] Loughnan, S. R., Baranski, M. D., Robinson, N. A., et al. 2004. Microsatellite loci for studies of wild and hatchery australian murray cod maccullochella peelii peelii (percichthyidae) [J]. Molecular Ecology Resources.

[157] Simon J. Nicol, Jason A. Lieschke, Jarod P. Lyon, et al. 2004. Observations on the distribution and abundance of carp and native fish, and their responses to a habitat restoration trial in the murray river, australia [J]. New Zealand Journal of Marine & Freshwater Research.

[158] Rourke, M. L., H. C. McPartlan †, Ingram, B. A., et al. 2011. Variable stocking effect and endemic population genetic structure in murray cod maccullochella peelii [J]. Journal of Fish Biology.

[159] Tonkin, Z., Kitchingman, A., Lyon, J., et al. 2017. Flow magnitude and variability influence growth of two freshwater fish species in a large regulated floodplain river [J]. Hydrobiologia.

[160] Bird, T., Lyon, J., Wotherspoon, S., King, R., et al. 2017. Accounting for false mortality in telemetry tag applications [J]. Ecological Modelling.

[161] Thiem, J. D., Wooden, I. J., Baumgartner, L. J., et al. 2017. Recovery from a fish kill in a semi-arid australian river: can stocking augment natural recruitment processes? [J]. Austral Ecology.

[162] Forbes, J., Watts, R. J., Robinson, W. A., et al. 2016. Assessment of

stocking effectiveness for murray cod (maccullochella peelii) and golden perch (macquaria ambigua) in rivers and impoundments of south-eastern australia[J]. Marine & Freshwater Research.

[163] Ingram, B. A., Hayes, B., †, M. L. R. 2011. Impacts of stock enhancement strategies on the effective population size of murray cod, maccullochella peelii, a threatened australian fish[J]. Fisheries Management & Ecology.

[164] Tetzlaff, J. C. 2011. Evaluating mark-recapture sampling designs for fish in an open riverine system[J]. Marine & Freshwater Research.

[165] Koehn, J. D., Todd, C. R. 2012. Balancing conservation and recreational fishery objectives for a threatened fish species, the murray cod, maccullochella peelii[J]. Fisheries Management & Ecology.

[166] Ingram, B. A., Hunt, T. L., Lieschke, J., et al. 2015. Field detection of calcein in marked fish-a cautionary note[J]. Fisheries Management & Ecology.

[167] 韩茂森. 2003. 澳洲虫纹鳕鲈的生物学特性及引养前景[J]. 淡水渔业.

[168] 翁伯琦，罗土炎. 2016. 着力农业科技创新服务新兴产业发展——写在《澳洲龙纹斑种苗繁育与养殖技术研究》完稿之际[J]. 福建农业科技.

[169] 刘怡. 2014. 墨瑞鳕在广东全人工繁育成功[J]. 海洋与渔业.

[170] 赵志玉. 2014. 墨瑞鳕成功养殖"五步曲"[J]. 海洋与渔业.

[171] 张善发，林昌华，李琼文，等. 2009. 澳洲鳕鲈养殖试验初探[J]. 科学养鱼.

[172] 陈永乐，朱新平，刘毅辉，等. 199. 澳洲淡水鳕鱼及其人工繁殖[J]. 淡水渔业.

[173] 吕华当，罗茵. 2017. 又一条养殖亩产值达20万的鱼问世 墨瑞鳕鱼苗东莞生产 更适合中国养殖环境[J]. 海洋与渔业.

[174] 杨宝琴. 2014. 澳大利亚"国宝鱼"在广东东莞全人工繁育成功[J]. 水产科技情报.

[175] 郑石勤. 2011. 来自澳大利亚的独特娇客-墨瑞鳕[J]. 养鱼世界.

[176] 中国农业新闻网. 2014. 澳大利亚"国宝鱼"在东莞全人工繁育养殖成功[J]. 江西饲料.

[177] 农中侃. 2002. 澳大利亚养殖鱼类落户青岛[J]. 现代渔业信息.

[178] 顾志敏，刘启文. 2005. 澳洲淡水养殖珍品——鳕鲈[J]. 农村百事通.

[179] 蔡乘成，陈鹏，张祖兴.2012.澳洲鳕鲈引种驯化养殖的报告[J].养殖技术顾问.

[180] 钟云.东莞成功繁育养殖澳大利亚"国宝鱼"[J].千里眼·顺风耳.

[181] 罗钦，罗土炎，饶秋华.2016.连城县水产养殖现状及发展对策[J].福建农业科技.

[182] 2014.名优新品种 墨瑞鳕[J].海洋与渔业.

[183] 刘怡.2014.墨瑞鳕在广东全人工繁育成功[J].海洋与渔业.

[184] 罗钦，李巍，罗土炎，等.2017.一种鱼类自动称重装置.CN201720131608.7.

[185] 张栋国.2017.一种墨瑞鳕礼品袋.CN201621432488.6.

[186] 罗钦，陈红珊，罗土炎，等.2018.一种低水分澳洲龙纹斑鱼籽酱的制作方法.CN201710939962.7.